COASTAL HISTORIES
SOCIETY AND ECOLOGY IN PRE-MODERN INDIA

Coastal Histories
Society and Ecology in pre-Modern India

Edited by

YOGESH SHARMA

PRIMUS BOOKS
An imprint of Ratna Sagar P. Ltd.
Virat Bhavan
Mukherjee Nagar Commercial Complex
Delhi 110 009

Offices at CHENNAI KOLKATA LUCKNOW
AGRA AHMEDABAD BANGALORE COIMBATORE DEHRADUN GUWAHATI
HYDERABAD JAIPUR KANPUR KOCHI MADURAI MUMBAI PATNA RANCHI

First published 2010
Reprinted 2013

ISBN 978-93-80607-00-9 (hardback)
ISBN 978-93-80607-84-9 (paperback)

Published by Primus Books

Laser typeset by Digigrafics
Gulmohar Park, New Delhi 110 049

Printed at Sanat Printers, Kundli, Haryana

Contents

Foreword

Michael N. Pearson

'REFLECTIONS ON MARITIME STUDIES AND THE SPACE FOR COASTAL HISTORIES'

Historians used to write histories which were constrained by the boundaries of a particular state. Thus, we had the history of the wool trade in England, or the life of the ruler of a particular state, maybe Louis XV of France. And most history was the history of the West. To the extent that anyone wrote the history of India, it was the history of the British or Portuguese or French in India. Even Sarvepalli Gopal, an adornment of Jawaharlal Nehru University for many years, started off with three 'colonial' books: two viceroyalties, and a general study of British policy in India from 1858 to 1905. Later of course he wrote the standard biography of Jawaharlal Nehru. In the US after World War II this situation was rectified to an extent by the funding of 'Area Studies', such as South or South-East Asia, the USSR, or Sub-Saharan Africa. While this had the commendable advantage of turning more attention to the non-European world (and even this terminology is objectionable, for it places 'Europe' in the default category!), it still imposed straitjackets on what could be studied. One could compare two areas within Africa, but not two related areas if one was in Africa and one in India. Casablanca and Zanzibar were considered to be capable of being compared because they were both in Africa, but Zanzibar and Calicut were not.

The newly revived field of world history aims to transcend these problems and restrictions. Its premier practitioner, Jerry H. Bentley, is clear that world history does not have to include the whole world. Rather it is 'a historical perspective that transcends national frontiers'. Scholars need 'a regional, continental, or global scale' to look at many important forces in history. The key is to get away from national, state-based histories.[1] World history investigates themes, trends, relationships which extend beyond the bounds of any particular 'state'. Examples of areas not constrained by the 'Myth of Continents' could include Eurasia, which would focus on the eastern Mediterranean and the Arabian Sea, or islands in South-East and East Asia, or any of the great oceans.

Here is where the Indian Ocean studies comes in, for it seems to me to be an excellent example of an area, the study of which will have a reciprocal benefit. On the one hand, the Indian Ocean studies will, I hope, contribute to World History; on the other, world history techniques and examples will enrich our studies of the Indian Ocean.

Oceanic studies have expanded greatly in the last few years. We all of course still find inspiration in Braudel's classic study of the Mediterranean, though we need also to keep in mind his cautionary note: 'A historical study centred on a stretch of water has all the charms but undoubtedly all the dangers of a new departure.'[2] The small maritime area of the Mediterranean still attracts eminent scholars, as seen in Horden and Purcell's important recent work.[3] The major publisher Routledge has under way a series called 'Seas in History', of which so far studies of the Atlantic, Baltic and North Seas, and the Indian Ocean have appeared.[4]

There have also been some attempts to assess the status of maritime history today. I will make no attempt to be comprehensive here, but merely note a few items which have struck me as useful or innovative. The most important journal, which from time to time publishes overviews and surveys, is *The International Journal of Maritime History*. In 1999 the *Geographical Review* published a very extensive analysis of maritime history, with a total of ten excellent articles.[5] A recent overview, called simply 'The Sea', aimed at an undergraduate audience, was put out on the English website 'History in Focus', which is sponsored by the Institute for Historical Research at the University of London.[6] A recent discussion in the very prestigious and widely read *American Historical Review* was guilty of a major sin of omission.[7] A long introduction by Karen Wigen was followed by analyses of the Mediterranean, the Atlantic and the Pacific. Curious indeed that the Indian Ocean was ignored. Could it be that the reason is that for most of its history the Indian Ocean was crossed and used by people from its littorals, not by Europeans, while the three examples chosen by Wigen are all dominated by Europeans for most or all of their histories? This complaint about a Eurocentric approach applies to an extent to a very recent book, *Seascapes*,[8] where again the Indian Ocean is largely absent and European and American controlled oceans and subjects are privileged. Indeed this Eurocentric bias goes back a long way: Braudel's study of the Mediterranean was notoriously weak on the southern, Islamic, shore of the sea. Even earlier than this, early in the twentieth century many European authorities considered the Indian Ocean was only a 'half ocean' as it did not extend far into the Northern Hemisphere!

Despite this neglect from the American academic mainstream, Indian Ocean studies are in fact flourishing. Very recently Markus P.M. Vink rectified the parochialism of the American Historical Review with an excellent overview of the state of Indian Ocean studies today.[9] Some years ago I attempted a general overview of its history, and more recently Sugata Bose wrote a stunning analysis of the last two centuries.[10] Conferences proliferate: as examples, one called 'The Maritime Heritage and Cultures of the Western Indian Ocean in Comparative Perspective' was held in Zanzibar in 2006, another in Leiden in 2006 called 'Culture and Commerce in the Indian Ocean' tried to introduce cultural studies notions to our ocean. In early 2007 Jawaharlal Nehru University hosted an innovative conference, 'Histories from the Sea: Multimedia for Understanding and Teaching Europe–South Asia Maritime Heritage'. Yet another, held in South Africa, again attempted a more 'theoretical', less empirical focus: 'Eyes Across the Water: Navigating the Indian Ocean', University of Witwatersrand, August 2007. A conference in Delhi in 2003, organized by H.P. Ray of JNU, produced a useful book.[11] Another very recent book takes a much more anthropological view of the Indian Ocean.[12] Oxford University Press in New Delhi recently republished in one volume classic works by Holden Furber, Sinappah Arasaratnam and Kenneth McPherson, though the introduction by Sanjay Subrahmanyam is a lamentable failure. As usual he indulges in vitriolic personal attacks and fails to engage in any academic or intellectual discussion.[13]

It is noticeable that the western Indian Ocean, or the incorrectly named 'Arabian Sea', seems to be more popular at present than the eastern ocean, that is the equally incorrectly named 'Bay of Bengal'. Indian Ocean information is increasingly to be found in cyberspace also. New and important sites include Gwyn Campbell's 'Indian Ocean World Centre',[14] and a very extensive site sponsored by the European Union, 'The Europe-India Maritime History Project'.[15]

This very sketchy survey does show that Indian Ocean studies are moving ahead. However, we need to think analytically about what we are doing. We have to demonstrate that the Indian Ocean has enough commonality to be a viable focus of analysis. If it is too discrete, if there are no connections around its shores, then it will not be integrated enough to make up a viable area of analysis. Thus, our task is to find links and connections and patterns all over and around the ocean. Many scholars have already produced excellent histories of trade in the Indian Ocean, and there is no doubt that this was an important mechanism for holding the ocean together. Others are beginning to study religious connections, especially Islamic ones. Anne Bang and David Parkin come to mind here.[16] Some are concerned to get beyond rather dry lists of trade items and introduce some sense of the sea in their work: a whiff of ozone perhaps. My own very preliminary investigations of the social worlds of European and Asian ships in the early modern period point to much more harmony and cooperation on board the latter, possibly a result of the presence of many more women and children on Asian ships as compared with European ones. Another possibility is to look at the coasts of the ocean, and consider if we can find elements of commonality here.[17] One could look at foods, houses, occupations, religious beliefs and kinship patterns to see if someone in western Malaysia had anything in common with a resident of Lamu. One of my on-going research projects looks at the role of cultural brokers. It is a matter of investigating how communication was achieved in the port cities all around the far-flung shores of the ocean despite cultural and linguistic differences. This in turn speaks to our continuing search for elements of commonality, to our attempt to show that there was in the early modern period something identifiable as an Indian Ocean World.

It is here that the present book makes a real contribution, for it advances considerably our knowledge of coastal societies in India. These are mostly micro-studies, all of which provide empirical data which can be fed into a more analytical and wide ranging investigation of the littoral of the whole ocean. Many of these chapters started life as presentations at the workshop 'Coastal Histories: Society and Ecology in Coastal India, Sixteenth–Eighteenth Centuries', initiated by Professor Yogesh Sharma and sponsored by the Centre for Historical Studies of Jawaharlal Nehru University. The workshop was held on 13 and 14 February 2007, and I was honoured by being invited to present an inaugural address and to act as General President for the entire proceedings. Several postgraduate students also presented papers, which are not included here, but it is most encouraging to see that there is an emerging critical mass of scholars at JNU, both staff and students, who are interested in maritime history. I have great pleasure in commending to the world of scholarship this excellent collection of research papers.

NOTES

1. *Journal of World History*, I (Spring 1990), 1: iii–v.
2. Fernand Braudel, *The Mediterranean and the Mediterranean World in the Age of Philip II*, London, 2 vols., 1972, p. 19.
3. Peregrine Horden and Nicholas Purcell, *The Corrupting Sea: A Study of Mediterranean History*, vol. I, Oxford, 2000.
4. Paul Butel, *The Atlantic*, London, 1999; David Kirby and Merja-Liisa Hinkkanen, *The Baltic and North Seas*, London, 2000; Michael Pearson, *The Indian Ocean*, London, 2003, and paperback edition November 2007.
5. Karen Wigen and Jessica Harland-Jacobs, eds., 'Oceans Connect', *Geographical Review*, 89, 2 April 1999, pp. 161–313.
6. http://www.history.ac.uk/ihr/Focus/Sea/index.html
7. 'Oceans of History', *American Historical Review*, 111, 3, June 2006, pp. 717–80.
8. Jerry H. Bentley, Renate Bridenthal and Karen Wigen, eds., *Seascapes: Maritime History, Littoral Cultures and Transoceanic Exchanges*, Honolulu, 2007.
9. Markus P.M. Vink, 'Indian Ocean Studies and the "new Thalassology" ', *Journal of Global History*, II. 1. 2007, pp. 41–62.
10. Sugata Bose, *A Hundred Horizons: The Indian Ocean in the Age of Global Empire*, Cambridge, 2006.
11. H.P. Ray and Edward Alpers, eds., *Cross Currents and Community Networks: Encapsulating the History of the Indian Ocean World*, New Delhi, 2007.
12. Edward Simpson and Kai Kresse, eds., *Struggling with History: Islam and Cosmopolitanism in the western Indian Ocean*, London, 2007.

13. *Maritime India*, New Delhi, 2004, which contains *Rival Empires of Trade in the Orient, 1600-1800* by Holden Furber, *Maritime India in the Seventeenth Century* by Sinappah Arasaratnam and *The Indian Ocean: a History of People and the Sea* by Kenneth McPherson.

14. http://indianoceanworldcentre.com/index.html

15. http://www.edumaritime.org/

16. Anne Bang, *Sufis and Scholars of the Sea: Family Networks in East Africa*, 1860–1925, London and New York, 2003; David Parkin and Stephen C. Headley, eds., *Islamic Prayer across the Indian Ocean: Inside and Outside the Mosque*, London, 2000.

17. For a preliminary discussion of the littoral see my 'Littoral Society: the Concept and the Problems', *Journal of World History*, 17, 4, December 2006, pp. 353-73.

Acknowledgements

This book is based on the proceedings of the seminar workshop organized by the Centre for Historical Studies at Jawaharlal Nehru University in February 2007. I, along with other organizers—Arvind Sinha, Pius Malekandathil and Joy Pachuau—express our sincere gratitude to Professor Michael Pearson for performing the role of general president of the seminar. His constant encouragement and extensive suggestions to scholars during and after the seminar proved very valuable in the making of this book.

Thanks are also due to Professors Dilbagh Singh and Aditya Mukherjee for granting funds from U.G.C. supported schemes sanctioned to the Centre for Historical Studies. Aditya Mukherjee provided space for the seminar held at the Jawaharlal Nehru Institute of Advanced Studies of which he is the Director. Sincere thanks to several of the History Centre scholars working on themes related to maritime history who also participated in the seminar. Most of them—Anirudh Deshpande, Joy Pachuau, Radhika Chadha, Rashmi, Mahesh Gopalan, Sonali Mishra, Smarika Nawani, Vaibhav Sharma, James Guite, Vikram Harijan, Uma Shankar Pandey and others—are teaching in prestigious academic institutions.

I would also like to acknowledge and communicate my gratitude to my father, B.K. Sharma, I.A.S. (Retd.) and mother Chandramukhi Sharma, who have been constant teachers to me over a lifetime.

YOGESH SHARMA

Introduction

Facets of Ecology and Society
in Coastal India in the pre-Modern Phase

Yogesh Sharma

IN STUDYING COASTAL histories it is important to understand that the coast does not merely denote or refer to the band of territory facing the sea and its wider component, as one moves inland. It concerns and encompasses the deep mass of territory reaching far inland, which was impacted upon by the reach of the sea in many ways ranging from climate, vegetation, patterns of socio-economic formation, and several other facets. As Fernand Braudel pointed out in the context of the Mediterranean that 'The vital force of the sea . . . has repercussions reaching far into the land mass, effortlessly drawing into its orbit all regions that look seawards.'[1] These regions inevitably looked seawards for their environment and change of seasons; for their commercial networks and movement of manufactures; for the arrival of the overseas merchants, priests, travellers, all of which cumulatively helped in turning the wheels of society and the economy.

What gives a coastal region its particular identity and specific regionality occurs or evolves over time, in general historical circumstances. Fernand Braudel in his monumental study of the Mediterranean has a section titled 'Seas and Coasts', which gives attention to the significance of the several smaller seas which together conjoined to form the larger corpus of the Mediterranean Sea.[2] Braudel however does not provide an explanation of what constitutes or characterizes a specific coastal zone. But he refers to what he calls the 'coherence of historical areas', or a sense of identity and self-identity that characterized specific regions. By and large coastal areas and sub-regions had their own distinct character and identity which evolved due to various factors such as commonality in the politico-cultural experience and in the socio-economic formation of the region. Besides, other significant underlying factors such as homogeneity of language and cultural outlook, communicational proximity and topographical bindings were important in forging an identity.

One of the earlier writings pertaining to this important historical zone, particularly in the context of India was by J.C. Heesterman. He pointed out that, 'the littoral forms a frontier zone that is not there to separate and enclose, but which rather finds its meaning in its permeability.'[3] M.N. Pearson's study, drawn from several writings pertaining to different regions of the world, constitutes an authoritative survey and reflection, on the theme.[4] He indicates that the primary importance of the coast arose from the fact that it served as the interface region between the land and the sea, between overseas and hinterland societies, ranging from commerce to culture. Philip Steinberg makes an important categorization by distinguishing the coastal sea corridor from the vast expanse of the ocean, as the primary area of maritime-commercial activity that had a consistent bearing on the coastal regions.[5]

An important observation by Braudel concerning the influence of the sea upon coastal societies and beyond was the dynamism felt by the peninsular regions in particular. He pointed out that the seas tended to have a more pronounced impact on 'the vast peninsular blocks, particularly since they present to the intervening seas coasts of exceptional ability'.[6] Braudel had in mind regions such as the Italian, the Greek and the Iberian peninsulas which thrived with maritime and commercial activity in Europe, across all ages.

The land spaces of the regions referred to by Braudel were considerably less than the continental sweep of the land mass in peninsular India, between the Arabian Sea and the Bay of Bengal. Nonetheless in a similar manner, peninsular India witnessed a surcharge of commercialization which enriched the coastal societies as well as the inland regions, due to the reach of the sea. Several contemporaries made observations about the prosperity of the region and of its potentates who managed to establish control across both the coasts of the great Indian peninsula. The Italian Ludovico di Varthema who came to India in the early sixteenth century (1502-8) was greatly impressed by the power and riches of the Vijaynagar rulers whose kingdom was spread across peninsular India. Describing these ocean rimmed, coast-endowed dominions he noted, 'The King of Narsinga [i.e. Vijaynagar] is the richest king I have ever heard spoken of . . . his realms are placed as it might be the realm of Naples and also Venice, so that he has the sea on both sides.'[7]

Varthema partly attributed the magnificence of Vijayanagar due to its territorial command over the coasts on both sides of the empire which gave it significant leverage in the Indian Ocean world. So immensely fabled were the riches and power commanded by its rulers that 'His horse is worth more than some of our [Italian] cities on account of the ornaments he wears. When he rides for his pleasure he is always accompanied by three or four kings and many other lords, and five or six thousand horse.'[8] The great political influence and wealth wielded by the Vijaynagar rulers brings to mind Braudel's observation about ancient Greece, Rome and the Iberian Hapsburgs, that 'Every time the political unity of the peninsula has been achieved it has announced some momentous change.'[9]

The importance of Vijaynagar as a peninsular kingdom endured for long and Bijapur was viewed as a kind of successor state to this legacy. Thevenot who visited the kingdom in 1666 noted 'Bisnagar which was formerly called

Narsingue . . . the provinces thereof crossed from the coast of Coromandel to the coast of Malabar, reaching a great way towards the Cape of Comory.'[10] Thus, control over the sea and the coasts was viewed as a key factor in conferring riches, revenues and power upon the rulers who made use of these elements.

The majesty of the sea was such that entire regions and peoples came to be referred to in the context of the seas and the sea-coasts, which became the principal markers in defining the regions. For example, the Dutch chronicler Van Linschoten wrote at the end of the sixteenth century, 'The Malabares are those that dwell on the sea coast between Goa and Cape Comorin, southward from Goa where the pepper growth.'[11] Though the description was not accurate as it incorporated the Canara Coast along with the Malabar, nonetheless it highlighted the strong identification of the Malabar region as coastal territory, and its inhabitants as coastal people.

For the Europeans who came to India the coast became the yardstick for describing principalities and peoples. The interior was more frequently plagued by shifting boundaries than the coastal regions which had a sense of permanence, and thus enabled descriptions to be simple and effective. For example, the Portuguese chronicler Tome Pires noted, 'The kingdom of Cambay [i.e. Gujarat] must have seventy or eighty leagues of coast, and inland it is not large but noble and rich and civilized.'[12] While the dimensions of the coast were somewhat specific, those of the interior tended to be less precise. Tome Pires similarly described Malabar as being 'a hundred and ten to a hundred and twenty leagues along the coast, and inland it is about five leagues wide in some places and fifteen in others'.[13] The European trading organizations which had a strong maritime orientation, focussed on the coastal regions as the primary zone of activity. For the Portuguese seaborne empire, based at Goa, the conception of its territory in western India was entirely coastal in perspective. Van Linschoten noted 'The coast from Goa to Daman or the turning into Cambaia is called by those of Goa, the Northerne Coast and from Goa to the Cape de Comorin is called the southern coast.'[14]

In the vocabulary and terminology of the contemporary writers the coast figured constantly, and territory was frequently referred to from a coastal perspective. The English doctor John Fryer, writing in the 1670s referring to the politico-territorial ambitions of Shivaji noted that his extensive coastalized possessions were no longer adequate for him. He referred to the Maratha proto-state under Shivaji as, 'a diseased Limb of Duccan, impostulated and swoln too big for the Body . . . all the Conchon [i.e. Konkan] being little enough for him extending along the sea coasts 250 Leagues that is from Balsore [Bulsar] Hills to the river Gangole [in Dharwar]'.[15]

THE RICH COAST

There were several facets about coastal societies that gave them a somewhat different orientation and identity from the continental interior. The Frenchman Thevenot remarked perceptively about the strong interface between human society and the sea, thus indicating the significance of the coastal regions. He noted, 'The most considerable [or important] part of Gujarat is towards the Sea, on which the towns of Surat and Cambay stand, whose ports are the best of all Mogulistan.'[16] European contemporaries had a strong coastal perspective in evaluating the financial and commercial strength of the Indian principalities. The French director Francois Martin was unequivocal in his assessment that Bengal was the most important 'government' or 'subah' of the Mughal dominions, while Gujarat was the second most significant province.[17] Thevenot, however, was of the opinion that Gujarat was 'the pleasantest province of Indostan though it be not the largest'.[18]

The pull of the sea was such that it exercised a major influence upon the lives of the ordinary inhabitants in the coastal regions. Referring to the Canara Coast around Goa, Van Linschoten observed 'their chief dwelling places [or settlements] are on the Sea side in the countries bordering upon Goa for that the palme trees doe grow upon the sea coasts or upon the banks on the river side'.[19] The English doctor Fryer was able to perceive a kind of interface between ecology, economy and coastal societies which gave them a distinct advantage and specific identity. He noted 'On the backside of the town of Bambain and Maijm [Mahim] are woods of Cocos . . . these Hortoes being the greatest Purchase and Estates on the Island for some miles together.'[20] Fryer recorded that a little further 'lies Messegoung [i.e. Mazgaon] a great Fishing Town, peculiarly notable for fish called "Bumbelo", the sustenance of the poorer sort, who live on them and "Batty", a coarse sort of rice, and the wine of the Cocoe called Toddy'.[21] This convenient affordability of rice, fish and wine was rendered further palatable by 'the Washes of the Sea [which] produce a Lunary Tribute of Salt, Left in Pans or Pits'.[22]

Settlements along the sea-coasts were usually considered blessed and relatively safer from the devastating effects of famine, which struck with crippling severity every other decade or so. The Jesuit priest John Cabral described the flourishing Portugese settlement of Hughli as one 'one of the most populous in the East and as it was a refuge against famine'.[23] The alternative food resources provided by the Sea coastal region saved them from famine which was an ever recurring reality. Pietro Della Valle who spent some time in the western coastal regions, remarked perceptively about the common living standards: 'their dyet for the most part is nothing but rice . . . of which is found everywhere in abundance, so that everybody even of mean fortune keeps a great family.'[24]

THE PROSPEROUS COAST

The coastal regions of India were by and large endowed with high levels of fertility, having the capacity to generate considerable food surpluses. The English sailing Master Thomas Bowrey described the Gingelli coast (comprising northern Coromandel and the southern Orissa coast) as, 'the most pleasant and commodious sea-coast that India affordeth, beinge a most delicate Champion

Land and one of the most fertile lands in the universe'.[25] Elaborating on the high agricultural productivity of the coast zone Bowrey noted: 'They have annually 3 crops each yieldinge great Encrease, and nothwithstandinge the sea cost, as also the inland be extraordinarily populous yett they transport above 10,000 gorse[26] of graine yearly [to other regions].'[27] Undoubtedly the capacity to generate surpluses for food deficit regions was considerable.

Similarly, most of the other coastal regions were also acknowledged to be richly productive. Francois Bernier was unambiguous in his assessment of Bengal as the most fruitful country in the whole world. Its tremendous output of rice and sugar which catered to several other regions and overseas countries, made it far surpass Egypt as a producer of food stocks.[28] Thomas Bowrey noted that Bengal's exceptional fertility was augmented by its deltaic character, 'the great River of Ganges and the many large arms thereof'.[29] The ecological interface between the riverine and coastal region created very 'Fruitful lands affording great plenty'.[30] An indication of high food productivity in some coastal regions was the considerable export of food and grain from these areas to other regions. Fryer noted that during his voyage to the Canara coast he came across 'the Portuguese Armado from the south with two hundred paddy-boats with their convoys', which was on it way to Goa and to the Portuguese settlements in Gujarat.[31]

The Englishman Thomas Herbert similarly remarked about the Malabar region that 'no part is without abundance of fruits and provisions.'[32] Tome Pires was likewise struck by extraordinary profusion of trees: 'There are countless palm trees and arecas along the coast of Malabar',[33] which gave the region commercial vitality.

The coastal regions frequently had a higher concentration of population than the interior, as they provided livelihood to several socio-vocational groups—fishermen and seamen, agriculturists, those engaged in horticulture, manufacturers, weavers and merchandisers. The fishing communities and their settlements were a vital part of the habitational space and demographic profile in all parts of coastal India. For example, the French doctor Jaques Dellon described Calicut as a sprawling city of thatched houses with a grand bazar spread over five streets, and a village of Machua fishermen towards the coast.[34] The English doctor Fryer described the growing settlement of Bombay where 'Confusedly live the English, Portuguese, Topasses, Gentues, Moors,[35] Cooly Christians [Kolis who were] most fishermen'. Thomas Salmon who served at Madras in 1699 remarked about its social topography: 'There is a little suburb to the southward of the "White Town", inhabited only by black watermen and fishermen.'[36] Such a sociological pattern was replicated in almost all the coastal settlements in every region of India. In fact, many of the new Indo-European towns that grew up in the seventeenth century, such as Madras or Pondicherry, were originally nothing more than fishing hamlets.

The fishing communities were a conspicuous and visible socio-vocational segment of coastal societies. Their importance was enhanced by the fact that they provided the sailors and seamen for the maritime fleets. To indicate their significance in coastal society, in January 1686 twenty-two heads of the principal working castes at Madras, such as weavers, washermen, painters, Chettis, Komathis, etc., gave a representation to the English Council not to impose a house tax on the 'Black Town' residents. Of these, two caste leaders represented the Muckwa and Patnavar castes,[37] which were principally fisher communities, many of whom also served as boatmen and worked at the dockyards for the English company. Thomas Lockyer wrote that these people who received poor wages, and were not infrequently subjected to the fury of nature and cyclonic storms had a simple philosophy of life—to search for happiness and joy in the routine ordinariness of life. He remembered them as 'merry birds' who sang aloud in energetic chorus to the rhythm of their stroking paddles in choppy waters.[38]

A conspicuous aspect of coastal societies in India was that they supported a vast manufacturing sector where large segments of the population were engaged in the manufacturing sector as weavers, painters, washermen, etc. To indicate the scale of the non-agricultural sector in coastal areas, a French Company report of February 1688 from Pondicherry stated that the proximate 'Canton' or small district was inhabited by over 30,000 textile workers who were exclusively engaged in cloth manufacturing.[39] Both Francois Martin and Streynsham Master recorded during their travels in the Coromandel coastal region that several townships and villages had a very large concentration of artisan population which catered to the extensive demands of the European Companies and Asian buyers.[40] Similarly, near Surat the Englishman Salbancke noted that there were villages like Sobay that 'consisteth altogether of spinners and weavers, and there is much cotton cloth made.'[41]

COASTAL ENVIRONMENT AND MORTALITY

In studying coastal history it is important to bear in mind environmental factors were particularly important in shaping material life. Several factors such as the shape of the landmass, the topography, the wind systems, the 'monsoons' were all important in determining the region's climate, seasonality, sailing conditions and the structure of everyday life. The shipping traffic on the coasts of India, and indeed throughout maritime Asia was determined by the wind pattern which blew in a predictable manner. Throughout maritime Asia the sailing people, merchants and the coastal inhabitants were quite familiar with this, as their socio-vocational activities were determined by these factors. The French voyager Jean Baptiste Tavernier remarked about the strong impact of the wind system on navigation in Asia. He noted: 'Navigation in the Indian seas is not carried on at all seasons, as it is in our European seas, it being necessary to take the proper season, outside which no one ventures to put to sea.'[42]

The Englishman William Finch who spent some time in India between 1608-11 observed the navigational pattern on the west coast of India. He recorded 'Monsoons [winds] here serve for the south [towards Indonesia] in April and September; and for Mokha [towards West Asia] in February and March.'[43] Finch further noted about the pattern of inward bound vessels towards the west coast of India: 'from the South [i.e. Indonesian zone] ships come hither in December, January and February, and from Mocha about the fifth of September, after the raines: from Ormus for the coast of India in November'.[44] The rainy season brought shipping activity to a halt as the seas and particularly the coasts received violent thrashing from the rains and storms that lashed the region. As the English chaplain Edward Terry noted: 'In the long season of the raine [May-August] and a little before and after it, the winds are commonly so violent that there is no coming but with great hazard into the Indian sea.'[45]

The topography and the shape of the land mass was also important in determining coastal life. For example, the elevated 'ghats' or hill ranges along the coasts of peninsular India partly obstructed the wind and navigational conditions. Tome Pires noted with reference to the Coromandel coast that the high rising southern hills and plateau prevented the 'South-west and west winds [to] blow in the kingdom of Narsinga [i.e. Vijaynagar]—that is if you are coming from Ceylon to the coast of India with fresh winds blowing from the above, they cease blowing when you reach the Choromandel coast.'[46] In fact, the Coromandel Coast did not receive the south-west monsoonal rains; and it was the retreating north-east monsoons that brought winter rains to the Coromandel.[47]

The Portuguese chronicler Tome Pires remarked about the variation in climate and rainfall between the two coasts on either side of the peninsula—Coromandel and Malabar. He observed 'Since Malabar is lacking in dry [land] winds it is fresh [i.e. fertile] and gracious, while the province of Choromandel which lacks wet [sea] winds is sterile.'[48] While the Coromandel was not sterile, nonetheless the inadequate south-west monsoons it received made it relatively dry, compared to wet western coastal regions, that is the Konkan, Canara and Malabar coasts.

The climate of the tropical coastal areas in India could be harshly oppressive and hazardous in the hot summer months. In fact the sharp heat combined with high humidity could at times be fatal in the hot season, particularly for the Europeans who were not accustomed to such a climate. The English company official Peter Floris who was at Masulipattam in May 1614, received disconcerting news about his subordinate Thomas Essington, who was posted at the proximate port town of Narsapur. The Young Englishman fell seriously ill 'of a Suddayne heate. . . He had some byles [i.e. boils] about him which at this time of the yeare are very common, among which one verye great one uppon his shoulder, which would not breake, and this was thought to be the cause of the heate.'[49] By evening he was dead.

It were not only the Europeans who were affected by the harshness of the climate. The English factor William Methwold who served long tenures in

India, remarked about the hazards of daytime travel during the hot season that 'many of the Natives are in their travel suffocated or perish [by the heat]'.[50] Methwold further narrated about the vulnerability of the Europeans during his tenure at Masulipattam: 'And of the Christians, a Dutchman as he was carried in his palamkeen and Englishman, walking but from the towne to the barre [i.e. harbour], little above an English mile dyed both in the way.'[51]

The stark glare of the sun and the heat felt on the open beach front, in the tropical coastal regions, could in fact be used as an instrument of physical torture. The French doctor Jacques Dellon remarked about the peculiarly harsh method of punishment administered to criminals in the Canara coastal region, through the harsh facet of mother nature. He noted: 'The manner in which criminals are punished in the Canara appeared to me appropriate to be remarked upon: They are exposed naked, feet and hands tied together upon the sand under the harsh sun, to be consumed little by little by the heat and by the insects.'[52]

Several of the coastal habilitational areas were notorious for their aggravated rates of mortality, particularly during the monsoonal season, and Bombay was reputed to have a damaging climate. The English Chaplain Ovington who spent four months at this new English settlement during the rainy season in the late 1680s recorded that 20 out of the 24 of their ship's passsengrs were buried in Bombay as were 15 of the ship's crew, thus amounting to a total 35 deceased from a single ship in four months.[53]

It was a tragic fact of coastal and maritime history that large numbers of people died on voyage, and many more lay sick and dying. Many of them travelled vast distances only to be buried on distant shores, soon after arrival. The Frenchman Francois Pyrard who spent some time at the famous large Jesuit Hospital at Goa in 1608 noted, 'the arrival of the fleet from Portugal sometimes brings more than 300 [sick persons];' many of whom died subsequently.[54] The Portuguese fleet of 5 ships which arrived at Goa in 1685, had sailed with 2500 persons. Of these 720 persons died during the passage and 600 died after reaching the Indian coast.[55]

Many coastal settlements were situated in marshy malarial terrain, having brackish waters, which made conditions of existence somewhat precarious, leading to high rates of mortality. For example, the grand port of Masulipattam was surrounded by low lying swamps which rendered the site unhealthy. While the Indians were relatively more hardy and more accustomed to the local climate, the Europeans tended to be severely vulnerable to adverse effects of the climate and environment. This factor led to the coinage of poignant phraseology concerning life and death amongst the contemporaries that was remarked upon by the chaplain Ovington. He noted that due to the high incidence of mortality, 'which common fatality has created a proverb among the English there that two monsoons are the age of a man'.[56] In fact, Theon Wilkinson's perceptive study of the social history of mortality amongst the Europeans in India, which focusses on the nineteenth century derives its title

Two Monsoons from the social vocabulary prevalent in seventeenth century coastal India.

The sea coast at times also witnessed facets of human trauma and pathos particularly during times of calamity. Pre-modern societies such as in India frequently faced famines of varying intensity, as it was mainly dependent upon the monsoons for its annual rainfall which was vital for its agricultural productivity. The inadequacy or absence of rainfall for over two seasons could lead to severe drought and shortage leading to large-scale mortality. It led to human trafficking, particularly that of children in very sad circumstances, in the east coast region. The English factor William Methwold, who was posted at Masulipattam noted that in the 1630s combination of civil war and famine caused 'such extreme want . . . that parents have brought thousands of their young children to the sea-side, selling there a child for five fanoms [small gold coins] worth of rice, transported from thence into other parts of India, and sold again to good advantage'.[57] The parents sold their children, not mainly in expectation of money, but with the hope that the children would survive elsewhere, rather than perish in the famine.

The hugely destructive famine of the 1680s highlighted the problem of export of slaves from the Coromandel coast to other regions of Asia. In 1687 the French director Francois Martin who was located at Pondicherry recorded: 'The English and the Danish have been doing large-scale trade of slaves whom they send to Achin [in Sumatra] for considerable profit. They can be purchased on the [Coromandel] coast for very little because of the widespread misery.'[58] It was estimated that about twenty thousand persons were sent as slaves to Sumatra during this famine phase by the English and Danish companies.[59]

This slave trade from coast to coast involved deep pathos and misery, but also carried with it the hope of survival and a new lease of life. The slaves arrived in miserable condition on the shore of the country where they landed. Maxmillan Bowman, an Englishman who witnessed the arrival of Indian slaves at Achin during the earlier famine, recorded in October 1646, 'There came uppon this small vessel upward of 400 slaves so hunger starved that they were scarcely able to crawl when they brought them ashoare.'[60] They were sold at 5 to 6 taels per head, at a discount for half the usual price, due to their wretched physical condition.[61]

THE BRACKISH COASTAL WATERS

The seas conferred innumerable bounties upon the coast by bringing food, wealth and prosperity to its inhabitants. But the nature and character of the seas was so overwhelming that it added more than a pinch of salt to the fresh waters along the coast, making it brackish and difficult to drink, which posed a severe challenge to coastal societies. The ingress of the sea water, deep into the mainland, and its tidal push into the rivers created a problem in most regions. The English doctor John Fryer who spent a long time in the west coast regions remarked that at Bombay the wells were all affected by sea water, with their water 'hardly leaving its brackish taste'.[62]

It was not only the wells in the coastal regions that were touched by the seas. Above all, the rivers which served as the vital lifeline to these regions and several cities, succumbed to the salty push from the seas. The river Tapti on whose banks was located the grand port city of Surat was known to be brackish. The Frenchman Jean de Thevenot who visited Surat in 1666 took notice of this aspect. 'The River of Tapty is always brackish at Surrat, and therefore the inhabitants make no use of it, neither for Drink nor watering of their grounds but only for washing their bodies.'[63] Similarly, John Fryer, who was quite environment conscious, noted about the poor quality of the waters on the east coast at Masulipattam. He noted: 'The water they drink they dig for, not that they are without Rivers, but they are brackish.'[64]

A comprehensive explanation about the problem of brackishness in the areas near the coastal regions was provided by Pietro Della Valle who studied this aspect during his travels, along the west coast. He joined a caravan of people going from Broach to Cambay across the wide marshy creeks, in February 1623. He found that the river water was saltish 'in four places where we forded the running water of the river, which nevertheless is salt there, the great strength of the sea overcoming that of the river'.[65] During his subsequent travel to Daman and further along the coast, Della Valle was struck by the physical and environmental impact of the sea. He recorded that 'the tide of the sea at ebbing and flowing being here [west coast] very strong and overcoming that of the rivers, hence it comes to pass that 'tis hardly perceived that they have any stream or no; and the water going very far into the land comes likewise to be salt'.[66]

It was this push of the sea's salt water into the river channels deep inland that made the rivers very brackish, and also created salinity in the adjoining areas of land and the underground water. Della Valle further pointed that the first appearance of the salty waters deceived the wisest as it deceived him in his 'judgement of the rivers making me take them for arms of the Sea'.[67] The problem was not confined to a few pockets and regions but was extremely widespread, as the entire western coast was densely intersected by rivers and rivulets. As Della Valle pointed out, 'They are almost innumerable upon all the coast of India,'[68] which gave a brackish character to the waters in the coastal areas. Thevenot noted at the port of Coulam in the Malabar region, 'The tide runs a great way up in the River.'[69] The mighty seas thus carried along with their waters a minor curse for the coastal people. They urged them to find other sources of drinking water to satiate their thirst, because the embrace of the sea upon land could be bitter for the waters. The answer to the problem was found in building huge artificial reservoirs and dams to store fresh rain water on a massive scale. It led to much architectural creativity and excellence in the construction of beautifully embanked water tanks and step wells in all parts of the country, which became vitally important components of all urban centers, particularly in the coastal regions.

RELIGION AND CULTURE

Coastal societies were particularly amenable to cross cultural influences and the impact of religious traditions that were transmitted along with the movement of peoples across distant regions. The sea-routes carried along not only trade goods and commodities, but also ideas, military ambitions, new technologies, and above all missionaries, belief systems and religio-cultural traditions, which created a sharp impression across distant lands. It is in this context J.C. Van Leur argues that oceanic routes served to lay down the material foundations of cultural history.[70]

Due to its geographical construct peninsular India reached across the seas, towards the Islamic countries in the west and towards the extensive South-East Asian regions and beyond up to the Far East. Over centuries a multiplicity of religious traditions became implanted in the Indian coastal regions as well as the interior. In these processes, trans-oceanic movements of peoples played an important role. An important aspect concerning the Indian coastal regions was that they were both transmittors and receivers of these different major influences and faith systems. They were visible in the form of the splendid religious architectural monuments, as well as the huge religious congregations, festivities and ritual activities that the sea coasts, the beaches, and the riverside near the coasts were witness to.

The Hindu places of worship were found everywhere in the coastal regions, as if the influence of the seas inspired religiosity. For the coastal people the sea was a divine, mythological and spiritual space where the gods dwelt, just as the mountains and the skies were the abode of the gods in interior societies. The Jesuit visitor Nicholas Pimenta who travelled on the coastal route from Negapattam to St. Thome over twelve days, in 1599 noted, 'so prodigious and innumerable were their Idols in many fair temples, and other lesser Oratories [were] almost without number'.[71] On the way he witnessed huge religious processions where colossal idols were moved around in 'Chariots as high as [Church] steeples by thousands of men setting their shoulders to the wheels.'[72]

Auspicious activities began and ended with a visit to the temple. Even barbarians from distant overseas regions were brought here, without being hindered by ritualistic taboos. The diarist of Vasco da Gama's expedition, which arrived at Calicut in 1498 recorded: 'They were no sooner landed . . . were brought into a sumptuous temple esteemed of great sanctitie.'[73] It was a matter of honour as well as ritual in the coastal kingdom of Calicut, to escort a foreign dignitary to the sacred places, to express their welcome, and to obtain blessings that the visit be cordial and fruitful. Vasco da Gama's, diarist further noted,

'Thence they passed to another temple of like magnificence, and after that to the king's palace.'[74] Divine visitations were appropriately given precedence before a royal audience. But apparently subsequent developments concerning Portuguese presence on the west coast, which were characterized by severe hostilities, indicated that Calicut's temple deities and their priests were not able to ward off the aggressive designs of the Portuguese.

The sea coast was an inspirational space where temples, seminaries and institutions of learning were frequently located. Writing about the Gingelly and Orissa coasts Thomas Bowrey noted, 'They have many . . . large fabricks of stone called pagodas more large and of greater Antiquitie then all the land of the Hindoos doth afford . . . theire most holy and esteemable Pagoda Jno Gernaet [i.e. Jagannath].'[75] The wonderous celebration of religiouisity of the Jagannath cult on the Orissa coast was remarked upon by several contemporaries.[76] Manrique wrote about the 'sumptuous and magnificent processions' involving 'Vast crowds of Pagamism', on the sea coast.[77] The Juggernaut in fact entered the lexicon and vocabulary of the Western world as a symbol of the awesome magnificence of congregational devotion with frenzy, on a massive scale, as practiced on the east coast.

The English doctor John Fryer was much impressed by the religious centre of Gokaran on the Canara coast, where he visited the well endowed 'University of Brachmins'. But he noted somewhat sceptically: 'This though a Principal University can boast of no Bodlean or Vatican; their Libraries being old manuscripts of their own Cabalas or Myteries understood only by Brachmins.'[78] He visited several temples 'coasting along the sea side', and participated in the grand religious processions, since it was the pilgrimage season.[79] Thus, all the coastal regions of India had a strong presence of Hindu religious institutions and places of worship.

Buddhism spread throughout East Asia, by sea and land, though its influence diminished in India, which was its country of origin. Interestingly the Chinese built a great Buddhist pagoda at Nagipattnam to cater to the spiritual, and diplomatic and commercial requirements of the Chinese colony which flourished there from the eleventh till the fourteenth centuries.[80] The Buddhist Chinese legacy survived a long time. In 1688 a Chinese delegation came from Amoy to visit the Black Pagoda and report about it to the imperial court in China. The English governor Elihu Yale provided them with all amenities for their voyage from Madras to Nagipattnam in the hope that the English Company would get favourable treatment at the sea-ports of China.[81] The French governor of Pondicherry, Francois Martin recorded that 'the Chinese in earlier times inhabited Nagipattnam and were known to have been masters of the coast . . . there still exists a [Chinese] Pagoda.'[82] Thus, Buddhists and Buddhism kept travelling back to its country of origin, via the oceanic routes.

The expansion of Islam from the seventh century onwards rapidly carried it across the seas to the Indian coastal regions, leading to the growth of colonies of expatriate Muslim merchants, in several parts of coastal India. Over time the process of assimilation, conversion and inter-mixing led to the emergence of several Indian Muslim communities such as the Mappilas, Navayats, Bohras, Memons, Chulias, Labbais, Marraikkars, in different regions of India. The Italian traveller Varthema who visited the west coast regions at the beginning of the sixteenth century noted 'in Calicut there are at least fifteen thousand Moors, who are for the greater part natives of the country'.[83] Varthema who visited several ports cities such as Chaul, Dobhol, Goa, Bhatkal noted the strong presence of Muslim merchants and inhabitants in all these regions.

To indicate the strong imprint of Islam, the young French doctor Jacques Dellon who visited Surat in 1670, was surprised to notice that the metropolitan port of the Mughals had over two hundred mosques. Many of these were splendidly magnificent which were visible on the sky-line of the imperial port city.[84] The Italian Pietro Della Valle noted the constant presence of Islamic places of worship in the coastal towns of Gujarat particularly at Cambay. Travelling along the coast inland, from Cambay to Ahmedabad, Della Valle came across several 'Mahometan Meschita or Temple, unroofed and without walls about, saving a little wall at the front and a place markt where prayers are to be made; of which sort Meschitas many are seen in India, especially in the country.'[85]

The rapid expansion of Islam, led to the creation of places of public worship everywhere. The austere, simple nature of Islam facilitated the emergence of places of worship without necessarily being ostentatious. The new settlement of Pondicherry, which was founded in 1674, already had a mosque, indicating that a small populace of Muslims was living in the tiny township.[86] The French priest Abbé Carré noted that the port town of Gogha was inhabited 'half by Moors and half by Hindus'.[87] Similarly, the Dutch chaplian Baldaeus noted at the Kerala coast in the mid seventeenth century 'Cananor is a populous city, inhabited chiefly by rich Mahometan merchants.'[88] Islam gathered many adherents, and also the spiritual following of many from the other faiths. Pietro Della Valle who visited a grand Muslim saintly shrine at Cambay noted; 'there is a great concourse of people not only Mohametrans but likewise Gentiles.' He also pointed out that in this process of worship, 'Some Rite of theirs [i.e. Gentiles] hath adhered to the Mohametans.'[89] By and large Islam in the coastal regions tended to be more adaptive than in the interior regions where the stronger presence of state agencies and theological personnel tended to favour a more strict form of doctrinaire Islam.

Christianity established itself in India after the founding of the Syrian Christian Church tradition on the Kerala coast, from the first century AD onwards. The arrival of the Portuguese along with the Catholic missionary orders in the sixteenth century strengthened the process of the implantation and expansion of the Christian faith, particularly in coastal India. Caesar Frederick who visited India from 1563 to 1581 took notice of the major impact of Christianity in southern India. He recorded: 'alongst this coast [south Malabar] neere to the water side, and also to Cao Comorin down to the lowland of Chialao [i.e. the Fishery coast] which is about two hundred miles, the people are as it were all turned to the Christian faith'.[90] The fishery coast in southern Coromandel experienced the strongest impact of Christianity following the dedicated efforts of St. Francis Xavier and the Jesuit missionaries, who converted Paravan and Muckwa communites to the Christian faith. When the missionary Nicolas Pimenta visited the coast in 1599 he recorded: 'There came from the [nearby] regions about sixtie thousand [persons] into these tents of Fishermen bringing all their families. Our Priests say Masse in the Churches ereted on the shore, appease tumults and have care of good order.'[91] Thus, the sea and the coast inspired religiouisity and several Christian communities grew along the littoral. Pimenta noted that in the coastal dominions of the nayak of Madurai there were thirty-three churches which he visited. To shepherd these neo-Christians the Portuguese priests functioned as preacher, teacher, caretaker and community elder all rolled into one.

The phenomenal omnipresence of churches in the Portuguese territories in western India served to integrate the countryside to the Portuguese coastal cities, bringing them within the ambit of Portuguese control. Pyrard Laval pointed out that Goa itself had over fifty churches. 'The number of churches there is a marvel: there is no square, street or cross-roads without one.'[92] John Fryer recorded during his travels in the countryside around Bombay, 'We found everywhere provided with churches.'[93] He noted that Thane alone had seven churches,[94] while Bassein had 'six churches, four convents and two colleges'.[95] The Italian Gemeli Careri noted that the small town of Casabo which served as a country retreat for the 'fidalgos' of Bassein had nine churches and monasteries belonging to different missionary orders.[96]

Thus, all religious traditions flourished in the coastal regions, in proximity of the oceans and seas, which acted as carriers of these belief systems to distant shores.

WORSHIP OF THE SEAS

For the coastal people the seas were a sacred space, just as the mountains and the skies were the home of the Gods in interior societies. In 1670 the Spanish priest Domingo Navarette spent a night at a Hindu temple dedicated to three gods, on the sea coast near Masulipattam. He was informed by the temple priests that these deities had come from far away realms, 'upon the waters of the sea',[97] which were thus a divine space sanctified by the Gods. For the Gentile community whose polytheistic representation of religion comprised hundreds of Gods and deities, the mighty seas were certainly an object of devotion and reverence. According to religious tradition, as narrated in the 'Ram Charitra Manas' by Tulsi Das, when Lord Rama sought to traverse the sea, with his army, to go to Sri Lanka, he prayed for three days to the Sea God and sought his permission to cross the waters. In a society where worship of nature gods such as Varuna (God of the Winds), Agni (Fire Lord), Indra (Rain

God) was a way of life, it was logical that there be a tradition of prayers to the seas. In fact, adherents of the monotheitic Islamic faith also retained a sense of devotion for the patron saint of the sailors, and guardian of the seas, the mythological Khwaja Khizr, who was revered in coastal India.[98]

Prayers and ritual offerings were necessary to seek the blessings of the mighty waters which governed the destinies of individuals, communities and particularly the coastal societies in all regions. The Icelander, Jon Olaffson who served at the Danish Company's port town at Tranquebar in southern Coromandel in the early seventeenth century noted a religious ritual observed by these coastal people. He recorded 'The last week before Easter, three of their gods are conveyed down to the shore with much drumming and trumpeting and water is carried up from the sea in splendid gold vessels and poured over the gods.' As per the second part of the ritual, 'the same water is scattered over the sea by the priest . . . it is now consecrated and [becomes] holy water so that there shall be better luck with fishing and greater safely for fisherman.'[99]

This popular religious ceremony was addressed to appease the gods to confer upon the coastal people the benevolence of the seas and of the forces of nature. The English captain Peter Floris recorded a similar ceremony he observed in the northern Coromandel region at Masulipattam in 1614: 'The Gentives had a feast which feast is three times a yeare when . . . both men and women washe themselves in the sea, which they take for a great indulgence, among whome also the Bramenes and the Cometis [i.e. merchant caste] are comprehended.'[100] The observation by Peter Floris indicated that this concourse of festivities and prayers also involved the priests and the trading castes, as well as the usual common people, such as the sailors, boatmen, fisher folk and others who were part of coastal society.

Not all prayers were always addressed to the sea gods, though the site of the seashore or the river bank was usually the commonly accepted sacred space for the conservation of the festivities and worship. The ceremony described by Peter Floris was part of the cycle of Hindu religious practices which addressed different gods and all gods as per tradition, seasonality and ritual norms. Such prayers were observed in all regions of India. A distinguishing facet of these public prayers and festivities, particularly in coastal regions, was the particular sense of devotion attached to the waters—oceanic, riverine and in the sacred tanks. For example, during his stay at Goa, Pietro Della Valle recorded in 1623 that in the month of August 'the Gentile Indians kept a kind of festival to which a great number came to a place in Goa which they call Nave . . . in pilgrimage from countries to wash their bodies here . . . and casting Fruits, Perfumes and other things into the water as it were in Oblation to the Deity of the Water.'[101]

Prayers to the seas were performed by individuals or by groups, and by the community in different ways. Prayers by individuals and families were person specific. In other words, they were meant to seek the blessings of the seas for certain individuals and their ship which were in passage to or from distant lands. Second, prayers by groups of people were performed to obtain the favour of the elements for specific ventures, such as the launching of a ship, or a voyage in group partnership. Third, there were the community prayers by masses of people which were generally a part of community festivity and celebration. It also symbolized the manifestation and expression of the coastal culture, in the form of ships and boats being decorated and raced in the waters, to the joy of the people on the shore, and in the vessels. These festivities also had a secularist dimension as they involved the participation of different communities—Gentues, Muslims and others who lived by the seas and flourished in consequence.

RITUAL OFFERINGS TO THE SEAS

The paradox about the seas was that they were a major provider of wealth prosperity and livelihood, but they were fiercely unpredictable and brought much destruction, misery and loss of life to coastal societies. From the mighty merchant magnate to the poor fisherman the only answer lay in prayers, and hope that the seas would confer their blessings upon them.

The Frenchman Jean de Thevenot who visited western India in the 1660's noted: 'All the people of that coast [Gujarat, Konkan, Canara regions] are much given to seafaring, so that the Gentiles offer many times sacrifices to the sea, especially when any of their kindred or friends are aboard upon a voyage.'[102] Thevenot who witnessed this sacrifice or prayer being performed recorded that the ritual was ornate yet quite simple. He saw a woman carrying a basketful of food and fruit; 'three men playing upon the Pipe and Drum accompanied her, and two others had each on their head a basketful of meat and fruits; being come to the sea side they threw into the sea the vessel of straw after they had made some prayers and left the meat they brought with them upon the shoar, that the poor and others might come and eat it'.[103] Thevenot's description regarding these ritual prayers was quite interesting, though, by and large, Gentile' religious observations did not involve offerings of meat, but it was not unknown. Thevenot added that such offerings were not performed exclusively by the Gentiles, and he had 'seen the same sacrifice performed by Mahometans'.[104]

There was something simplistically beautiful and naive about the ritual of prayer to the huge and mighty ocean. A group of believers standing on the edge of the sea with their flowers, offerings and music, chanting hymns and prayers in small voices, seeking kindness and favours from the deep waters. It happened regularly in every region of coastal India.

The maritime rituals of prayers, festivity and feasting were colourful, yet poignant, because they concerned the safety of ships and passengers going to distant lands. Elaborate prayers and fanfare was witnessed before the launching of a ship. Van Linschoten noted that the heathen merchants of Goa who were very rich and frequently went out to sea: 'When they will make a voyage to sea they use at least fourteene days before, sounding of trumpets, and to make fiers (also beat pots) that may be heard and seen both by night and day, the ship being hanged about with flagges wherewith (the say) they feast their Pagod [i.e. Gods] that they may have a good voyage.'[105]

Members of all communities—shippers, officials, merchants—were involved in celebration and prayers in coastal India. The English captain William Hawkins during his stay at Surat in 1608 recorded: 'I was invited by Hagio Nazam [Haji Nizam] to the fraughting of his ship to Mocha, as the custome is, they make at the fraughting of their ships great feasts for the principallest of the Towne.'[106] The feast was held on the river front in appropriate symbolism to the waters. Hawkins noted that on the occasion two Portuguese vessels sailed in to collect their refreshments, and also their taxes, as the controlling maritime authority in coastal India.[107] The maritime and trading ethos was so important in coastal India, that celebrations were also held when the ships returned to port. Van Linschoten noted that 'The like [prayers] doe they at theire [ships] returne for a thanksgiving fourteen dayes long.'[108]

The strong monsoonal season made sailing out of the Indian ports completely hazardous and difficult. John Fryer noted the impact of the monsoon across the west Indian coastal regions: 'And now the Rains are set in all hostilities and commerce cease . . . here being no stirring out to sea or travelling in the country.'[109] The ferocity of the rains and storms brought maritime activity to a standstill; while the coastal interior also experienced a slackening of the commercial circuits, as well as military and naval campaigning.

The end of the monsoon signalled a return to maritime activity, which was inaugurated with prayer and festivity by August end. Thevenot noted: 'The Gentiles offer another [prayer] . . . that they call to open the sea, because nobody can sail upon their seas from May till that time; but that sacrifise is performed with no great ceremonies, they only throw Cocos into the sea and every one throws one.'[110] Thevenot witnessed this ceremony, known as the 'Nariali Purnima' observed on the day of the full moon, in a small port in the south Konkan region, where it was observed as a simple religious ritual and not as a major festive celebration. Dr. Fryer also referred to this significant event of coastal India, during his stay in the Bombay region. He observed that at the end of August at the time of the full moon, 'The banyans [trading community] assisted by their Brachmins go in a procession to the sea-shore and offer coconuts . . ., when they make preparations to go to sea, and about their Business of Trade.'[111]

The festival of the coconuts was an important event in the life of the imperial port city of Surat which was celebrated with gusto, involving riverine sport, the promenade of boats, and much rejoicing. The shippers and merchants opened their coffers for festivity and fanfare. Like the harvest festival (Baisakhi) or the kite festival in interior societies, the coconut festival was an event to remember, and to look forward to. The director of the French Company Francois Martin who was posted at Surat in 1680s recorded in his *Memoires*: 'There is a custom among the Gentiles. . . in the month of August to promenade in the river on the boats and to throw a quantity of coconuts in the water;

the ships which are in the river are decorated with pavilions and banners as are the boats; it is an ancient tradition; they believe thereby that the vessels which go to sea will have a happy voyage.'[112]

This public celebration of Gentile rituals at the blessed port of the Mughal empire invited the disapproval of the rigidly austere emperor Aurangzeb, particularly since Muslims were also known to participate in these festivities. His strong inclination towards doctrinaire Islam, and his personal antipathy towards music, and such festivity, led to the policy assertion that a ban be placed on these activities. Francois Martin recorded in August 1683, 'Since the Moors have been much involved in many of these rituals there were orders from the Mogol [emperor] to put a stop to these type of festivities. Consequently, the governor [of Surat] put a ban on them.'[113]

The Mughal's directives and Surat's reaction to these promulgations reflected the kind of dichotomy that could exist between the imperial decree and the concerns of the distant coastal society. The coastal elites which comprised the powerful Muslim officials, shippers and maritime merchants were sensitive to the traditions of local society, and the social mores of its dynamic 'Bania' community. The overwhelming bulk of the merchants, brokers, money-changers and accountants belonged to the trading castes of the Hindu and Jain communities, whose commercial acumen was significantly instrumental in turning the wheels of commerce of the grand port, and of the coastal region. Martin recorded that 'the banians and people of other castes protested against the ban.'[114] The coastal Muslim elite on their part, did not view the maritime rituals as being disturbing or worrisomely contrary to their own religious beliefs. The response towards the imperial decree was lukewarm, and no ban was possible without their support or action on the matter. As Martin recorded briefly, 'but these [orders] were not observed',[115] and simply allowed to lapse, letting the port city continue with it simple festive ways. Martin's diary recording in the following year 1684 indicated that coastal Surat continued with its usual practice: 'The ceremony normally held every year to offer coconuts to the river was held on the 25th [August] . . . it is an ancient tradition of the Gentiles, banned by the Mogol, but nonetehles is also observed by some Mahometans who have their ship going to the sea.'[116] Coastal society had in fact ignored the orders of the grand emperor of the vast subcontinental lands of India.

'JIZIYA AND THE PORT CITY OF SURAT'

Variations between imperial policy the coastal world's own concerns, or between official decree and local compulsions surfaced in other matters as well. In 1679, as Martin recorded 'the Mongol had established the tax of "gisia" on all subjects, who were not of his religion, but also on all [Christian] foreigners'.[117] The levying of the *jiziya* by Aurangzeb signified a further shift towards a more orthodox position in the functioning of the state, which was facing turbulence in the form of civic agrarian disturbances in north India, the Maratha revolt in the Deccan, Rajput dissatisfaction, and other similar problems. The *jiziya* was more in the nature of ideological assertion, rather than a revenue enhancement measure, to rally together some important constituents of the status together. Quite naturally the prosperous port of Surat, and the trading communities of the entire region were logical addressees of this tax enactment.

Subsequent events however indicated that the imperial enactment was one matter, while its implementation and collection in the distant regions such as coastal Gujarat was another matter. The Mughal state perhaps soft pedalled on the collection of *jiziya* which was not done rigorously by the regional and local authorities. Francois Martin recorded in October 1683, more than three years after the imposition of *jiziya*: 'The principal bannia merchants assembled at the beginning of the month upon orders from the court to the governor to augment the "giria" [i.e. *jiziya*]. . . there were many meetings held during the entire month with little success.'[118] The local governor possibly thought of making the principal bannias or their *jati panchayat* known as the *Mahajan* which regulated the community's social affairs, involved in the collection of *jiziya*. There is no clarity on this issue, but it appeared that no effective effort or mechanism was made for this purpose. Martin in fact stated that: 'From the banias and other Gentils revenue from this cess alone, if all of its came to the Mughal's coffers, would be more considerable.'[119] Quite evidently the state machinery was slightly slack or somewhat indifferent about pushing the collection of *jiziya* with severity in these coastal areas, which brought several advantages and riches to the Mughal state and officials.

CHRISTIAN WORSHIP

Prayers at the onset of a voyage and upon returning and reverence towards the seas and forces of nature were expressed and observed by sailing people belonging to all faiths. At Goa the simple tradition of prayers and salutations to the Church of *Nossa Senora do Cabo*, overlooking the bay, was observed by all passing ships to seek benediction. The Frenchman Francois Pyrard de Laval who spent almost two years (1608-1610) at Goa remarked about this simple maritime ritual observed by all passing vessels. He noted: 'All ships, whether galleys or other vessels, entering or departing and whether for the wars or for traffic, whether Christian or other, salute this monastery in passing with their guns.'[120] The holy 'Lady of the Cape' at Goa had come to be acknowledged as the patron deity of the ships, mariners, and maritime passengers, who sailed up and down the bay leading to Goa.

Religion in fact became closely associated with the destiny of ships, as they crossed a thousand dangers to reach distant destinations. At Goa, there was a church mainly consecrated to maritime matters and specifically to a ship which reached Goa, following very perilous circumstances. John Fryer who was quite impressed by the architectural splendour of the river front leading to Goa started: 'adorned all along with stately churches and palaces'. He noted that halfway to the city was 'depainted on a church, a story of a ship brought from the *Cape Bon Esperanzo* [Cape of Good Hope]. . . and fixed where the church is now built, and by that means helping them with the Timber for the roof and two crosses set up as far off as the Ship was in length'.[121] This church symbolized and commemorated the glory and perils of the maritime world, particularly ships which were perceived as divine instruments, on which vested many hopes and fortunes.

THE SEASHORE

The seashore was a very cherished congregational space where the elements met, and people came together. The confluence of the forces of nature—land, ocean and the skies—conferred a sense of sacredness upon the sea coasts that was valued by coastal societies. It was serene as well as tumultuous; attracting and sometimes forbidding. But always available as a vast unspoilt common space, where the natural elements by and large prevailed to prevent human society from spoiling it, and rendering it unusable. The rains, storms and pounding surf cleansed the beaches and conferred upon them a propriety that was always inviting.

The shore and the beach front was nature's gift to people in common which was difficult to carve out into individualized property spaces. Nature's fury could easily nullify such claims and particularistic pretentions. No doubt the different political entities would claim them as part of their territory. But the privileged and the elite found it difficult to create their spaces of exclusive activity in perpetuity at the seafront, other than some strongholds at vantage points. Like the vast forests, the various faces of the sea-face were difficult to aggrandize on a vast scale and hold on to. So community activities such as festivities and religious congregations, involving masses of people took place on the seashore. They could come in large numbers, occupy themselves with their socio-cultural purposes, and leave. Nature would, to some extent take care, sweeping away the debris and leave the place anew for other people and groups. However, the considerable needs of society and the push of commercialism did lead to the polluting of these wonderful coastal fronts. John Fryer during his travel on the west coast remarked about 'Salset a Fishing Town where lay several boats to carry off mountains of fish salted on the beach, the scent whereof was very noysom.'[122]

The serene yet busy ambience of the beach was described by Fryer, upon his landing at the shore at Marmagoun near Goa. He noted: 'At our landing the Sea bestowed a kind of Murmur on yielding Sand and cast us ashore in a place quadrated more for still Retirement than noisy Commerce, there lying before its banks Canooses [i.e. boats] belonging to fishermen, and Baloons [rowing boats] of pleasure only; the Seignioros minding nothing less than merchandizing and the Povo [i.e. common people] employing their fish-hooks, and knitting needles to get a livelihood.'[123] The beach served as a commercial strand for the local merchants and gentry, particularly on designated days, to conduct business. The fisher folk and common people used it as a vocational space, and as a functional area for buying, selling and other activities of simple daily life. An important and interesting facet of all such activities at the beach, and particularly their infrastructure, was that they were impermanent; yet they were regular and routinized.

The beaches were also used as pathways for travelling, particularly by the porters and palanquin carriers, who found it preferable to walk on the soft sand while carrying their heavy burdens. Abbé Carré made the journey from Daman to Tarapur a distance of about sixty miles, mainly along the beach, in the company of the Portuguese *fidalgo* Dom Francisco. He was escorted by his black *caffre* slaves who were armed with matchlocks and heavy blunderbuss 'following on foot their masters' palanquin, which go more quickly than a horse would'.[124] The sight of such a travelling party trotting at a brisk pace, on the picturesque beaches was quite enchanting, though it was hugely laborious for the porters, as well as the heavily armed escorts. Abbé Carré noted that they started at daybreak, but by morning, 'The sand on the shore was so hot that our poor coolies, though quite accustomed to fatigue, were obliged to stop from time to time to rest under the shade of trees which grow along this coast.'[125]

Francois Martin who voyaged along the sea coast from Srikakulam to Krishnapatam in June 1681 remarked: 'we marched on the sands across several riverbeds and courses of streams which hardly had any water.'[126] Such coastal journeys were pleasant when the sun was down. The twin port cities of Madras and San Thame lay about five miles apart; and a brisk walk along the beach was the preferred option to the road routes. Abbé Carré recorded that after reaching Madras in April 1673, 'we marched the whole time along the seashore and arrived at St. Thome'.[127]

The seashore belonged to everyone in common; such was nature's way. But more particularly it was the socio-vocational sphere of engagement for the ordinary fisherfolk and the boatmen who lived by the sea. But their interest and focus was on the seas and not so much on the beach, which was only a base, and could shift from time to time. A telling facet about the sense of temporariness concerning such activities were the pearl fishing hut clusters or seasonal hamlets that were put up on the Fishery Coast beaches, to gather pearls from oysters. They were set up at particular times of the year, under licence from the Portuguese captains of the region to generate revenues, and also to prevent excessive pearl diving, which could denude the oyster and pearl stocks. Caesar Fredericke who visited the region in the 1570s recorded that every year in the month of March or April: 'they make or plant a village with houses and a Bazaro, all of stone, which standeth as long as the fishing time

Lasteth, and it is furnished with all things necessarie'.[128] After about fifty days or so, the pearl season over, the village was abandoned. Moreover, 'one yeere it is in one place, and another yeere in another place of the same sea'.[129]

It was an unusual and remarkable aspect of coastal life in India that fairly large seasonal villages or marketing centres were established every year to cater to the commercial requirements of the shippers and overseas buyers, in different parts of coastal India. Caesar Fredericke noted about one such well known temporary commercial establishment near Satgaon in Bengal. He observed: 'Every yeere at Buttor they make and unmake a village, with houses and shoppes made of strawe, and with all things necessarie to their uses, and this village standeth as long as the ships ride there, and till they depart for the Indies [South-East Asia], and when they are departed, every man goeth to his plot of houses, and there setteth fire on them.'[130]

Caesar Frederick was quite overawed by this phenomenon wherein a dramatic ritual end was put to the entire habitational-commercial structure by setting fire to it at the end of the trading season. He noted: 'For as I passed up to Satagon I sawe this village standing with a great number of people, with an infinite number of ships and bazars and at my returne I was amazed to see such a place soone razed and burnt, and nothing left but the sign of burnt houses.'[131] Perhaps these people dreaded to permanently colonize these deltaic coastal stretches, which were vulnerable to the depredations of the Magh and Arakan pirates, and the ferocious Portuguese slave raiders, who carried away people and goods with impunity.[132]

EUROPEANS IN COASTAL REGIONS

The coastal world witnessed a surcharge of commercialization and urban formation due to the fact it was the vital zone of interaction between overseas peoples and local society, and the vast hinterland interior which was linked up with the coast. Several of the contemporaries commented about the pronounced facet of trade, manufacturing activity and the general prosperity of the coastal cities. Jean de Thevenot who travelled around Gujarat noted that apart from important centres like Surat, Cambay and Ahmedabed, 'There are above thirty others on which depend a great many Bourgs and villages, but those [cities] which lie near the sea are the most considerable.'[133] By and large the coastal regions and the urban centres located near the sea acquired greater dynamism due to the multiplicity of influences and opportunities that the overseas connectivity created in the littoral areas.

Metropolitan ports such as Surat, Calicut, Masulipattam were major hubs of international trade which brought a congregation of people from several countries. Abbé Carré was amazed by the vast floating population of foreign merchants who were active in the Indian ports such as Surat. He recorded 'There seem to be in the streets and in Public places mainly Persians, Arabs, Turks, Armenians, Chinese and merchants from all European nations.'[134] The truly international ethos of Surat impressed the young French lieutenant De Lespinay who wrote: 'One sees here all kinds of nationalities . . . and each one can live without any constraint to his religion.'[135]

The European states and their merchant representatives, who sought to establish trading links with India in the late sixteenth century, were mainly maritime powers who were basically interested in the coastal regions of India to set up their establishments. Their perception about the states and dominions in India tended to be coast centric. Accordingly in addressing the mighty Mughal emperor Akbar, they exhibited a lack of clarity about the land-based power or the continental strength of the Mughal monarch, and concerned themselves with a marginal coastalized perception of his immense dominions. This was reflected in a letter sent by Queen Elizabeth to Akbar dated February 1583, which was carried by the merchant emissaries John Newbury and Ralph Fitch who came to the Mughal court to establish diplomatic and commercial links.

Queen Elizabeth's letter to Emperor Akbar addressed him, as follow, 'To the most invincible and most mighty prince Lord Zelabdin Echebar, King of Cambay, Invincible Emperor, etc.'[136] From the maritime, mercantilist perspective of the English merchants, who would have been involved in framing the Queen's letter, the primary perception of the grand Mughal monarch was purely that of a ruler who exercised command over the coast around the major port city of Cambay. It did not matter to them that the 'Kingdom of Cambay' was a small though rich coastal province, in the very vast dominions and territories of which Emperor Akbar was the mighty sovereign overlord.

The Indian coastal rulers, naiks and governors, on their part also had a hazy perspective about the important European kingdoms and maritime powers who were interested in opening trade with India. Their understanding and formulation about the Christian kingdoms was obtained from their overseas merchants, who had linkages and connections with people from the Christian world. This interesting insight emerges from the account of the young French lieutenant Bellanger de Lespinay when he came to India with the expedition of the French Admiral de la Haye (1670-5), who captured the Portuguese settlement of San Thome in the southern Coromandel region. In 1673 Admiral de la Haye sent de Lespinay to Sher Khan Lodi the Bijapuri governor of the Carnatik principality in southern Coromandel. To pre-empt the French mission, the Dutch Company sent its own representatives to Sher Khan who tried to persuade him that France was not an important country in maritime-commercial matters, and would not create valuable trading ties with the Indian coastal regions.[137]

Sher Khan's response to the Dutch illustrates the kind of information networks the Indian coastal rulers relied upon to manage their maritime policies. He told the Dutch ambassador that he had information about France and the Netherlands from a Bijapuri ship captain who knew a famous Armenian merchant who had travelled extensively in Europe and was quite knowledgeable about Christendom. Sher Khan had learnt from these merchants, that 'the king of France was the most powerful king amongst the Christians, and that the Hollanders themselves had only a small country full of water, and their livelihood only depended upon trade and commerce'.[138]

There was considerable curiosity amongst the Indian coastal elites about the distant lands from where the Europeans came. The Portuguese presence in the Indian coastal regions for over a century, and the strong impact of the Christian missionaries, had created an awareness about the European-Christian world in India. There was a wish to know more. The Frenchman Francois Pyrard de Laval who spent some time in Malabar in 1608, remarked about this when he met the Raja of Muttangal, a port city based principality in northern Malabar. He recorded: 'The king questioned me much about France when I told him that I was of that country, and asked me the difference between the English, the Hollanders and us. Next he enquired of the estate and greatness of the king of France.'[139]

Quite clearly the Indian coastal elites had an interest in Europe, and wanted to understand the configuration of power politics amongst European rulers. They sought to understand who possessed greater territories, and who had greater leverage in financial and military matters, in order to determine the policies to be formulated vis-à-vis the Europeans. But by and large it appears that the knowledge about the European countries was rudimentary, and not based upon report, and travel accounts of emissaries, merchants or other Indians or Asians who may have visited these countries. This lack of awareness and interaction with distant overseas countries and coastal principalities of the 'other' was detrimental to Indian interests in a long-term sense. It deprived the Indians of the possibility of obtaining knowledge and direct interaction with the European regions, that could have been translated into influence when dealing with the aggressive European mercantilist states.

An important and visible factor about the coastal regions and ports in India was that they were highly receptive to overseas traders and foreigners, belonging to different nationalities and backgrounds. The Spanish priest Domingo Navarette recorded that when their travelling party reached Masulipattam in 1670, 'a wonderful crowd of people came to see us English, Dutch, Persians, Armenians, Portugueses, Mungrels, Mahometans, Gentiles, Blacks and Natives were all spectators'.[140] But the city and its people made a favourable impression upon the friar who observed, 'I liked the natives who were all good men it seemed to me.'[141]

The overseas visitors frequently complained about the harshness and the extortionist attitude of the customs officials. But there was considerable appreciation and a feeling of being overwhelmed by the Indians in the coastal regions. The English captain William Hawkins upon his arrival at Surat: 'At my coming on shore, after their barbarous manner I was kindly received, and multitudes of people following me, all desirous to see a new come people.'[142] Apart from the curiosity about the Europeans, the coastal people also understood the significance of overseas traders and trade. Another English Captain Martin Pring recorded that when he landed at the small coastal town of Coranga in northern Coromandel in 1619, 'the governor sent his horse for mee . . . then caused his Palankeen to be made readie to conveigh me unto lodging'.[143]

The European contemporaries were also aware that the coastal inhabitants, particularly those from the ports, tended to be more familiar and receptive to foreigners, than people from the interior. The French director Francois Martin who undertook a long overland journey from Pondicherry to Surat in May–July 1681 also remarked about this factor. When he was about four days travel time short of Surat he recorded: 'Since the place where we were was not too far from Surat and because the Europeans are much better known there than in the interior regions, the governor of the place, sent his people to me with his compliments, and offered to do service.'[144]

The Indian milieu, particularly in the coastal regions provided the foreigners considerable security, advantages, and freedom of worship enabling them to pursue their vocations and live in favourable circumstances. As Pietro Della Valle remarked: 'any private person whatever of whatever country or religion may in these ports live with as much grandeur and equipage as he pleases; and such is the liberty here'.[145] In fact, the Europeans were able to acquire a fairly privileged position in several regions of India, which was remarked upon by their contemporaries. The English sailing master Thomas Bowrey recorded his observations about their influence in the Gingelli coast. He noted: 'Forraigners, more especially the English and the Durch have great freedome here, the same we have in other parts of this King's Dominions and live very pleasantly upon the fatt of the land.'[146] It was perhaps this advantage of being able to live 'very pleasantly upon the fatt of land', that whetted the appetite of the Europeans towards pursuing larger ambitions in India.

The Europeans came to exercise considerable autonomy and freedom in conducting their commercial affairs in the Indian principalities. The charter of agreement signed between the Surat authorities in November 1623 stipulated that the English shall 'freelie exercise their own religion, weare armes for theire owne defense and execute justice on theire owne people'.[147] The sweeping range of trading privileges included the authority to issue passes to Indian shipping in different coastal regions of India. For example, the English doctor John Fryer recorded the upon his arrival in India in 1672 when their ship rounded the coast of Ceylon, towards the Coromandel coast they came across three country junks; 'they produced English passes, the masters of the two [junks] being Portugals, the other a Moor'. By the seventeenth century the Indian coastal region had acquired a slightly cosmopolitan character. Apart from the Asian merchants sailing across, there were Portuguese, Armenians, and a cross section of European shipping personnel found in all the coastal regions of India.

In coastal India the Europeans operated from two kinds of establishments, which influenced the nature of their mercantilist presence, and their trade structures. The first category comprised the European fort towns such as Portuguese Goa, English Madras, Dutch Pulicat, French Pondicherry, which were autonomous, fortified, and where the Europeans enjoyed considerable extra-territorial privileges. European fire-power combined with naval support, maritime access, and shrewd diplomacy made these fortified centres relatively secure from the threatening overtures of local potentates and officials. For example, Francois Martin recorded that in February 1675 the Danish Company at Tranquebar became entangled in conflict with the local ruler, who stopped the supply of water and provisions to the Danish establishment, by mounting a blockade to pressurize the Danes to seek accommodation over some contentions matters. Martin remarked: 'This is almost the only method which is used by these petty princes to insult the Europeans who have their fortresses in their territories. They [local rulers] are not strong enough nor adequately vigorous to come and attack.'[148]

The second category of European establishments were their factory houses or commercial quarters located at all major ports and coastal regions, within the dominions of the Indian rulers. The Europeans took considerable precautions to safeguard their establishments against skirmishes or acts of hostilities by local officials. Writing about the Dutch Company's factory house at Masulipattam, the English governor of Madras Streynsham Master noted that 'They keep 30 or 40 European soldiers always in their factory and 15 or 20 horses. They have severall peeces of cannon, which they keep close and mount sometimes upon occasion of quarrells, and they inlarge their factory by buying in it at great rates what homes and ground lyes near them, soe that it is thought they design to fortifie themselves.'[149] Besides the deployment of several Indian peons and soldiers gave the European factories a certain measure of security to resist hostilities.

The growing militarization of European mercantilism in all regions of India gradually transformed the simple commercial nature of their trading establishments. The Companies had to contend with friction and warfare in several regions due to regional circumstances. For example, in October 1670 Shivaji raided and sacked the commercial metropolis of Surat, lasting over four days, in order to commandeer finances from its rich traders, and also to undermine the legitimacy of Mughal authority in western India. In the pillage that followed, the English Company pressed into service forty well-armed sailors from the company's ships at Swally port, in addition to the English factory personnel and soldiers at Surat factory, under the command of Streynsham Master. When a Maratha detachment surrounded and attacked the English factory house: 'The enemy found such hott service from our house, having lost severall men that they left us and fell on the Tartar quarter fiercely.'[150] Increasingly recurring tendencies of friction in the coastal regions, during the phase of Mughal decline, led to Europeans to fortify their trading establishments wherever possible, as in the case of Calcutta and Bombay, which greatly augmented their autonomous stature, within the body politic of Indian principalities, thus constituting an incipient threat to these polities.

FOOD FOR THE VOYAGE

Coastal societies prepared for shipping voyages with care and concern. They were well aware that proper provisioning for the long maritime voyages could make the difference between life and death. Food had to be well preserved and salted. In fact, several coastal settlements specialized in provisioning for voyages and functioned as stapling points of certain merchant groups. For example Van Linshoten recorded that the Portuguese town of Div particularly was renowned for preparing salted fish, meat and cheese for the ships and sailing people.[151] The Dutch chaplain Philip Baldaeus recorded that the port city of Gogha close to Surat similarly undertook the task of provisioning. He noted: 'Here all the ships designed for Arabia and the southern parts (by the Merchants of Cambaja and Amdabath) are careened and victualized, there being a safe road here.'[152]

The taboos, dietary codes and susceptibilities concerning the kinds of food eaten by different communities and people were recognized and observed. For the Europeans all foods particularly meats were edible. For the Muslims—merchants and sailors—pig meat or pork was taboo, just as for the Gentoos or Hindus, beef was usually avoidable. Swally harbour which catered to Surat's shipping fleets was a major provisiong centre for Asian ships. Nicolas Downton reported in 1611 that at Swally they went to the 'market there kept upon the strand divers sorts of provisions, to wit, Meale, Bread, Bullocks, Goats, Sheepe, hennes, Butter and Cheese, Sugar, Limes, Plantans, Water-Mellons, Gourds Onions, Radishes, Pallingeries, Cucumbers, Milk. . . and Tobacco, also Salt-fish dryed and Prawnes and Palmita wine which they call Taddy. All these aforementioned things were at reasonable cheape rates.'[153] Though palm wine was freely available, Downton did not mention the availability of pig meat, which was disapproved of as food in the Islamic dominions.

Most of the European ships stopped at Goa to provision themselves for their voyage. Abbé Carré recorded on his trip from Madras to Surat that when their ships stopped at Goa in December 1673 there came 'several boats filled with refreshments, viz pigs, oxen, sheep, bread, fowles, fruits, nipes [liquor] and Portuguese wines. These got a ready sale from the ships'.[154] Abbé Carré noted that the palm wine of Goa was very popular 'which serves as the ration for the crew on a voyage'.[155] The Goan elite society on its part had considerable requirement for goods brought by the European ships such as cheese, olives, hats, silk stockings, and European liquors.[156]

The European companies made large scale provisioning of food for their ships involved in the Asian trading circuits. Streynsham Master recorded during his voyage to Hughli in September 1676 that the Dutch Company maintained a 'Hogg factory' at Baranagar where 'they kill 3000 hogges in a yeare, and salt them up for their shipping'.[157] The Dutch were thus able to reduce their dependence upon the markets in Mughal dominions, and also cut down upon their expenses. In a similar manner the French Company's council at Pondicherry ordered the purchase of buffaloes on a large scale to be slaughtered and salted for the requirements of their ships and crews in February 1689. Since Pondicherry was located within the larger territorial ambit of the Hindu Maratha principality, there was a kind of ban, or strong disapproval related to cow slaughter. Consequently, as the director Francois Martin recorded, the French found it expedient not to antagonize the regional authorities, and use buffalo meat for their sailing crews.[158]

The sea could confer unusual surprises and benefits to its sailing persons and to the coastal people in different ways. The English Chaplain Ovington recorded in 1689: 'When the [ship] Benjamin had unloaded her cargo here [at Swally] after a long voyage at sea, the commander ordered her to be cleaned; and thereupon he found a multitude of large well tasted Oysters which grew upon the bottom of the ship, with which he feasted his sea-men and all the English at Suratt.'[159] This unexpected bonus was happily accepted by the English community at the port town. The coast on its part offered relief and a happy diversion for sailors who had been at sea for a while. Tavernier mentioned that when he arrived at Vengurla from Goa in March 1649, 'on landing we met some Dutchmen with the commander, who has come down to the seashore to eat oysters and drink Spanish wine'.[160] By and large the ports in India were adequately cosmopolitan and well provided to cater to the requirement of people from diverse backgrounds. For example, at Surat a Frenchman called Lafleur, who was married to a local Muslim woman, owned a tavern which catered to an international clientele.[161]

Coastal society, which usually commanded greater resources, due to its extensive linkages in several directions, was quite hospitable to visiting strangers, and important personages. The Frenchman Jean Baptiste Tavernier who visited San Thome on his way to Madras in August 1652 mentioned the generous food packages they received from the Portuguese governor and others. He noted, that they 'sent us a quantity of presents—hams, ox tongues, sausages, fish, water melons and other fruit of the country. It took nine to ten men to carry these presents'.[162]

The coastal establishments of the European companies frequently had to make urgent food provisions for their incoming ships, which had long been on voyage. Since the ship's crews were numerically large, the scales of provisions required even for short periods, were quite considerable. The English captain Martin Pring recorded that when their fleet of three ships arrived at Masulipattam in June 1619, they overshot their landing point by two leagues and had to stay there for a period of four days, due to the strong current and winds. To cater to the crews immediate requirement the English factory despatched 'Master Methwould [who] came from Messulapattam in one of the country boats and brought with him twentie Hogs, two great Jarres of Racke [or Arrack country liquor] six goats and two baskets of bread.'[163] The Company's coastal establishments kept their sailors and soldiers in good humour to prevent desertions, which were frequent due to the demands of the rival Europeans as well as Indian rulers for personnel from Europe.

The Europeans at times also encountered the problem of non-availability of beef in certain coastal cities within the Mughal dominions. This was due to the fact that at times and in certain places Mughal administration made accommodatory arrangements with the local people concerning their sensibilities in dietary matters, in keeping with their religio-customary sensibilities regarding cow slaughter. In this context, the English captain Nicholas Downton recorded that in 1614 the English Company's representatives petitioned the local governor 'at Baroch to have a bazar or Market by the water side, that we might have Beef for the [English] peoples eating in regard that other flesh was not good for them'.[164] The governor immediately agreed to the setting up of a market place on the strand for the provisioning of the English ships and personnel, but expressed his inability to make beef available in the market. He replied: 'A bazar we should have but for bullocks and kine, the king had granted his Firma to the Bannians for a mighty summe yeerely to save their lives.'[165] Apparently the Gentile community of Broach had been able to obtain a concession from Emperor Jehangir to prevent the slaughter and consumption of cows and bullocks, upon payment of a certain sum annually to the local administration.

The Mughal authorities were willing to heed the socio-religious susceptibilities of the Hindu subjects and accept a ban on cow slaughter. However, such concessions it appears were applicable or observed in some places only, and not everywhere. In fact, Nicolas Downton mentioned that in December 1614 the Nabob sent to the English, 'great store of provision, Goats, Bread, Plantans etc with a banquet of sweet meates. Cogenozan [Khwaja Nazam] sent me a present of five Bullocks.'[166] These animals were evidently not transport animals, but for the food requirements of the English. In fact, while describing the provision market at the Swally harbour Nicolas Downton mentioned that bullocks were available, along with other animals, for the provisioning of ships going on voyage.[167]

There appears slight ambiguity on this matter of the ban on cow slaughter by the Mughal state. But it is quite clear that the state during emperor Jehangir's reign did follow a policy which sought to address the religio-customary sensibilities of the Hindu populace in such matters. The Dutch factor Francisco Pelsaert who resided at Surat in the 1620s noted: 'Oxen and cows are not slaughtered, . . . their slaughter is strictly forbidden by the King.'[168] Pelsaert explained the rationale concerning this enactment. 'The king maintains this rule to please the Hindu rajas and banians, who regard the cow as one of the most veritable gods or sacred things.'[169] Pelsaert further pointed out that the Hindus from time to time obtained orders from governors of cities to put periodic bans, even on the slaughter of goats, sheep, buffalos in these places. 'Such orders are extremely inconvenient for ordinary people, but the rich slaughter daily in their own houses.'[170]

The strictness of adherence to these laws concerning cow slaughter is not altogether clear or certain. What is evident and significant is that the Mughal state framed the policy and tried to address the sociological concerns of the Hindu community on issues that were marginal and important at the same time. It sought to make a statement that for the state the community mattered; implicit was the connotation that for the community the state mattered in similar measure. The construct of state and community ties occurred in several ways, at several levels.

CROSSING THE FORD

The sea was a great facilitor for travel and movement across the waters, or along the coast, since it provided unhindered passage. But at the same time it could be a deterrent for travel along coastal land, due to the vast expanse of marshes and swampy land, particularly at the mouth of the rivers, which were numerous in the coastal areas. The topographical irregularity and the fluctuating ingress of the sea, caused by changing tides, created serious difficulties in travelling on the coastal routes. For example during Francois Martin's voyage along the Coromandel coast in June 1681, his travel party hurried to cross the Chipler rivulet near Rampatnam, before the rising of the tide. But by the time they reached, the sea had swollen considerably; 'we had to unload our wagons and place the baggages on the heads of the porters', which made the crossing tedious, difficult and time-consuming.[171]

The crossing of the vast marshy coastal tract on the Surat–Cambay circuit, where the river channels and the ocean impacted with raging fury, was an amazing feature in the lives of the coastal people of Gujarat, and of the several travellers who frequented this hazardous route. In fact the entire coastal region around the Gulf of Cambay was vulnerable to the fury of the on-rushing tide or bore, particularly at the time of the full moon or the new moon. The Italian Pietro Della Valle who was quite fascinated by this phenomenon noted at Cambay seafront: 'We saw the sea come roaring afar off like a most rapid river and in a moment overflow a great space of land, rushing with such fury that nothing could have withstood its force; and I think it would have overtaken the swiftest race-horse in the world.'[172] Della Valle was completely surprised that the sea rose sharply to its full height in less than half an hour, which would normally take around six hours in other countries.[173]

The Frenchman Jean de Thevenot who visited Cambay in 1666 pointed out that it took about 24 hours to go from Surat to Cambay in a 'Almadie' or fast boat. But the journey had to be undertaken at night due to the fear of the Malabar pirates. Moreover this sea journey was hazardous: 'Many of the Almadies are lost in the Gulf of Cambaye where the tides are troublesome and the banks numerous.'[174] Consequently, the most frequent, though dangerous method of journey was to cross the marsh at low tide across three and half leagues or about 9 miles of water, which was three to four feet deep, in ox-wagons or on foot.[175] But Thevenot was deterred by the reported risk of this voyage which he confessed: 'I was told that the waves beat so rudely sometimes against the

Chariot, that it required a great many hands to keep it from falling, and that some mischance always happened.'[176]

The French director Francois Martin who took this route during his inspection of the French establishments in Gujarat in March 1683, similarly remarked about the lethal unpredictability of the Mahi river crossing which he traversed. He noted: 'It is quite dangerous and if one is not careful about the rush of the tidal waters; the sea rises with surprising rapidity similar to a flash flood, and very often people are carried away by the rushing waters.'[177] Pietro Della Valle was awestruck by this crossing of the Mahi river from Jambusar onwards towards the Gulf of Cambay, in February 1623. He recorded that, 'the flux and reflux of the sea is more impetuous and violent and with more rapid current than perhaps in any part of the world'.[178] In fact, the high spring tides were known to rise as much as 33 feet.[179]

The passage across this extensive coastal tract involved considerable drama, adventure and the assertion of the human spirit to encounter the perilous forces of nature, despite these people's vulnerability to the caprice of the tidal waters. The young Englishman John Jourdain who crossed this ford on his return from Ahmedabad to Surat in October 1611 recorded, 'goinge over the river of Cambaia we wett all our stuffe, the water beinge high and the stream so swifte that itt is very dangerous goinge over, because it is att least half a mile over, and deepe to the arme pitts;. . . itt hath manie, both horses, coatches and men drawned'.[180] It was amazing that men and women belonging to the genteel 'baniya' community and other similar simple folk took this dangerous coastal route, instead of the much longer inland route, as if the marshy passage were a kind of pilgrimage. Pietro Della Valle who undertook this journey twice recorded that they met up with a large special caravan at Jambusar (on 23 February 1623), and the following day after a journey of 5 kos reached the gulf an hour after sunshine.

Considerable expertise about the timing and the place of crossing was necessary to minimize the possibility of accidents. 'So being united in a great troop the better to break the stream, we passed over all that space of five coss[181] [about 12 miles], which was moist yet firm ground saving in four places where we forded the running water of the river.'[182] The crossing of the river streams, which were thrashed by sea tides, was daunting and intimidating, particularly since the water 'came higher than the belly of the oxen which drew the coaches'. A number of men were hired to steadfastly hold the coaches on both sides to prevent them from being carried off in the waters, and to carry the bundles on their heads.[183]

The Cambay crossing was an unusual phenomenon because large numbers of people routinely traversed this treacherous coastal stretch, disregarding the dangers and unmindful of concerns of the urbane civility in their dress codes. 'The men who go on foot in this passage either strip themselves naked, covering only their privities with a little cloth or pulling up their coats.'[184] The travelling men and women learnt to adjust to the elements by abandoning sartorial propriety and traditional mores. Pietro Della Valle noted with understanding. 'Tis certainly an odd thing to behold in this passage, which is very much frequented, abundance of people go everyday in this manner, some in coaches and chariots, others in horseback and on foot, men and also women naked, without being shie who sees them.'[185]

The dread of the forces of nature, concern for safety, and purposefulness about the journey, made men and women oblivious towards concerns of appropriate clothing. The eagerness to save time and effort led people to take this direct route across the swamp. Della Valle recorded that following the wet passage, a walk of only two coss along the seashore brought the travelling party to Cambay before dinner time 'having travelled that day in all twelve coss'.[186] The inland journey was almost four times longer.

The Cambay ford undoubtedly cast a strong impression upon the mind of Della Valle who described its perils again following his return journey to Surat a fortnight later on 7 March 1623. The water was much higher; 'The oxen and horses could scarce keep their Heads above Water, and the coaches being light, if men hired purposedly had not gone along in the water to hold them steady, and break the course [i.e. current] thereof by holding great Stumps of Wood on that side the tide came furiously in, without doubt the water would have swept them away.'[187] Della Valle and others had to stand upright on the wagons with the sea water coming upto the middle of their legs, the rest being submerged.[188]

Jean de Thevenot had perhaps been wise in avoiding this perilous journey which involved a stiff struggle with the sea, the tides and the river currents over a distance of 5 kos. Della Valle who twice experienced the fury of these elements was informed by the doughty regular travellers: 'it often carries away people, and sometimes with such violence that an elephant cannot bear up against it, but is swept away by the Water. Therefore, they wait certain fit hours to pass this ford, namely when the sea is at the lowest ebb.'[189] The only distraction they received from this elephantine turbulence was the beautiful sight of flocks of colourful flamingo birds skimming on the waters close to the land. By nightfall the travelling party reached Jambusar, and after three days reached Surat after crossing many rivers and stretches of coastal expanse.[190]

In general crossing the tidal rivers and the swampy marshy terrain along the coastal routes, was hazardous, strenuous and time consuming under all circumstances. Even the routine process of crossing rivers was quite cumbersome for the travellers, the porters and the transport animals who comprised the travelling party. As noted by Pietro Della Valle during his travels in western India that several rivers emptied into the sea: 'They are almost innumerable upon the coast of India.'[191] This factor made coastal travel quite difficult particularly for those who had large ox-carts and carriages. The Frenchman Thevenot recorded in 1666 about such river crossings during his voyage from Surat to Ahmedabad. He noted: 'I crossed river Tapty in a boat big enough, but very incommodious for taking in of the chariots because the sides of it were two foot high.'[192] It was an extremely tedious process, as a number of porters manually lifted and placed the wagons onto the boats, and again unloaded it on the other side. Thevenot noted that crossing the rivers Kim and Narmada further along the route was a similarly difficult process.

But the to and fro passage from Cambay to Broach, described by several contemporaries constituted a formidable testimony of human spirit and endurance, manifested by ordinary men and women, particularly from the coastal regions, to face the savage elements in their fury. The route was very short in distance compared to the interior route, which was very time consuming. What was astonishing was that simple genteel folk—the sheltered urbane trading people, and their more sheltered delicate women folk—were willing to pick up the gauntlet and wade across 5 kos of marsh lands, where the buffeting tidal waters in the streams, reached up to the shoulders of these vulnerable people. Pietro Della Valle recorded with trepidation that the ox cart in which he stood upon, holding its ceiling, the water reached up to the waist, while the cart itself shook precariously, threatening to go with the swirling waters any moment.[193] What was baffling was that these dangerous journeys had come to be accepted as a routine normal way of life, by these coastal people.

PORTUGUESE ARAB CONFLICT

One of the enduring sagas of maritime hostility and recurrent ravages on the coastal establishments of each other was the bitter conflict between the Portuguese and the Arabs of Muscat which simmered throughout the sixteenth and seventeenth centuries. This pernicious feud was a conspicuous and long enduring feature in the western part of the Indian Ocean which centred on the coasts of western India and the Arabian Peninsula. It was the legacy of a kind of 'civilizational conflict' that had its genesis dating from the Moorish conflict of the Iberian Peninsula (tenth-twelfth centuries); and the subsequent feats of aggression by the Portuguese at Ceuta, Morocco, along the north African coastline in the late fifteenth and early sixteenth centuries.[194]

Hostilities were aggravated in 1507 when the Portuguese raided the port of Muscat, set the city on fire, burnt 34 ships and 'put all the Moors, with their women and children found in the houses, to the sword, without giving any quarter'.[195] Further, after burning the grand mosque they rounded up many of the Moors who had fled; 'Dalbuquerque gave orders that their ears and noses be cut off and they should then be liberated.'[196] In later years, the macabre magnanimity of not killing the hostages but carrying them away as prizes became a pattern in the conflict between the two adversaries.

This pernicious and debilitating warfare which dragged on for the next two centuries was remarked upon by several contemporaries. By the mid-seventeenth century the Portuguese power had declined considerably and the Arabs were able to inflict severe damage upon the Portuguese homeland zone in coastal western India. In January 1669, the Muscat Arabs pillaged the Portuguese stronghold of Div. The French priest Abbé Carré who visited the town lamented that the Arabs 'took the town with all its riches and treasures and carried away 6000 Christian women as slaves. They left sad traces which will ever remain, having destroyed the roads and demolished the principal houses.' The priest was given an exaggerated numbers of the women who were carried away, but nonetheless the raids were chillingly brutal born out of generational vitriol, agony and thirst for revenge.

Francois Martin who was in the Persian Gulf region in June 1669 similarly noted that 'the Portuguese and the Arabs have been involved in a cruel war without giving any quarter to the other'.[197] He recorded the fierce engagement that took place on sea in which the Arabs lost one of their generals and a large number of men.[198] The Portuguese coastal cities had been functioning in the manner of a frontier society since their inception. As the English doctor Fryer noted: 'To check these incursions of the Arabs, the Portuguese every year are at the charge of a lusty squadron in these seas . . . who are no sooner gone than the Arabs sent their fleet to do this mischief here.'[199] Such was the intensity of desire for revenge that in November 1672 Admiral Antonio de Mello de Castro was imprisoned by the Portuguese viceroy for concluding a truce with the ruler of Muscat. Further his estate was confiscated for returning with his fleet, contrary to his orders to wage a punitive war against the Arabs. In fact, the king of Portugal had also issued stern instructions to wage war, and appointed a new Portuguese admiral to inflict harsh reprisals on the offending Arabs.[200]

Next the Arabs completely worsted the Portuguese in their homeland zone in the west coast, in this ruthless dragging warfare. In 1674, the Arabs raided the Portuguese settlements around the Bombay region, 'carrying away their Fidalgo prisoners together with their wives and families, butchering the Padres, and robbing the Churches without resistance'.[201] Fryer noted with a sense of historical awareness that the bitter hostility was 'conceived on a deadly Feud partly out of revenge of the Portuguese cruelties at Mushat; but chiefly out of detestation of each other's religion, insomuch that quarter is denied on either side'.[202] The problems of the Portuguese kept mounting in many coastal regions. In October 1679, the Portuguese at Goa received news that their viceroy Dom Pedro de Almeida attacked the port of Pata and took possession of it. Some days later an Arab fleet arrived there, and in the ensuing battle the Portuguese vessels were severely damaged and they were forced out of the town. The viceroy retreated to the Portuguese colony of Mozambique where he died out of grief and remorse.[203]

By the end of the seventeenth century the Portuguese ability to hold command over the western seas and the coasts had declined significantly. The English chaplain John Ovington who was posted at Surat in the 1680s noted with some sadness and some disdain that the Portuguese were 'now as low and declining as they formerly were powerful'.[204] On the other hand, 'the Muscatters daily increase their naval forces and prevail against them [the Portuguese] incontestably at sea'.[205] Ovington in fact wrote scathingly, without concealing his Protestant disapproved of the Catholic Portuguese power. He noted: 'The Arabians are a stout, couragious and hardy nation. . .The Portuguese are mighty sunk as well in their courage as in their fame and fortune.'[206] The renowned *Estado da India*, by this phase had ceased to be a seaborne empire.

PIRACY: THE MALABAR CORSAIRS

If the sea was the source of profit and power it could also be a source of menace and disaster. Since it was an open territory without barriers or physical zones of exclusivity, it was capable of bringing fort unpredictable and unexpected problems. Banditry of the seas, or piracy, was a major problem that the seafaring people and trading communities had to live with. John Fryer who spent a long time in coastal India observed 'No part of these seas were without these Vermin, the Bay of Bengal being as much infested as the coast of Coromandel.'[207] He made particular mention of 'the most accursedly base of all mankind', the Arrakan and Magh pirates who infested deltaic Bengal raiding with impunity and carrying away people as slaves.[208] Francois Bernier similarly referred to the 'cruel and incessant devastations of these barbarians', which afflicted coastal Bengal due to 'the unjust and violent proceedings of the pirates established in Rakan'.[209]

The west coast of India was far more seriously affected by the activities of pirates from Gujarat down to Malabar. The dreaded Sanganian pirates of Gujarat and the Sind coast were particularly notorious during the sixteenth and seventeenth centuries.[210] In fact, the slight decline of Indian shipping during this phase led many India's seafaring communities to resort to unlawful piratical activities, alternating with their sailing vocation. But it was the terror and harsh striking capability of the notorious Malabar corsairs for over two centuries from about 1530 onwards that led to the Portuguese *Estado da India* to start the 'caphila'-armada or convoy system of escorting European and Indian ships along the coasts of India, as a prominent visible feature of maritime activity.

The Malabar coastal polity was fragmented among a large number of principalities and potentates during the medieval phase. The Frenchman Francois Pyrard who visited the northern Malabar region at the beginning of the seventeenth century noted, 'Each of these port hath its own king, but all are in some way, under the Samory [the Samudri Raja of Calicut].'[211] This fragmentation of power in Malabar enabled the Portuguese to establish a shadowy suzerainty in the coastal regions and impose their system of maritime control. This fluid polity also permitted the rise of the Kunhali Marakkar Corsair also known as Malabars, who established their parallel influence in the northern Malabar region.[212] They acquired a legendary status by their daring deeds, which found expression in several ballads in the folklore of the region. They were viewed as an important part of coastal society, who offered resistance to the Portuguese aggressors and contested their system of maritime control.

The Kunhali Marakkars shifted from Kollam to the Kottakal Badgara region after their destruction by Henrique de Menezes in 1525, and inaugurated a policy of piratical aggression.[213] They created a kind of coastal confederacy of corsairs who commanded several galleys and fleets, of *paros* (country vessels), and exerted a strong influence over the western seas. Francois Pyrard who visited several of their strongholds at Muttangal, Badara, and Kanyirottu in the sourthern Kanara region was impressed by 'the corsairs who rule a great extent of Countries and peoples'.[214] The French doctor Jacques Dellon who served at the French company establishments in Malabar in the 1670s, wrote a century and half after the initial founding of the Marakkar power about their strong influence. 'The Seigneur Kunhali is a famous pirate, who has ten or twelve *paros* at sea, each carrying five to six hundred men; his subjects following his example have become corsairs and are all rich and haughty to the point of being insolent.'[215]

The entire west coast was subjected to the maritime threat of the Malbaris, whose reach extended up to the Gujarat coast. Thevenot mentioned that maritime traffic between the two rich commercial cities of Cambay and Surat was vulnerable to their piratical activities. He noted: 'Vessels go not commonly but in the night time that they might not be discovered by the Malabars . . . they skulk behind some Rock and fall upon it in its passage.'[216]

The Englishman Thomas Herbert recorded witnessing a sea fight in which his own ship was involved in 1626, off the west coast in the Konkan region, with a Malabar corsair vessel. Herbert recorded that their ship launched a cannonade and then resorted to the bravado of chasing the Malabar ship 'with barges manned by fifty Musquetiers'. The fight that ensued, and the response of the Malbaris was daringly impressive. Herbert noted: 'Wee made too much haste in boarding her, being entertained with such store of fire-works and Granados, a volley of cruell shafts; in a word we were opposed with so much desperate courage that after small hurt to them, wee retreated with shame, the better halfe slaine, hurt and scalded, our ships all the while being made an unwilling Theater of this affront.'[217] Such stout courage, martial prowess and maritime capability enabled the Malbaris to challenge European domination for two centuries and be dreaded by the Asian as well as European shippers who crossed their path. They belonged to a strong seafaring mercantilist tradition, 'almost all being merchants, robbers or sea warriors', and 'such as are not brave enough for soldiers, they make mariners.'[218]

The Malabars lived under several kings, as law abiding people who paid customs, duties, and gratifications to their lords. They enjoyed strong support due to their legendary status, and due to their favourable role in local societies. Pyrard noted: 'All the merchants of the coast when they hear that the galliots of the pirates are about to come in, hold themselves in readiness to buy their goods cheap.'[219] It completely suited their merchant ethics to obtain goods at rock bottom prices, without feeling disturbed that they belonged to other merchants whose goods had been stolen and several amongst whom faced bankruptcy. Nor did these corsair people feel any serious moral dilemma that their piratical activities involved plunder of Asian, including Muslim merchant ships, who were brought to ruin by their action. In an interesting revelation about their self perception, the Malabars felt that by vocation they 'were all robber and pirates, a trade that was no dishonour to them, seeing they practiced it from father to son'.[220] For them, it was merely a vocational hazard that they were aggressors, plunderers and aggrandizers whose activities were severely damaging for others.

In was perhaps to alleviate their conscience and to seek divine exoneration that corsairs gave away a significant part of their booty to the priestly and poor segments of society. Pyrard who spent some time with the Malabars observed: 'the priests of their religion too, and the poor are on the lookout and come distances of thirty leagues to get their share; for they know well that these Malabars have made vows, in case they make a good prize, to give so much to the poor, and never fail to acquit themselves therein'.[221] Thus, there was a kind of Robin Hood virtue associated with the Malabar corsairs, where the priestly persons and the poor shared in the proceeds from sea brigandage. It tended to confer an unusual coating of piety on acts of maritime brigandage. Plunder and piracy thus acquired a garb of respectability, by finding community support and acceptance amongst these coastal people. To a considerable extent, it had its genesis in the coercive actions of the Portuguese who had uprooted centuries old tradition of peaceful trading system, based on participation by coastal peoples across the Indian Ocean belonging to several nationalities and religions.

At the end of the sixteenth century, the legend surrounding the fourth Kunjali Marakkar was so considerable that it was said, 'He was the grandest corsair ever seen in those parts.'[222] Pyrard extolled his phenomenal reputation in unusually expansive terms: 'His fame and terror were spread abroad from Cape of Good Hope even to China',[223] due to his command over the western seas. The Kunjali revolted against his overlord the Zamorin who sought the help of the Portuguese, who sent a powerful squadron under Andre Furtado de Mendoca in AD 1600. With the Zamorin's help, the Kunjali was captured and carried away in chains to Goa, where he was ordered to be stoned by children to die a slow and painful death.[224]

The tradition of the Marakkar however continued and flourished throughout the seventeenth century. In fact, Alexander Hamilton's testimony as well as the English Company's records indicated that the Malabars continued to impose their hegemony well into the eighteenth century, obliging ships to carry their passes.[225]

RITUALS FOR THE SEAS IN THE MALDIVES

The tradition of prayers and offerings to the seas was also prevalent outside the Indian mainland in the proximate region of the Maldives where the dominant religion was Islam. The Maldive Islands were truly maritime–coastal in their way of life and constantly interacted with the elements, with ships and sailing people, with voyages across the turbulent waters, storms and calms as a routine facet of life. Consequently, they evolved a strong tradition of worship to the forces of nature, which was not in conflict with the tenets of Islam. The French voyager Pyrard de Laval who visited these islands at the beginning of the seventeenth century noted: 'when at sea, if they are caught by contrary winds, by calms or by storms, they make vows to him who rules the winds, who is called not God but King; and there is no island but has a "siare" [i.e. *ziarat*][226] as they call it which is a place dedicated to the king of the winds, in a corner of the island remote from the world, where those who have escaped from danger come to make offering daily of little boats and ships fashioned on purpose and filled with perfumes, gums, flowers and odiferous woods. The perfumes are set on fire, the little boats are cast upon the sea, and float till they are burned.'[227]

This solemn ceremony and prayer to nature was religious, secular and pagan at the same time. It had its own ritual of prayers which were addressed to a 'King' rather than to a 'God', so as not to be in contravention or violation of the doctrinaire position of Islam, which was primarily observed in the Maldives. But other than the persona, or the divine surrogate, to whom these prayers were addressed there was not much to distinguish this form of worship from the ceremonies observed by the Gentiles in several parts of coastal India, which were referred to by many contemporaries. Perhaps, the coastal communities and the ship people needed some solace or protecting force to help them face the remarkable uncertainties in their journeys across the seas, and in their lives on the coast which were more severely exposed to the fury of nature. So if there was to be no God of the seas for them then they would invent one in another form, that was equally venerable, but not located in religion.

Pyrard de Laval recorded the form of ceremony, ritual and prayer addressed to another such mythical persona, different from the 'king of the winds' referred to earlier. He noted: 'Likewise they believe in a king of the sea, to whom in like sort they make prayers and ceremonies while on voyages; or when the go a-fishing.'[228] The omnipresent and savagely furious seas around the Maldives led to the evolution of several taboos, rituals and observances to communicate a constant sense of reverence to the lord of the seas. He noted: 'they dread above all things to offend the kings of the winds and of the sea. So too when they are at sea they durst not spit nor throw anything to windward for fear lest he should be offended and with like intent they never look abaft.'[229]

The sacrifices to the seas were also made at the time of launching of ships or during their transit process, and were similar to the coconut festivities observed in coastal India. Pyrard noted: 'so when they have any difficulty in launching their ship or galleys, they kill some cocks and hens and cast them in the sea, in face of the ship or boat they desire to launch'.[230] It was not merely to remove obstructions or maritime problems that prayers and sacrifices were offered. Worship of the seas and seagoing crafts was constant. 'All boats, barques and vessels are dedicated to the powers of the winds and the sea; and indeed they treat their boats with as much respect as their temples, keeping them exceedingly clean and abstaining from all filthy or indecent actions on board.'[231]

The French priest Abbé Carré also observed that a sense of 'purity' was desirable during voyages on the sea. He recorded that during his return voyage from Surat to Bandar Abbas in February 1674, he boarded an Asian ship, whose crew and passengers were mainly Asians, mostly Muslims. During the journey the ship was caught in terrible 'calms which gave us much trouble. This induced the passengers and rich Muslim merchants to try all kinds of sorcery and witch-craft to raise the wind. . . They also obliged all the followers of their faith, headed by their priest, to bathe their bodies in the sea in order (they said) to wash away the dirty impunities they commit with their young slaves of which there were more than 200 in the ship.'[232] Thus, homosexuality, though widely prevalent, was frowned upon as a perverse practice that brought misfortune upon ships on voyage. And the rites of purifications were necessary to mitigate its evil effects. Only then could one safely reach the safety of the coasts.

A.J. Qaisar points out in the context of coastal India the importance of the Islamic mythological figure Khawaja Khizr, who was regarded as the guardian of the seas. This patron saint of the sailors to whom prayers were addressed was located within the general structure of the custom and tradition of maritime India and was integral part of Islamic societies. In the case of Maldives, Laval's observations indicated that its coastalized society had created forms of surrogate divinity that appeared non-Islamic and seemed somewhat pagan. He noted: 'Likewise they hold in honour the kings of other elements (as they call them) as him of war, and pay them all great ceremony. These Kings of the elements were in the mould of para-religious entities who were powerful, magnificent and benign, to whom prayers and sacrifices were made routinely. Perhaps they addressed society's problems and concerns beyond the format of structured religion, appealing directly to the lords of the seas, the winds, of war, of life and death, who mattered immensely. It was a kind of cult worship that existed, without coming into conflict with the doctrines of orthodox religion. It focussed on the facets of human frailty and social trauma vis-à-vis the elements, that formal religion was not completely able to take care of or answer. The desire for safety and survival made them willing to accept plural traditions of worship.

CONCLUSION

The coastal people endured several kinds of problems and calamities that were specific to their environment and location. Problems such as famine, plagues, floods, warfare, bandity, afflicted all regions with equal severity, both interior as well as coastal. But the impact of cyclones, storms, shipwrecks, piracy and loss of vessels, specifically affected coastal regions. They could severely depress the fortunes of several mercantile establishments and commercial people, just as favourable voyages and opportunities brought wealth and prosperity.

The seas were not always benign. The eastern coast of India, particularly the Coromandel, was vulnerable to severe storms and cyclones. On it were located several important coastal cities such as Masulipattam, Petapoli, English Madras, Portuguese San Thome, French Pondicherry, Dutch Negapattam and Pulicat. Francois Martin recorded that the savage cyclone of November 1679, which hammered Masulipattam, broke throgh the coastal embankments and drowned 20,000 persons. It dragged away the large wooden bridge in the city, which extended about three-fourths of a mile, and had cost over 200,000 *pagodas*[233] to build.[234]

The storm of November 1681 struck seventy miles across the coast in the southern Coromandel region, affecting several coastal towns, particularly Negapattam, Porto Novo and Tranquebar. About 2000 fisherfolk took shelter in a grand temple built on high ground in Negapattam. But the entire structure crashed upon them, killing them all, unable to withstand the fury of the seas. Apparently the prayers of the people were not always answered by the Gods and the seas. The total deaths in Negapattam area alone were recorded to be above 14,000 persons.[235]

How did the coastal people, particularly the ordinary fisherfolk, boatmen, dock workers and all the others deal with these coastal problems that occurred periodically and disturbed their lives? The ritual of prayers and offerings was more suited to the richer segments of society who had the services of the priests, and the almanacs. But the ordinary working people, who usually had a low caste status, had to seek strength in their commonsense philosophy of life and community caste bindings. Most contemporaries were quite impressed that these unprivileged people lived quite contentedly, and adapted well to their tough lifestyles. Van Linschoten writing about the Muslim sailors in western India took notice of their carefree attitude on ships: 'when they doe anything such as hayling [ropes] they sing and answere each other [sweetly] that it seemeth to be Musick'.[236]

The Englishman Charles Lockyer, similarly noticed at Madras that the boatmen who performed the tough job of carrying goods in their catamarans and other coastal crafts, on the very choppy surf, in fact received very low wages. He felt that it was 'Money dearly earned . . . however they are merry birds, howling out a Ela, Yela, as chorus to their songs, at almost every stroke [of the oars].'[237] Apart from the seafolk, other coastal communities such as farmers, porters, the ordinary working people, the 'pariahs' had a spirited attitude towards life that was simple and appreciable. Thomas Bowrey who was based at Madras for 19 years and well understood their attitude of life remarked perceptively: 'they seemingly live very contentedly . . . as cheerful in poverty as any mortal can be.. they sing and dance very frequently as men secured from all cares and feares that accompany the wealth of the universe'.[238] Perhaps, these simple coastal people well understood that joy and suffering, life and death were both a part of nature and life, just like the land and the sea.

NOTES

1. F. Braudel, *The Mediterranean and the Mediterranean World in the Age of Philip II*, trans. Sian Reynold, vol. I, New York, 1976, p. 162.
2. Ibid., pp. 103-67.
3. J.C. Heesterman, 'Littoral et Interior de l'Inde' Itinererio 1 (1980): 89, cited in M.N. Pearson, 'Littoral Society: The Concept and Problems', *Journal of World History*, vol. XVII, no. 4, 2006, pp. 358-73.
4. M.N. Pearson, 'Littoral Society: The Concept and Problems', *Journal of World History*, vol. XVII, no. 4, 2006, pp. 358-73.
5. Philip Steinberg, *The Social Construction of the Ocean*, New York, 2000, pp. 138-9, cited in M.N. Pearson, 'Littoral Society'.
6. F. Braudel, *Mediterranean World*, vol. I, p. 164.
7. R.C. Temple, ed., *The Itinerary of Ludovico di Varthema of Bologna from 1502 to 1508*, London, 1928, p. 53.
8. Ibid.
9. F. Braudel, *Mediterranean World*, vol. I, p. 164.
10. S.N. Sen, ed., *Indian Travels of Thevenot and Careri*, Delhi, 1949, p. 127.
11. A.C. Burnell, ed., *The Voyage of John Huyghen Van Linschoten*, vol. I, London, 1935, rpt Delhi, 1997, p. 277.
12. A. Cortesão, ed., *The Suma Oriental of Tome Pires*, vol. I, London, 1944, rpt Delhi, 1990, p. 40.
13. Ibid., p. 66.
14. A.C. Burnell, ed., *Van Linschoten*, vol. I, p. 63.
15. William Crooke, ed., *John Fryer's A New Account of East India and Persia, Being Nine Years' Travels, 1672-1681*, vol. II, London, 1909, rpt Delhi, 1992, p. 58.
16. S.N. Sen, ed., *Thevenot and Careri*, p. 8.
17. A. Martineau, ed., *Memoires de Francois Martin*, vol. II, Paris, 1932, p. 312.
18. S.N. Sen, ed., *Thevenot and Careri*, p. 8.
19. A.C. Burnell, ed., *Van Linschoten*, vol. I, p. 261.
20. William Crooke, ed., *John Fryer*, vol. I, rpt Delhi, 1992, p. 173.
21. Ibid., pp. 173-4.
22. Ibid., p. 175.
23. C.E. Luard, ed., *Travels of Frey Sebastian Manrique 1629-1643*, vol. II, London, 1928, pp. 329-30.
24. Edward Grey, ed., *The Travels of Pietro Della Valle in India*, London, 1892, p. 42.
25. R.C. Temple, ed., *Thomas Bowrey's A Geographical Account of the Countries Round the Bay of Bengal 1669 to 1679*, London, 1905, rpt Delhi, 1997, p. 120.
26. A gorse equals roughly 4 tons.
27. R.C. Temple, ed., *Thomas Bowrey*, p. 121.
28. Francois Bernier, *Travels in the Moghul Empire, 1656-1668*, Paris, 1670, trans. Archibald Constable, 1891, rpt Delhi, 1997, p. 120.
29. R.C. Temple, ed., *Thomas Bowrey*, p. 132.
30. Ibid., p. 132.
31. William Crooke, ed., *John Fryer*, vol. III, rpt Delhi, 1992, p. 14.
32. Thomas Herbert, *Some Years Travels into Africa and Asia*, London, 1638, rpt Chennai, 1998, p. 313.
33. A. Cortesão, ed., *The Suma Oriental of Tome Pires*, vol. I, p. 120.

34. Jacques Dellon, *Relation d'un Voyage des Indes Orientales*, vol. II, Paris, 1685, pp. 9-10.
35. William Crooke, ed., *John Fryer*, vol. I, p. 172.
36. Thomas Salmon, *Modern History or the Present State of All Nations*, London, 1739, vol. I, p. 274.
37. Madras State Archives, Records of Fort St. George, *Diary and Consultation Book, 1686*, 1911, pp. 5-6.
38. Charles Lockyer, *An Account of Trade in India*. London, 1711, p. 10.
39. C²63, Archives des Colonies, Archives Nationales, Paris, lettre du comptoir de Pondicherry, fol. 108.
40. A. Martineau, ed., *Memoires de Francois Martin*, vol. II, pp. 226-43; R.C. Temple, ed., *The Diaries of Streynsham Master 1675-1680: And Other Contemporary Papers Relating Thereto*, vol. I, London, 1911.
41. Samuel Purchas, ed., 'Observations of William Finch', in *Purchas His Pilgrimes*, vol. 3, Glasgow, 1905, p. 82.
42. W. Crooke, ed., *Travels in India by Jean Baptiste Tavernier*, trans. V. Ball, vol. I, London, 1889, rpt Delhi, 2000, p. 4.
43. Samuel Purchas, ed., 'Observations of William Finch', in *Purchas His Pilgrimes*, vol. 3, p. 29.
44. Ibid.
45. W. Foster, ed., *Early Travels in India, 1583-1619*. London, 1921, rpt Delhi, 1999, p. 302.
46. A. Cortesão, ed., *The Suma Oriental of Tome Pires*, vol. I, p. 66.
47. Jacques Dupuis, *Madras et le Nord du Coromandel*, Paris, 1960, pp. 11-12.
48. A. Cortesão, ed., *The Suma Oriental of Tome Pires*, vol. I, p. 66.
49. W.H. Moreland, ed., *Peter Floris: His Voyage to the East Indies in the Globe 1611-1615*, London, 1934, p. 121.
50. W.H. Moreland, ed., *'William Methwold's Account'*, in *Relations of Golconda in the Early Seventeenth Century*, London, 1931, p. 7.
51. Ibid.
52. Jacques Dellon, *Voyage des Indes Orientales*, vol. II, p. 38.
53. J. Ovington, *A Voyage to Surat in the Year 1689*, ed. H.G. Rawlinson, London, 1929, p. 86.
54. Albert Gray, ed. and trans., *The Voyage of Francois Pyrard of Laval*, vol. II, London, 1887, rpt Delhi, 2000, p. 12.
55. A. Martineau, ed., *Memoires de Francois Martin*, vol. II, p. 412.
56. J. Ovington, *A Voyage to Surat*, p. 87.
57. W.H. Moreland, ed., *Relations of Golconda in the Early Seventeenth Century*, London, 1931, p. 3.
58. A. Martineau, ed., *Memoires de Francois Martin*, vol. II, p. 493.
59. Ibid.
60. William Foster, ed., *English Factories in India 1646-50*, London, 1914, pp. 54-5.
61. Ibid.
62. William Crooke, ed., *John Fryer*, vol. I, p. 173.
63. S.N. Sen, ed., *Thevenot and Careri*, p. 37.
64. William Crooke, ed., *John Fryer*, vol. I, p. 99.
65. Edward Grey, ed., *The Travels of Pietro Della Valle*, p. 65.
66. Ibid., p. 131.

67. Ibid.

68. Ibid.

69. S.N. Sen, ed., *Thevenot and Careri*, p. 124.

70. J.C. Van Leur, *Indonesian Trade and Society*, The Hague, 1956, pp. 5-6.

71. Samuel Purchas, ed., *Hakluytus Posthumus*, or *Purchas His Pilgrimes*, vol. 10, Glasgow, 1906, p. 210.

72. Ibid.

73. Samuel Purchas, ed., *Hakluytus Posthumus*, or *Purchas His Pilgrimes*, vol. 2, Glasgow, 1905, p. 70.

74. Ibid., p. 71.

75. R.C. Temple, ed., *Thomas Bowrey*, pp. 126-7.

76. See W. Crooke, ed., *Travels in India by Jean Baptiste Tavernier*, vol. II, pp. 175-8.

77. C.E. Luard, ed., *Travels of Frey Sebastian Manrique*, vol. I, pp. 69-70.

78. William Crooke, ed., *John Fryer*, vol. II, pp. 37-8.

79. Ibid., pp. 35-8.

80. Harprasad Ray, 'Indian Settlements in China', in *Indian Ocean and Cultural Interaction 1400-1800*, ed. K.S. Mathew, Pondichery, 1996, pp. 52-81.

81. Madras State Archives, Records of Fort St. George, *Diary and Consultation Book, 1688*, 1908-16, pp. 49-50.

82. A. Martineau, ed., *Memoires de Francois Martin*, vol. II, p. 167.

83. R.C. Temple, ed., *The Itinerary of Ludovico di Varthema*, p. 61.

84. Jacques Dellon, *Relation d'un Voyage des Indes Orientales*, vol. I, Paris, 1685, p.113.

85. Edward Grey, ed., *The Travels of Pietro Della Valle*, pp. 69, 92.

86. A. Martineau, ed., *Memoires de Francois Martin*, vol. III, Paris, 1932, p.137.

87. Abbé Carré, *Voyage des Indes Orientales*, vol. I, Paris, 1699, p. 138.

88. Philip Baldaeus, *A True and Exact Description of the Most Celebrated East Indian Coasts of Malabar and Coromandel and also of the Isle of Ceylon, AD 1640*, rpt Delhi, 2000, p. 622.

89. Edward Grey, ed., *The Travels of Pietro Della Valle*, p. 69.

90. Richard Hakluyt, *The Principal Navigations, Voyages, Traffiques and Discoveries of the English Nation,* ed. Edmund Goldsmid, vol. 5, Glasgow, 1906, p. 395.

91. Samuel Purchas, ed., *Hakluytus Posthumus*, or *Purchas His Pilgrimes*, vol. 10, Glasgow, 1906, p. 206.

92. Albert Gray, trans. and ed., *The Voyage of Francois Pyrard of Laval*, vol. II, p. 57.

93. W. Crooke, ed., *John Fryer*, vol. I, pp. 188-9.

94. Ibid.

95. Ibid., p. 192.

96. S.N. Sen, ed., *Thevenot and Careri*, pp. 168-9.

97. J.S. Cummins, ed., *The Travels and Controversies of Friar Domingo Navarrete, 1618-1686*, vol. II, Cambridge, 1962, p. 320.

98. A.J. Qaisar, 'From Port to Port: Life on Indian ships in the 16th and 17th centuries', in *The Indian Ocean*, ed. Ashin Dasgupta and M.N. Pearson, Calcutta, 1987, pp. 331-50.

99. R.C. Temple, ed., *The Life of Icelander Jon Olafsson Travellor to India*, London, 1931, p. 143.

100. W.H. Moreland, ed., *Peter Floris*, p. 134.

101. Edward Grey, ed., *The Travels of Pietro Della Valle*, p. 187.

102. S.N. Sen, ed., *Thevenot and Careri*, p. 117.

103. Ibid.

104. Ibid.

105. A.C. Burnell, ed., *Van Linschoten*, vol. I, pp. 227-8.

106. Samuel Purchas, ed., 'Captain William Hawkins Account', in *Hakluytus Posthumus, or Purchas His Pilgrimes*, vol. 3, Glasgow, 1905, p. 7.

107. Ibid.

108. A.C. Burnell, ed., *Van Linschoten*, vol. I, p. 228.

109. William Crooke, ed., *John Fryer*, vol. I, pp. 196-7.

110. S.N. Sen, ed., *Thevenot and Careri*, p. 117.

111. Ibid., p. 197.

112. A. Martineau, ed., *Memoires de Francois Martin*, vol. II, p. 325.

113. Ibid., p. 325.

114. Ibid.

115. Ibid.

116. Ibid., pp. 359-60.

117. Ibid.

118. Ibid., p. 328.

119. Ibid.

120. Albert Gray, ed. and trans., *The Voyage of Francois Pyrard of Laval*, vol. I, London, 1887, rpt Delhi, 2000, p. 31.

121. William Crooke, ed., *John Fryer*, vol. II, p. 9.

122. Ibid., p. 41.

123. William Crooke, ed., *John Fryer*, vol. I, pp. 21-2.

124. Abbé Carré, *Voyage des Indes Orientales*, vol. I, p. 172.

125. Ibid.

126. A. Martineau, ed., *Memoires de Francois Martin*, vol. II, p. 233.

127. Abbé Carré, *Voyage des Indes Orientales*, vol. II, p. 377.

128. Richard Hakluyt, *The Voyage and Travell of M. Caesar Fredericke, Marchant of Venice into the East India, and beyond the Indies (1563)*, vol. 5 of *The Principal Navigations, Voyages, Traffiques and Discoveries of the English Nation*, London, 1907.

129. Ibid.

130. Ibid., p. 410.

131. Ibid., p. 411.

132. See J.J.A. Campos, *History of the Portuguese in Bengal*, Calcutta, 1919.

133. S.N. Sen, ed., *Thevenot and Careri*, p. 44.

134. Abbé Carré, *Voyage des Indes Orientales*, vol. I, p. 21.

135. Henri Froidevaux, ed., *Memoires de L.A. Bellinger de Lespinay, Vendomois, Sur son Voyage Aux Indes, Orinetales 1670-75*, Vendone, 1895, p. 72.

136. Richard Hakluyt, *The Voyage and Travell of M. Caesar Fredericke, Marchant of Venice into the East India, and beyond the Indies (1563)*, vol. 5 of *The Principal Navigations, Voyages, Traffiques and Discoveries of the English Nation*, p. 450.

137. Henri Froidevaux, ed., *Memoires de Belangar de Lespinay*, p. 195.

138. Ibid., p. 196.

139. Albert Gray, ed. and trans., *The Voyage of Francois Pyrard of Laval*, vol. I, p. 337.

140. J.S. Cummins, ed., *The Travels and Controversies of Friar Domingo Navarrete, 1618-1686*, vol. II, p. 321.

141. Ibid., p. 325.

142. Samuel Purchas, ed., *Hakluytus Posthumus, or Purchas His Pilgrimes*, vol. 5, Glasgow, 1904, p. 20.

143. W. Foster, ed., *Early Travels in India, 1583-1619*, p. 71.

144. A. Martineau, ed., *Memoires de Francois Martin*, vol. II, p. 266.

145. Edward Grey, ed., *The Travels of Pietro Della Valle*, pp. 41-2.

146. R.C. Temple, ed., *Thomas Bowrey*, p. 127.

147. William Foster, ed., *English Factories in India 1646-50*, pp. 310-11.

148. A. Martineau, ed., *Memoires de Francois Martin*, Paris, 1932, vol. II, p. 11.

149. R.C. Temple, ed., *The Diaries of Streynsham Master 1675-1680: And Other Contemporary Papers Relating Thereto*, p. 297.

150. C. Fawcett, ed., *The Western Presidency 1670-77*, vol. I of *English Factories in India*, Oxford, 1952, pp. 195-7.

151. A.C. Burnell, ed., *Van Linschoten*, vol. I, p. 59.

152. Philip Baldaeus, *Description of the Most Celebrated East Indian Coasts of Malabar and Coromandel*, p. 513.

153. Samuel Purchas, ed., 'Journal of Nicolas Downton', in *Hakluytus Posthumus, or Purchas His Pilgrimes*, vol. 3, p. 361.

154. Abbé Carré, *Voyage des Indes Orientales*, vol. III, p. 714.

155. Ibid.

156. Ibid.

157. R.C. Temple, ed., *The Diaries of Streynsham Master 1675-1680: And Other Contemporary Papers Relating Thereto*, vol. I, p. 324.

158. A. Martineau, ed., *Memoires de François Martin*, vol. III, p. 28.

159. J. Ovington, *A Voyage to Surat*, p. 100.

160. W. Crooke, ed., *Travels in India by Jean Baptiste Tavernier*, vol. I, p. 168.

161. A. Martineau, ed., *Memoires de Francois Martin*, vol. II, p. 423.

162. W. Crooke, ed., *Travels in India by Jean Baptiste Tavernier*, vol. I, p. 215.

163. Samuel Purchas, ed., 'Captain Pring's Journal', in *Hakluytus Posthumus, or Purchas His Pilgrimes*, vol. 5, Glasgow, 1905, p. 20.

164. Samuel Purchas, ed., 'Nicolas Downton's Account', in *Hakluytus Posthumus, or Purchas His Pilgrimes*, vol. 4, p. 219.

165. Ibid., p. 220.

166. Samuel Purchas, ed., *Hakluytus Posthumus, or Purchas His Pilgrimes*, vol. 3, p. 235.

167. Ibid., p. 361.

168. W.H. Moreland and P. Geyl, trans., *F. Pelsaert's Jahangir's India*, Cambridge, 1925, p. 49.

169. Ibid.

170. Ibid.

171. A. Martineau, ed., *Memoires de Francois Martin*, vol. II, p. 234.

172. Edward Grey, ed., *The Travels of Pietro Della Valle*, pp. 103-4.

173. Ibid.

174. S.N. Sen, ed., *Thevenot and Careri*, p. 19.

175. Ibid.

176. Ibid., p. 19.

177. A. Martineau, ed., *Memoires de Francois Martin*, vol. II, pp. 313-14.

178. Edward Grey, ed., *The Travels of Pietro Della Valle*, p. 62.

179. Ibid., p. 65n, cited in Hunter's Gazetteer.

180. William Foster, ed., *The Journal of John Jourdain 1608-1617*, Cambridge, 1905, p. 175.

181. 'coss' or 'kos', Indian unit of measuring distances. One kos equals about 2¼ miles.

182. Ibid.

183. Ibid.

184. Edward Grey, ed., *The Travels of Pietro Della Valle*, pp. 65-6.

185. Ibid., p. 66.

186. Ibid.

187. Ibid, pp. 118-19.

188. Ibid.

189. Ibid., p. 64.

190. Ibid., p. 119.

191. Ibid., p. 131.

192. S.N. Sen, ed., *Thevenot and Careri*, p. 8.

193. Edward Grey, ed., *The Travels of Pietro Della Valle*, pp. 103-4.

194. See C.R. Boxer, *The Portuguese Seaborne Empire*, New York, 1969, pp. 1-38; Sanjay Subrahmanyam, *The Career and Legend of Vasco da Gama*, Cambridge, 1997, Chap. 2, pp. 74-5.

195. Walter de Gray Birch, ed. and trans., *The Commentaries of the Great Afonso Dalbuquerque*, London, 1875-84, rpt Delhi, 2000, pp. 81-2.

196. Ibid.

197. Abbé Carré, *Voyage des Indes Orientales*, vol. I, p. 132.

198. A. Martineau, ed., *Memoires de Francois Martin*, vol. I, p. 202.

199. William Crooke, ed., *John Fryer*, vol. I, p. 193.

200. Abbé Carré, *Voyage des Indes Orientales*, vol. I, pp. 176-7.

201. William Crooke, ed., *John Fryer*, vol. I, pp. 192-3.

202. Ibid.

203. A. Martineau, ed., *Memoires de Francois Martin*, vol. II, p. 176.

204. J. Ovington, *A Voyage to Surat*, p. 125.

205. Ibid., pp. 125-6.

206. Ibid., p. 254.

207. William Crooke, ed., *John Fryer*, vol. II, pp. 152-3.

208. Ibid.

209. Francois Bernier, *Travels in the Moghul Empire, 1656-1668*, p. 179.

210. William Crooke, ed., *John Fryer*, vol. II, p. 152; Abbé Carré, *Voyage des Indes Orientales*, vol. I, pp. 130-1.

211. Albert Gray, ed. and trans., *The Voyage of Francois Pyrard of Laval*, vol. I, p. 338.

212. For an authoritative study on Malabar and the Portuguese see Genevieve Bouchon, *Regent of the Sea*, Delhi, 1988.

213. W. Logan, *Madras District Gazetteers, Malabar*, Madras, 1951, vol. I, p. 462.

214. Albert Gray, ed. and trans., *The Voyage of Francois Pyrard of Laval*, vol. I, p. 344.

215. Jacques Dellon, *Voyage des Indes Orientales*, vol. II, pp. 4-5.

216. S.N. Sen, ed., *Thevenot and Careri*, pp. 18-19.

217. Thomas Herbert, *Some Years Travels into Africa and Asia*, p. 34.

218. Albert Gray, ed. and trans., *The Voyage of Francois Pyrard of Laval*, vol. I, p. 337.

219. Ibid., p. 342.

220. Ibid., p. 347.

221. Ibid., p. 342.

222. Ibid., p. 350.

223. Ibid., p. 352.

224. Jacques Dellon, *Voyage des Indes Orientales*, vol. II, pp. 4–5; Albert Gray, ed. and trans., *The Voyage of Francois Pyrard of Laval*, vol. I, pp. 350–3; W. Logan, *Madras District Gazetteers*, vol. I, p. 462.

225. W. Logan, *Madras District Gazetteers*, Malabar, vol. I, pp. 462–3.

226. *Ziarat* (Arabic) refers to holy pilgrimage.

227. Albert Gray, ed. and trans., *The Voyage of Francois Pyrard of Laval*, vol. I, pp. 175–8.

228. Ibid., p. 178.

229. Ibid.

230. Ibid.

231. Ibid.

232. Abbé Carré, *Voyage des Indes Orientales*, vol. III, p. 795.

233. Small gold coin prevalent in southern India.

234. A. Martineau, ed., *Memoires de Francois Martin*, vol. II, pp. 176–7.

235. Madras State Archives, Records of Fort St. George, *Diary and Consultation Book, 1681*.

236. A.C. Burnell, ed., *Van Linschoten*, vol. II, p. 269.

237. Charles Lockyer, *An Account of Trade in India 1711*, London, 1711, p. 11.

238. R.C. Temple, ed., *Thomas Bowrey*, pp. 79–98.

Implantation of Commercial Crops: Cochineal Culture and the Regional Ecology in the Eighteenth Century Coromandel

Arvind Sinha

I N RECENT YEARS, there is a growing recognition for the importance of preserving and promoting heritage linked to the maritime past. It includes not only the sea itself but also the coast and manmade activities that affected different strata of society. Maritime heritage had an ecological component to which commercial activities gave a distinct identity. Fernand Braudel's masterpiece, *La Méditerrenée et Le Monde Mediterranéen' à l'Epoque de Philippe II* (1949) had brought out finer aspects of human geography and its relationship with the sea. An attempt has been made in this chapter to recount English attempts in the late eighteenth century to promote in the newly acquired territories along the Coromandel coast in the south by experimenting and transplanting new varieties of crops and plants like pepper, tobacco, indigo, cotton, sugar, mulberry and teak.[1] But the most remarkable of all the experiments was the transplanting of a commercial product in the form of cochineal, an insect found in the Spanish colony of Mexico and used for dyeing cloth, to the coastal lands of the Coromandel. This was done after identifying the climatic and environmental suitability of such an exercise.

The beginnings of large-scale production in England caused by the Industrial Revolution resulted in some discernible shift in the English colonial policy. The focus now shifted to control the major centres of production and to utilize the colonial lands to promote cultivation of those products that had commercial value in the rapidly changing home economy.

From the late eighteenth century, British capitalism was steadily assuming a global character through an imperial state structure with a global trading network. This development had been noted by Immanuel Wallerstein, who described it as the 'globalization of British foreign trade in the eighteenth century'.[2] According to him, Britain was successful in unifying the world market in two ways—the first involved the exercise of political and military force which created the colonial empire. The second was achieved through substantive exercise of politico-economic hegemony—the integration of producing regions into the world market. They thus placed colonial control within a general economic context of a need for markets, raw materials and investment outlets.

After the acquisition of Diwani in Bengal and the rich territories in the Deccan and the 'Northern Circars' along the Coromandel, India was becoming an important region for Britain. The Seven Years War had made the East India Company grow from a commercial enterprise into a military and territorial power and thereby laid the foundation of the British Empire. The importance of colonies was realized during this era. Several Englishmen, including the elder Pitt stressed the value of the colonies; they saw the colonies as primarily raw material producing regions to promote the British industrial interests. The last two decades of the eighteenth century held a significant place in the economic history of India because of her steadily turning into a raw material producing region. One of the possible reasons behind the shift in the English policy was the financial difficulties of the company and the growing criticism levelled at it for failing to improve its resources. This was accompanied by the loss of American colonies. It was felt by a large number of people that England must strive to regain the territorial losses of America elsewhere. Several historians on the subject suggest that this transformation in the English policy of making India the raw material producing region to sustain the English manufacturing sector took place from about the 1820s. Friedrich Engels seems to have set 1800 as the dividing line between the 'import' and the 'export' phases of British colonialism, that is, the British conquest of India helped in transforming the metropolitan country from an exporter of manufactured goods to its importer. In Marx's perception the change came after 1813.[3] It was expected that the commercialization of agriculture would compel the Indian peasants to raise raw materials for the world market instead of domestic crops for village-level consumption. However, a number of documents and company despatches suggest that this change had been attempted much earlier.[4]

The English adopted numerous steps to encourage cultivation of such crops which possessed commercial value and which could promote the company's exports to Europe or elsewhere. The Court of Directors instructed the English authorities to give priority to new commercial products which could be grown on Indian soil and provide financial benefits to the authorities. The present Andhra and Orissa coastal region was particularly favoured for their experiments. It included the prized possessions of the northern Circars, centre of trade and manufacturing. These formed the districts of the peninsula situated between Ganjam and river Krishna. The river disembogues itself into the sea south of, the famous port town, Masulipatam. The Ganjam district was under the Madras Presidency. It was the northern-most district and was mostly mountainous. It also had some large fertile plains, and was rich in iron deposits. To its south were the districts of Chicacole and Vizagapatam. This region was mountainous and intersected by hill ranges. The hills reached a height of 1500 feet near Vizagapatam and formed a kind of bay. The intersecting ranges of hills formed valleys; some of those were of considerable width. The soil was a mixture of loam, vegetable earth and gravel and was regarded fertile and good for cultivation.[5] The low country that lay between the hills and the sea and to the east of the Godavari river was Rajahmundry. Low hills, steep slopes and a forest clad region enclosed the valley of the river. It had extended rice fields and other crops included oil seeds and sugar. The English found this region useful from the point of view of experimenting with commercial products and trade. Bounded on the north by the river Krishna and the Bay of Bengal on the east was the Guntur Circar. This region could be divided into three parts: the delta region comprising the eastern coast, the stony uplands and hills consisting of the southern and western territories, and the open plains covering the rest of the circar, abounding in black cotton soils, caused by the weathering of the limestone. In places where moist climate was experienced, large and thick forests existed.[6] The two types of vegetation included the forests, mostly a scrub variety, and the plain vegetation that included paddy, jowar, cotton, tumeric, maize, sugarcane and sesame trees. These circars were particularly known for the manufacture of coarse and fine varieties of cotton that had wide markets in East Asia, Europe and Central Asia.

The East India Company records are full of information on the English attempts to promote and nurture mulberry, cotton and indigo cultivation on the coastal region and other newly administered territories in the south. Dr. Anderson had made several proposals to the company officials to order mulberry plantations at every village on the coast. The company government gave directions to the Chief and Council of Masulipatam, Guntur and many districts of the Northern Circars to instruct the collectors and *zamindars* to appropriate small portions of ground which were suitable for mulberry trees.[7] The Mulberry Plantation Report accepted the advice of Dr. Anderson and suggested the extension of mulberry plantation in the territories between Velout and Bocklore, the region intersected by cultivable lands and *nullas* (drains).

Serious endeavours were also made to encourage indigo plantation through Dr. Roxburgh. The Governor of Madras recommended those schemes. The Board of Trade communicated to the commercial resident that the 'Government desires that every encouragement should be given to the manufacture of indigo, a very valuable article of commerce. A copy of the process adopted by Mr. Roxburgh for the improvement of indigo is transmitted for dissemination to the natives. Musters of every description of indigo manufactured in the district may be collected and purchased at fixed rates. No effort should be spared in the collection and improvement of indigo culture.'[8]

The Board of the Trade also sought a report from the Chief and Council of Masulipatam regarding the various varieties of indigo produced in that district, the season of cultivation, the quantity produced and other information relating to its cultivation, manufacture and the prices of each variety.[9] In a letter sent to the Court of Directors with regard to indigo cultivation, it was clarified that:

The Chief and Council at Masulipatam offered it as their opinion that it would be most advisable to induce the natives, if possible, to undertake the culture of Indigo as well as the process of extracting dye, upon their own account, rather than to trust to the speculators of Europeans which would afford comparatively—with a view to excite an emulation among the natives, they proposed to ensure to them a ready sale for any quantity they could manufacture by receiving it as an article of your Investment of fixed prices according to its quality. . . .[10]

The Board of Trade also recommended that every description of indigo should be received, the better sort for the European market and the inferior quality to be resold to the local inhabitants. It was found that almost the entire quantity of indigo needed by the dyers came from Pettapore and Hyderabad. The crop was grown in the fertile track of land twice a year in August and in December, in periods of normal rains. The process observed for extracting indigo from the stems, branches and leaves was not considered appropriate and the manufacturers here made no use of oil in its preparation, as was the case in the West Indies. It was felt that there was urgent need to improve the technique of preparation of indigo to meet English requirements.[11]

Another commercial product that was found on the western coast of India but not yet accessible to the British was pepper. Some of the English officials were keen to grow it in the northern Circars. The Court of Directors were equally interested in this project which is evident from its letter, which stated:

We approve of the measures you mention to have taken in letters of 14 October and 23 December last, respecting assisting Mr. Roxburg in the cultivation of pepper plant. The sample you have transmitted of what was produced in the Rajamundry Circar under Masulipatam has been shown to some of the principal Buyers who are of opinion, it is of a quality fit for the market, but whether it is like to answer as a profitable article we are unable to determine as you have afforded us no information of the price at which it may be furnished or the quantity that may be probably be produced. . . .[12]

The English attempt to grow pepper in the Coromandel region lost momentum once they acquired the territories on the western coast after defeating Tipu Sultan.

The English showed even greater concern for cotton cultivation. The Court of Directors was anxious that India should contribute in the supply of cotton to England. Dr. Anderson was entrusted with this task. In 1790, a variety of foreign cotton seeds were obtained from Mauritius and Malta. Of the numerous experiments with different types of cotton, only one proved to be of commercial value. This was called the 'Bourbon cotton' (probably introduced in 1804). It was also called 'the shem parutti' or *seemai parutti* and is believed to have been introduced by the French who transplanted it from the West Indies. Its chief advantage was that it could be grown on red or gravel soil, especially calcareous and quartzite soils. This was found in many parts of the coastal south. The plant could survive five to six years. Dr. Anderson's personal interest and labour played a significant role in the early success of Bourbon cotton cultivation in Tinnevally, Salem and Coimbatore in the south, and Masulipatam and

Vizagapatam in the northern Circars.[13] Of these, the farm at Vizagapatam was the most successful.[14] The English Commercial Residents were instructed to promote this variety of cotton. This experiment did not last long as the quality of cotton could not be maintained due to the exhaustion of the soil resulting from continued cropping on unmanured land.[15]

Thus, in the newly acquired territories in the south, the English made specific efforts to try out different crops such as pepper, tobacco and indigo, encouraged gardening of mulberry plants for silk production and took steps to extend sugar and cotton cultivation. The English also carried out attempts to introduce teak planting in the low grounds of the Circars though teak trees were commonly seen in the hills of Rajahmundry Circar.[16] The English authorities instructed local officers that every vacant corner of the land should be planted with teak saplings.

Apart from these measures adopted by the English in the last quarter of the eighteenth century, the most noticeable attempt was to foster cochineal culture in southern India. In order to secure responsible and expert advice, the company appointed an experienced botanist to carry out experiments. Though most of these efforts did not succeed in the end and fell below the expectations of the English, these attempts clearly indicate the mind of the British authorities. An attempt has been made in this chapter to examine the motives and the nature of efforts in introducing cochineal culture in India as a commercial product in the late eighteenth century.

The English Company's keenness to promote this commercial product along with other marketable crops through agricultural experiments in the Carnatic and Andhra region is evident from the Company's report, which stated, 'efforts be made to foster the cochineal industry and to extend sugar and cotton cultivation and to make experiments such as the growing of vines'.[17]

Cochineal was an article which had been monopolized by Spain for a considerably long period. Very little study has been made on the trade of cochineal. It was a shield insect, native of Central and South America and was grown on the leaves of the nopal cactus. Hundreds of eggs were laid by the the female insect on the nopal plant which were hatched thirty-five to forty days later. Five months later the young insects were gathered and placed under the sun or heated over a low fire and later crushed to produce a dye of brilliant crimson red. Perhaps this method was discovered by the Spaniards in Mexico in 1518 and made known to Europe in 1523. The European sources mention the name of Hernan Cortés, the Spanish conquistador in Mexico, who was the first to know of it. Cochineal was used as a dyestuff by the Aztec and Mexican Indians since time immemorial. Its value was instantly recognized in Europe and Charles V sought further information on it from Cortés. The merchants of Antwerp, the well known centre of cloth dyeing, had started purchasing cochineal in raw and powdered form from the Spanish markets by 1540s. The cochineal trade in Europe came into existence when a few Spanish and Italian merchant bankers provided financial help to the Habsburg rulers of Spain in the purchase of this item. Their trade established linkages between

Seville or Cadiz, Genoa, Livorno and Florence.[18] Later its demand in Europe picked up. There is very little information on the cochineal trade during the seventeenth century but a number of sources are available which throw light on its increasing demand. According to the English sources, there were two forms or qualities of the cochineal insect: the *Grana fina* and *Grana Sylvestris*. The former was generally spoken of as the cultivated and the latter as wild cochineal. The cultivated one was larger and more valuable than the wild, but whether these two were distinct species was not very clear. The *Grana fina* was reported to be a native of Mexico while the *Grana sylvestris* came from South America. There is some possibility of the Portuguese making an attempt to introduce it into India in the seventeenth century. In 1786, Dr. James Anderson, Physician–General of the Madras Army, had found an insect, which he believed to be a kind of cochineal and with which he dyed pieces of flannel, shawls and satins. These specimens were forwarded to the Court of Directors. The Court ordered various experiments on the specimens sent to it, but these were found to be entirely useless for the purpose of dyeing. Sir Joseph Banks obtained similar results after having ascertained in May 1787, that the specimens sent to him were those of a real species of Coccus. He came to the conclusion that the true cochineal might easily be cultivated on the Coromandel Coast as the climate here was very similar to that of the West Indies.[19] Cochineal was used in dyeing wool, silk and cotton. It was employed as a colour ingredient of drugs and confectionery and as artists, pigments. The two different red colours obtained from it were the bluish red, called crimson and a fiery red called scarlet. Cochineal was comparatively rich in tinctorial matter compared with most of the other natural dyes. It contained between 10 and 20 per cent of the pure substance which was like a glucoside, from which carmine red was readily produced. The dyestuff required no preparation for the market. Before being supplied to the dyer the insect was beaten to a powder. The most valuable commercial form of it was the 'silver grey'. Cochineal was almost exclusively used for the production of scarlet shades on wool.

A French source also brings out the importance of this product in the following statement, 'Cochineal is an object of great importance for dyeing and colouring, the price of this insect is very high and has huge consumption throughout in all the nations of Europe, who receive it from the Spaniards in Mexico…. [T]he English are going to cultivate it in different parts of the Coromandel Coast. The English have encouraged the cochineal culture along the banks of Tamil country from Negapatam to Masulipatam.'[20]

Cochineal was found in several parts of the Western world but only the Spaniards had established possession of the true cultivated sort in their South and Central American settlements. Spain supplied this drug to Europe and Asia, which in point of value was regarded next to gold and silver. It had not yet been cultivated with success in other parts of the world. When the demand for cochineal was growing with the expansion of the textile industry, Spanish authorities were enforcing strict control over its production and its trade on the mercantilist lines. Many countries were trying to learn of its cultivation to circumvent Spanish monopoly. For example, in 1777 the French botanist Thiery de Menonville wrote his observations in a book.[21] He was sent to Oaxaca to study the production of cochineal. The French had made an unsuccessful attempt at Haiti to cultivate it. Around the same time the English were also making similar efforts to introduce the cochineal culture in their colonies. Dr. Anderson, the Chief of the Hospitals of Madras, having established it in 1787 on the Coromandel Coast, was encouraged to promote its culture in southern India. It is another matter that his initial finding was the case of mistaken identity.

In the Commercial Despatches[22] from England, the Court of Directors clearly stated:

Cochineal in addition to its medical qualities as a dyeing drug is an article of highest importance in many of our manufacture of the woolen in particular'. The cost of cochineal in 1788, as mentioned in the source, varied from 18 to 21 shillings per pound and the British were paying an increasing sum of money to Spain to import this commodity.[23] In 1736, the English sources suggest that no less than 880,000 pounds. of cochineal was imported into Europe[24] and the Spanish American was paid half a million pounds sterling annually at the old prices. To the Directors of the Company, it was 'a commerce certainly worth endeavouring to transfer at least a part of it to our settlement.[25]

The following table indicates the exact import of cochineal into England and the portion of its re-export from 1773 to 1786.[26] However, the export figures for the years 1783 to 1786 are not available.

IMPORT OF COCHINEAL

Year	*Imported quantity (in lbs)*	*Exported (in lbs)*
1773	169,245	44,153
1774	238,415	44,695
1775	198,053	60,136
1776	211,147	34,605
1777	194,159	19,283
1778	130,255	23,960
1779	100,891	13,522
1780	99,057	12,502
1781	120,566	18,020
1782	104,216	17,665
1783	270,935	
1784	368,575	
1785	306,623	
1786	208,457	

While emphasizing the reasons why India was ideally suited for growing cochineal, the Commercial Despatches to Madras pointed out that its cultivation required a country well stocked with industrious population. India fulfilled this condition. Second, the Spaniards had been the sole possessors of this drug. They imposed repeated duties on its export. Its production in India would make it relatively free of such duties because of British political domination over important regions in India. Furthermore, the crops in America were annually affected and damaged by certain insects which fed on the cochineal. It was assumed that some of these insects may be destroyed in the course of the long voyage as it was argued that India was admirably situated for the cultivation of real cochineal.[27]

The same source also mentioned why cochineal could not be profitably cultivated in the West Indies. It was felt that though climatically well suited for its production, west India had very high labour costs. The Madras region in India, on the other hand enjoyed as good a climate as the West Indies. The soil was suitable for the growth of the cactus and the labour cost was lower in India in comparison to Mexico or the West Indies. It was further hoped that if the cultivation of cochineal in Madras met with success, the English would be able to meet the Spaniards at an advantageous position in the market, especially as the English believed that this article was so light that the freight would bear a very small proportion to the real volume. Second, it was believed that there was an additional advantage of introducing cochineal culture in India. It was experienced that an insect of the fly kind always accompanied the cochineal in America which fed upon the young ones, sometimes desolated whole plantations and levied a grievous tax upon all. It was hoped that in the course of the long journey from America to India, the insect would remain on board for a protracted period. An attentive man employed to keep a close watch on it may succeed in destroying the whole race of flies which accompany the cochineal insects, which could render the cultivation free from the previous threat.

A consignment of cochineal insects was brought from Brazil in a ship under the care of Commodore Phillips and delivered to Dr. Anderson, one of the most important persons in charge of the English botanical experiments in India in 1788. The entire activity of procuring the insects from Mexico reflects a secret character. An English source reflects the sentiments of the local officials, 'As it appears the Hon'ble the Court of Directors did not intend the instructions relating to the care and preservation of the cochineal insect at sea should be made public. . . .'[28]

In the meantime, a careful study was carried out whether cochineal insects could be found within India. An insect was found which was believed to be of a cochineal type but careful examination by Dr. Anderson revealed that it had no similarity with the real cochineal in colour, figure, size or chemical properties, nor did it give turmeric, scarlet or crimson dye.[29]

In India, cochineal insects were reared on the Naga Kolli plants which grew generally in clumps or as hedges surrounding the little gardens of the local residents.[30] A few insects placed on a plant or on a bit of *Naga Kalli* leaf cut off with insects pinned on it by means of the thorns of the plant. These were sufficient to start a colony. Dr. Anderson reported that the cochineal insect operated as a complete flight to the indigenous *opuntia*.[31] It was capable of spreading over 16 miles of the country in a course of 12 months but no plant except *opuntia* had anything to fear from it.

How seriously the company was involved in the cochineal experiment is evident from the letter of Anderson to Campbell.[32]

I have devised another method of sheltering the plants from high winds by fences of such plants as no vermin can live . . . of this description there are many on the coast such as the great aloe in use in Mexico: but the thick Hedge and Mirgora Tree will prove sufficient; the acrid jeuces (juices) of the one and bitterness of the other, will not admit the generation of any insect that can prejudice the cochineal.

Anderson requested Campbell to immediately arrange for 100 *Mammoty* men to plant the fences, to mix manure in the soil and bring 20 cart loads of manure.

The Committee of Warehouses of the Court of Director also endorsed the view of the court that the true cochineal could be cultivated on the Coromandel coast as the soil was suitable for the production of cactus with few spines, and labour as cheap as in Mexico. The Committee stated in 1788,

The supposed discovery of Dr. Anderson in the environs of Madras, which, although unsuccessful in the issue, has nevertheless, led your committee to conceive that the insect may very successfully be introduced and propogated in the British settlements in India, to the advantage of the natives, the company, and the British nation, by giving to the former a new article of culture, to the second an additional article of commerce and to the latter a participation as a lucrative article of Trade, which has hitherto been enjoyed unrivalled by a neighbouring power.[33]

A thorough preparation was carried out on the hills of Rajahmundry Circar to introduce the Mexican insect in India. James Anderson wrote to Col. Kyd in 1789, 'The enclosures I now transmit you, are the continuation of my correspondence with government for the establishment of a Nopalya plantation on the proper ordering of which the preservation of our first American insect must depend.'[34]

Dr. Anderson was personated to establish the Company's Nopal nursery for the cultivation of the several kinds of cactus, on which alone the true cochineal feeds. Dr. Berry was placed under him as the superintendent of this garden. Plants of cactus were obtained with considerable difficulty from far distant lands like Canton, Manilla and the Isle of France as well as from Kew Gardens. In a short period of three years, there were over two thousand *Opuntia* plants in the Company's Nopal plantation.

In the Godavary district, the work of carrying out experiments of cochineal culture was to be carried out by Berry Heyne, a botanist who developed a large cactus plantation with the help of numerous coolies.[35] It was suggested to the Company authorities that in order to encourage the native Indians to promote cochineal culture the company should in the first instance give the lead by creating small plantations in each village and show how it could be done. The Indians should be acquainted with the advantages of this product as an article of trade so that it could be grown in the interior parts of the country.[36] At the same time the company was warned that the rapid and unregulated multiplication of the insect could cause more harm than good unless proper care and frequent examination of plants was carried out by the company experts.

That the cochineal experiment in southern India had succeeded initially is evident from the letter of Andrew Berry, who wrote, 'I might speak with some degree of certainly as to the success of their (cochineal) culture here and with pleasure I can say that the climate seems most congenial and that they (insects) thrive in all situations, both sheltered and exposed—even to the direct rays of sun'.[37]

The Company directed its officials to carry this experiment of cochineal culture with seriousness and great care as is revealed by the extract of the letter from the government. The Superintendent of the company while writing on the subject of the cochineal insect which had been lately introduced on the south-eastern coast, informed the Collectors under the Board of Revenue that he attached greatest importance to the object of promoting the breed of this insect as the early experiments indicated, that it was likely to become 'a source of essential benefit to the company'.[38]

The Madras Government, in order to encourage the production of a larger quantity of cochineal, offered to the cultivators of cochineal one pagoda per pound. The Madras Government had collected 21,744 pounds of cochineal in September 1797 at a price of nearly one pagoda per pound. The specimens of the insects sent from India to the Court of Directors for examination revealed that it was the *Sylvestre* or the wild specie and that there was little prospect of its being cultivated to any advantage for the supply of the European markets unless it was offered at about one-third the price of *Grana fina* or at about five to six shillings per pound covering all charges. The sale of Madras cochineal in England in the years 1797, 1798 and 1799 was about 55,196 pounds at an average of 8s. 8¼ d. per pound.[39] This was slightly more than its prime cost in India. In 1807, the management of the purchase of cochineal at Madras was transferred to the Board of Trade. Despite the lack of success of this experiment in terms of price factor and absence of profit, the Court of Directors was not disillusioned. The Court of Director declared in 1807, that, 'As the prices which we have obtained for the cochineal on sales have not been such as to reimburse the prime cost and charges, our sole reason for continuing to suffer a considerable annual loss upon this article, has been with a view to encourage the breeding of the insect, until it should become perfectly understood among the natives.'[40]

We do not have sufficient information regarding the results of these experiments in the long run. Perhaps the progress of the industrial revolution and some major breakthrough in the form of aniline dyes must have diminished the importance of cochineal as a commercial product. At the same time, it is worth mentioning that these early efforts of developing commercial products in the coastal belt of India by the English failed in the short run. Such measures could not have any significant impact on the rural population nor could these transform agrarian relations, but they do give us a glimpse of the subsequent policy of the commercialization of Indian agriculture and the English efforts of converting India into a raw material producing region to feed the English industries.

NOTES

1. J.F. Royle, 'Essay on the Productive Resources of India', London, 1940, pp. 59-64.

2. Immanuel Wallerstein, *The Modern World System, II, Mercantilism and the Consolidation of the European World Economy*, New York, 1980, p. 271.

3. Engels to Conrad Schmidt, 27 October 1890, Selected Correspondence, pp. 420-1, Karl Marx, *Tribune*, 11 July 1853, *On Colonialism*, p. 51, in Iqbal Husain, *Karl Marx on India*, from the *New York Daily Tribune* (including articles by Frederick Engels) and *Extracts from Marx-Engels Correspondence 1853-1862*, New Delhi, 2006.

4. Measures were adopted for introducing and establishing the filature of silk in the Carnatic and Northern Circars, *Madras Despatches*, E/4/878, vol. 18. 19 September 1792, India Office Records (hereafter IOR), London: Mulberry Plantation Report, 30 January 1793. *Factory Consultations, Vizagapatam District Records,* Andhra Pradesh State Archives (henceforth APSA). *Mulberry Plantation Report, Chingleput District Records*, correspondence regarding Jaghire lands 1792-4, vol. 445. Tamilnadu State Archives, (henceforth TSA). These papers also discuss measures regarding tobacco farming. Efforts of Roxburg in promoting pepper cultivation in Rajahmundry Circar on the eastern coast is mentioned in *Madras Despatches,* 31 July 1787, E/4/873, IOR.

5. Pennant while describing the Godavari and Kistnab deltas (River Krishna has been spelled differently by the local people and the Europeans) made a special mention of the Isle of Nagur. According to him, the circar of Rajahmundry was divided into three parts by the forks of the river which formed the famous Isle of Nagur. These deltas were very fertile as they were enriched by the soil brought down by the annual inundations. The banks included hills covered with immense forests of teak trees. Its wood was floated down and ships were built with it. Adjacent to Rajahmundry was the circar of Chicacole that extended up to Chilka Lake. It included parts of sandy wastes as well as a range of wooded mountains which bound the whole of western border. Pennant, *The View of Hindoostan*, vol. II, *Eastern Hindoostan*, London, 1818. Commercial Resident to Board of Trade, 18 April 1793, vol. 3323, *Nagore Factory Correspondence*, APSA, Board of Trade to Commercial Resident, 20 March 1793, vol. 3345, ibid.

6. Ibid.

7. Dr. Anderson's Proposals, 10 May 1791, *Masulipatam Consultation—Revenue*, 1791, vol. 2795, pp. 300-4. Letter received from the Board of Revenue, 26 April 1791, pp. 823-31, APSA, *Revenue Department Consultations*, pp. 591-5, 486-90. Roger Darvall, Collector to David Haliburton and Members of the Board of Revenue, 17 May 1793, Chingleput District Report, *Mulberry Plantation Report*, TSA.

8. Report on Indigo Cultivation, *Masulipatam Commercial Consultations*, 6 July 1790, vol. 2840, APSA.

9. Ibid.

10. Ibid.

11. Letters from the Court of Directors, 31 July 1787, E/4/873, *Madras Despatches*, India Office Library.

12. Despatch of the Court of Directors to the Governor General, 20 August 1788; Reports on cotton wool, 1836, National Archives of India; J. Taboys Wheeler, *Handbook of Cotton Cultivation in the Madras Presidency*, Madras, 1862.

13. Ibid.

14. Darvall, Commercial Resident of Vizagapatam to the Board of Trade, 26 June 1796 and 17 January 1798, Godavari District Records, *APSA*.

15. Letters from Madras Board of Trade, 27 May 1819, Tamilnadu State Archives.

16. Pennant, op. cit.

17. General Letter t the Court of Directors in England, 3 May 1793, Commercial Despatches to England, TSA.

18. Carlos Marichal, 'Forgotton Chapter of International Trade: Mexican Cochineal and the European Demand for American Dyes, 1550-1850', Paper presented to the 'Conference on Latin America Global Trade and International Commodity Chains in Historical Perspective', Stanford University, 2001. The dye came to be used for the finest fabrics used by princes, nobility, popes and the wealthy people in European cities. Its demand continued to grow from the mid-sixteenth century as the best crimson dyestuff for textiles. Within fifty years of its introduction in the European markets, cochineal as a dye virtually replaced kermes, the traditional dye material of Europe. James Watt, *The Commercial Products of India*, London, 1908, p. 347.

19. Joseph Bank to Dr. Anderson, 22 May 1787, James Anderson, *Correspondence for the Introduction of Cochineal Insects from America ...*, Madras, 1791.

20. M. Legoux de Flax, Essai Historique, geographique et politique sur L'Indoustan avec le tableaux de son commerce, Paris, 1807, p. 156.

21. Joseph Thiery de Menonville, Traite de la culture dun opal et de l'Education de la cochenille dans les Colonies Française de l'Amerique, Paris, 1787 in Rushika Hage, Cochineal, htpp//bell lib.umn.edu/Products/cochineal. html.20. Public Sundries Book, nos. 47 and 48, Record office nos. 1395-96, TSA.

22. Commercial Letters, 21 July 1787, *Madras Despatches*, E/4/873, IOR.

23. Ibid.

24. Ibid.

25. *Commercial Department Despatches from England to the President and Council of Fort St. George*, 10 May 1788, TSA.

26. Ibid.

27. Ibid.

28. Extract from the Minute, 22 July 1789, Public Consultations, 187A/42, *Commercial Department Despatches,* op. cit. TSA.

29. Letters from the Board, 29 August 1796, vol. 3707, Vizagapatam District Records, vol. 2728, APSA.

30. Ibid.

31. Dr. Anderson's Report on *opuntia*, 28 July 1796, APSA, *Letters from the Board*, vol. 3706, *Vizagapatam District Records*, APSA.

32. James Anderson to Archibald Campbell, 6 January 1789, *Public Department Consultations,* Madras 1538, TSA.
33. Letters to Sir Joseph Banks, J.F. Royal, op. cit.
34. Ibid.
35. Berry Heyne to Thomas Inodgrass, Collector of Masulipatam District, 3 November 1795, *Letters to the Board of Revenue,* vol. 884, *Godavari District Records,* APSA.
36. Andrew Berry on Cochineal, 23 March 1796. *Letters to Board of Revenue,* vol. 885, *Godavari District Records,* APSA.
37. Andrew Berry to Lord Hobart, Governor and Council Fort St. George, 26 August 1795, *Letters to the Board of Revenue,* op. cit.
38. Extract of Bluthenglon's letter, 29 August 1795, *Letters to the Board of Revenue,* op. cit.
39. J.F. Royal, op. cit.
40. Ibid.

Of Rivers and Roads:
Transport Networks and Economy
in Eighteenth-Century Bengal

Tilottama Mukherjee

TRAVEL AND MOVEMENT was the defining feature of eastern India's early modern landscape. The wandering trader, the itinerant bard, the mystic who travelled through the subcontinent untrammelled were integral components of the region, as was the immense array of commodities and services an inseparable part of the economic terrain. Transport systems were an area of vital importance, without which no economy can hope to survive—a truism that holds for the eighteenth century, as it does for the present. Markets and fairs did not arise in a vacuum, nor did they survive in isolation. The regional specialization of commodities was dependent not only on climate and geography but also on transport. These sites of exchange were connected together by land transport and rivers, carrying goods and people through a regular and reliable system. Studies on Bengal's transport history have been relatively few and they have not been fully integrated into the economic studies. K.K. Datta,[1] Sukumar Bhattacharya[2] and Sabyasachi Bhattacharya[3] have confined their observations to brief descriptions of certain routes. K.K. Datta devoted a small section to communications in his chapter on general economic conditions and began by remarking that the state of communications within a country greatly influences its economic state.[4] Some pathways were described and 'the usual statement that the want of communications made the villages economically independent and self-sufficient units' was discounted. The reason for this,

*This essay is part of my book-manuscript under preparation, 'Markets, transport and the state in the Bengal economy in the latter half of the eighteenth century'. I wish to thank C.A. Bayly, Rajat Datta, Yogesh Sharma and Raziuddin Aquil for their support and encouragement as well as valuable comments on my work. Thanks also to the organizers and all those who commented on an earlier version of the essay at the 'Coastal Histories: Society and Ecology in Coastal India, Sixteenth-Eighteenth Century', Workshop and Seminar, Centre for Historical Studies, Jawaharlal Nehru University, New Delhi, February 2007. The usual disclaimers apply.

according to him, was to be sought in other factors. The author did not attempt any sustained analysis of the networks. Sukumar Bhattacharya too included a few pages under the same head.[5] On the other hand, Sunil Kumar Munsi's writings concentrate on a later period with emphasis on steam navigation and the railways.[6] C. Oldham,[7] one of the earliest writers on this theme, described the old highways and 'by-ways' right from the Mauryan Empire to the nineteenth century and emphasized their strategic importance. B.K. Sarkar in his 1925 monograph on medieval transport covered many regions including Bengal in an eighty-seven page survey.[8] He posited that the facilities of internal transport were 'fairly adequate, according to the necessities of the age', and noted, erroneously, that 'service or employment was almost unknown' given the 'self-sufficing character of the different parts of the country'.[9] In recent years, Jean Deloche in his monumental two volumes[10] has attempted a wide survey of Indian transport from ancient times to mid-eighteenth century. Despite the richness of empirical data covering all regions and huge swath of time, there are some obvious problems. The transport system is seen in isolation, divorced from the economy and polity. The almost timelessness of the systems described, reduce the value of the work.

Most available secondary sources assume the relative frailty of networks[11]—a comment difficult to reconcile with the ground realties as presented through a diverse array of sources. In the period of this study, the problem of infrastructure has indeed been inadequately addressed.[12] There was an extraordinary movement of commodities and people and this was in part due to the presence of a dense communications network. This essay focuses on the basic constituent elements of transport systems of eighteenth-century Bengal, moving beyond what Sanjay Subrahmanyam critiques as 'such distressingly vague notions as "vibrancy", "crisis" or "decline"'.[13]

Transport and movement directly relate to distance and time. How far, how fast, how frequently and how safely the transport system could convey its carriage were the essential criteria of the extent of its efficiency. Bengal, given the nature of the terrain interspersed with rivers of varying dimensions, relied on water transport to a very large extent. The great river systems had always exercised a definitive influence on the natural environment, as well as on the economic, social and religious life of the region. David Arnold and Ramachandra Guha note that for South Asia, as for many other parts of the world, rivers have been more significant than the adjacent oceans and have 'run like a central silver thread through their history', while maritime trade has been of only secondary importance, at least 'until the opening of the European age'.[14] Bengal, however, was not an insular region, but Janus-faced looked both out as well as inwards. In this internal environment, with which we are concerned here, rivers acted as the connecting filaments, which wove together links between the villages, towns, and cities, the varied and multicoloured pieces of the pattern that constituted the region. It was the Ganges, the Brahmaputra and their various tributaries and distributaries, which shaped the life of eastern India.

Though rivers were the veritable lifelines of this region, overland routes too played important roles, albeit in a secondary capacity to the mighty sinuous channels.

Eighteenth-century Bengal was characterized by a hierarchy of urban and rural networks, a multi-layered strata stretching from the peasants and artisans and moneylenders, landed proprietors, through the spectrum of big and small merchants, to the upper echelons of political power. Though these elements were very much present in other parts of the subcontinent in this century, what set Bengal apart was its unique geographical or ecological terrain, which aided in providing a high degree of connectivity and fertility. Almost the entire area of the province of Bengal is deltaic.[15] The temperatures are moderate but humidity is excessive.[16] Her fertile plains—owing to the excessive humidity, the soil of Bengal has much 'power of sprouting' and most of the lands grew three crops in a year as contemporaries observed,[17]—congenial climate and the 'industry' of her inhabitants, were conducive for the production of immense quantity of goods. Bengal, thus, had an edge because of its fertility and its ease of communication. Changes of river courses determined the fortunes of agriculture and settlement patterns. Nature, in its various manifestations, is a constant presence in Bengal. This set it apart from any contemporary regional economy in India, prompting the author of *Riyazu-s-Salatin* to call the region 'Jinnat-bilad' or 'Paradise of Provinces'.[18] The various natural advantages of Bengal enabled her to develop widespread commercial relations early. In the eighteenth century, her trade and commerce brought within its net not only the different countries of Asia, but also of Europe and Africa.

The English East India Company records contain incidental references to these varied aspects. Our aim here is to sketch a picture of the transport networks, as refracted through the prism of the Company records and a wide array of travel accounts written in English and French.

River Systems

In Bengal, waterways formed more functional axes of communication and exchange than did land routes and a boat could be laden with cargo, which if carried on land would require an entire caravan of pack animals.[19] According to Rennell, the Ganges and Brahmaputra and their tertiary rivulets furrowed through the region to 'form the most complete and easy inland navigation that can be conceived'.[20] He observed that water transport conveyed all the salt and a major part of the food of ten million people. Further rivers carried exports and imports, to the tune of two million sterling per annum, the 'interchange of manufactures and products, throughout the whole country'.[21] They supported a fishing industry and provided livelihood to a large section of the population.

Deltaic Bengal was intersected by numerous navigable canals and rivers, which enabled people to travel in four days where 'no Moor land army can go in twenty'.[22] The Resident of the English East India Company's Dhaka factory reported that boat hire constituted a major part of travelling charges as,

this must necessarily be in a country so intersected with creeks and rivers, as this is for one part of the year, or in the other part, when creeks and rivers are scarcely to be distinguished but by their currents, for nine tenths of the country is under water, there is no passing from town to town but by boats and in many towns and villages it is necessary to have a boat or raft to go from house to house.[23]

The interiors of eastern Bengal, especially the tract lying to the east of Dhaka, did not have as many good roads as the western parts.[24] In south-eastern Bengal, goods were transported mainly by means of boats, or by human labour and through the minimal use of horses.[25] Lakes and rivulets communicating with rivers and becoming passable in the rainy season could bring boats to the peasant's door.[26] It was the rainy season, which induced the travellers, inland traders, as well as farmers to take to the boats, for, as Thomas Twining noted, the innumerable streams and natural canals that intersected the country and afforded a most extensive and convenient communication through the interior in every direction.[27]

More than 5000 boats—a conservative estimate given the volume of goods in transit—of various descriptions, of 10 to 100 tons each, were employed in this navigation with at least 50,000 boatmen earning their livelihood in this immense sector.[28] All these rivers flowed towards the south, one branch of the internal navigation, which was directed towards the interior—pursued its course in an opposite direction, and

by this providential order, boats are enabled to descend and ascend the streams of India at the same time and at the very time when they are best navigable. The current is strong enough to carry the boats down the river against the wind without sails and with sails the wind is strong enough to carry them up against the current.[29]

If boats moved in the direction of the tide, they could travel rapidly into the middle of the stream without using oars until the tide turned in the afternoon. Boatmen anchored their crafts a few yards from the bank, to prevent their being left dry as the water level fell. This could make tracking of boats necessary.[30] River navigation did not merely depend on the ebb and flow of tides and currents, but also on seasons and weather.[31] Storms could cause major disruptions and havoc.[32] Any change in the water level affected the movement of boats.[33]

Commentators had remarked that 'next to the abundance of the necessaries of life', the facility of transporting them to the market was the most vital element, which contributed to its cheapness.[34] The numerous rivers in the region enabled the peasants and traders to bring their produce to the market with little expense and trouble. Often, the records acknowledge that transporting by river worked out to be cheaper than transit on land.[35]

Though nature itself had been generous in providing an intricate network of arteries and capillaries that imparted such a boon to the region's economy, minor improvements through landed élite, the Company state, hybrid, or state-private initiatives were not lacking. For instance, the Company's government issued orders to the surveyor in Calcutta to clear the ditch as far as the first bridge, to provide boats coming to the custom house a secure place for docking.[36] Private individuals, as well as Company officials, proposed a number of such projects and the latter carried out numerous feasibility surveys.[37]

SEASONAL RIVERS

As some of the rivers were of a seasonal nature, traders and others had to use both land and water carriage.[38] Often pressing circumstances compelled Company employees to deploy multiple means to transport goods like saltpetre to Calcutta.[39] Streams or rivulets intersected the roadways requiring fording or ferrying, and hence, the regular use of both forms.

Although movement throughout the year was possible, merchants and companies preferred to keep goods for the next season in order to use river transport. This was due to the comparative cheapness of river navigation.[40] Though additional expenses were incurred for storing, yet the cost of sending by river was so much lower that it made it economical to hold stock until the rivers became navigable.[41] Merchants took advantage of the seasonal flow of the principal rivers on which the towns depended for their supplies. A 1791 reference shows how the merchants manipulated and raised prices from September to June: 'This is the season when rice usually becomes dear, not because a real dearth or scarcity reigns, but that here in Calcutta and the province of Bengal, the corn merchants in whole sale and retail raise the price of their grain; and this they always do till the communication with the upper country is opened again.'[42] The great 'aquatic movement'[43] according to another contemporary generally began from June and concluded in about December. The waters rose in May, were at their highest in July and August and lowest from February to April. With the snow melting at the higher reaches of the Himalayas, rivers began to rise in June.[44]

TYPES OF BOATS

An enormous range of boats existed in eighteenth century Bengal, reflecting local traditions, the depth of rivers and other navigational necessities, the variety of commodities and needs of the people. Travellers' accounts, reports and even Company documents, sketches and paintings often depict these boats of different sizes and shapes plying on the innumerable rivers of Bengal. These included the *palwars*, *bajras* (with both sail and row, having in general twelve to twenty oars), *bhurs*, 'paunchway', *morpankhi*, 'boliaskhasa', 'bolias choor', 'motwas', *ulak*, 'ash churas' and pinnaces (*pinas*).[45] The rates of hire varied.

The various boats mentioned provide only a glimpse of the immense range of barges plying on the rivers. Their employment, their differing carrying capacity, the established rates of hiring, sometimes fluctuating according to demand—all testify to the complexity and extent of the economy which required such a wide variety of boats to ply on the rivers. Some contemporary observers noted that most of these vessels were very cheaply constructed.[46] Whether many individuals, beside rich merchants, could have a number of these crafts given the cost is difficult to say, as the price in 1744 of a 'very good teake budgerow' was Rs. 2000,[47] while a *pansi* cost Rs. 62.[48] Evidence, however, attests to their wide ownership. There was trade in old and second-hand boats as records from Calcutta show, thereby enabling other less wealthy sections of society to own one. Even traders in somewhat reduced circumstances could hire out boats to augment their resources as instances from Dhaka demonstrate.[49] Many, called 'naiyas' who were fishermen or farmers let out boats to hire. It was possible for even economically weaker groups to possess smaller vessels. Possession was not entirely gender specific as even poorer women could hold ownership and their income sometimes solely depended on the money earned by hiring out boats.[50]

THE ORGANIZATION OF RIVER TRANSPORT

River transport appears to have been a highly organized sector of the Bengal economy, with an assortment of specialized persons functioning in various positions in a hierarchy. The *ghats* had their chiefs who charged duties on the goods passing on the rivers and could use force to realize their dues as petitions from merchants show.[51] Some, such as the *ghat majhi*s, wielded immense influence in the procurement of boats,[52] and, hence, the smooth operation of the system, forming small nodes of power whose function scholars have largely ignored. The appointment of a *ghat majhi* was essential as noted by contemporaries, especially in all big cities, for securing boats for the government, for the troops and traders.[53] They regularly registered the names, places of residence of all *majhis*, *dandie*s and owners of boats.[54] They fulfilled a wide array of functions. Buchanan Hamilton notes that at each landing place, there was the *ghat majhi*, 'who is very importunate to be employed and liberal in his promises'.[55] He took money to advance to the boatmen on the assurances that everything was ready 'but when the boat is wanted the credulous freighter will find that the procuring this was not of all in contemplation'.[56] However, other observers and Company proceedings paint a much more positive image of these groups and indeed, no vessel could be obtained without their intercession. When the demand was high during the season, the Company even granted permission to its employees to build their boats.[57]

Landed proprietors played an important facilitating role in the transport sector, as in other economic domains. *zamindar*s and *taluqdar*s owned *ghats* or landing places of ferries, for instance, in Nadia.[58] They farmed out ferries to the highest bidder.[59] Reports of mishaps, especially of boats laden with commodities, do not occur in the records as often as one might expect. On the whole, the system functioned smoothly and efficiently. Local society mobilized all kinds of resources in order to keep the channels of movement open and in working order.

TRANSPORT WORKERS

A considerable number of people were engaged in various activities related to the transport sector. The smooth operation of the economy depended to a large measure on the actual physical transference of huge volumes of goods. If the different participants in this economy had to maximize profits, there had to be a regular and reliable system of movement. Provided Rennell's estimates of 30,000 boatmen plying on the Ganges are anywhere near the true figures (as also the observations of other contemporary commentators and travellers), we have a segment of society that is numerically significant. For not only were *dandies*, *majhi*s and *ghat majhi*s intrinsic components of this sector, but other related service providers—carpenters for making carts and different kinds of boats, the 'rungsaj' or houseboat and palanquin painters and rope makers,[60] among others were present in large numbers.

Wages were certainly not static, but generally on the lower side. The rates were negotiable and employers had to provide incentives in the face of competition from other merchants and companies.[61] A vast section of the population gained employment in the transport sector in some capacity or the other, with varying remuneration.[62] They, however, managed to survive, in a commercialized economy where *ghats*,[63] ferries, boats' sales were farmed. Some might have taken to these vocations in the slack season and might have been farmers of land during the rest of the year. The wages increased over the course of the years. The evidence suggests that it was only with the coming of the railways that the industry and the various people involved in the sector suffered disastrous effects.

Majhis and Dandies

Although the rivers of Bengal transported a huge quantity of commodities, very little information exists about these boatmen—*majhi*s and *dandies*—in the records. Often documents mention that 'manjees' receipts' were taken. Whether they were individually responsible or some kind of organization and collective responsibility existed, it is difficult to say. However, a few references indicate that a person stood as security and was responsible for goods lost in transit.[64]

The *majhi*s appear to have belonged to the poorer sections of society.[65] Every boatman rented a small piece of ground and paid his rent. His wages and his wife's spinning took care of clothing, religious or festival expenses. He had the produce of his land for food and raw materials for clothing.[66] Natural calamities could severely hit these groups, as they did not have adequate means for keeping aside sufficient stores of grain.[67] Any reduction in food supplies hiked up prices while wages remained constant.[68] As they and their families also had small farms and worked in the textile sector, albeit on a very limited scale, this might have helped them at least during minor scarcities.[69] When sick and unable to work, they usually borrowed money by pawning silver ornaments, articles of dress or copper and brass utensils.[70] This suggests that under normal circumstances, they could accumulate a few material possessions, which could later help them tide over difficult times. Absence of several 'key' consumer goods amongst the poor of this period does not necessarily signal a lack of knowledge of such goods, where to purchase them and how to use them as Sara Pennell[71] notes in a different context. There was no association among the workers for mutual relief during times of ill health or other misfortunes, though they did have some kind of community network for resolving professional problems. They assembled to adjust disputes, to settle the rates of wages and to subscribe for pujas.[72] The rates of wages varied according to seasons of the year, the highest being at the harvesting season and during the fairs.[73]

RIVERINE NETWORKS

The sinews of the transport organization bolstered and sustained the integrated yet decentralized commercial economy. The existence of a number of direct routes meant a high degree of connectivity. Routes radiating out of cities and towns formed an extensive and complex maze.[74]

TABLE 1: NUMBER OF DIRECT RIVER ROUTES
BETWEEN THE FOUR MAJOR CITIES

City	Calcutta	Dhaka	Murshidabad	Patna
Calcutta		4	4	3
Dhaka	4		2	2
Murshidabad	4	2		4
Patna	3	2	4	

Source: Based on James Rennell, *The Bengal Atlas* (*Memoirs of a Map of Hindostan or Mughal Empire and the Bengal Atlas*, ed. B.P. Ambashthya, Patna, 1975).

From the number of direct routes for each of the major towns in the region, it appears there was a kind of hierarchy where places like Chittagong, Dinajpur, Lakshmipur, Rangpur, Kumarkhali and Sylhet occupied a higher rung as compared to the rest. Though some were better serviced by way of roads and others through rivers, only a few such as Chittagong and Rangpur were equally well connected by both land and river.

FIGURE 1. RIVER CONNECTIVITY OF CALCUTTA, DHAKA AND MURSHIDABAD

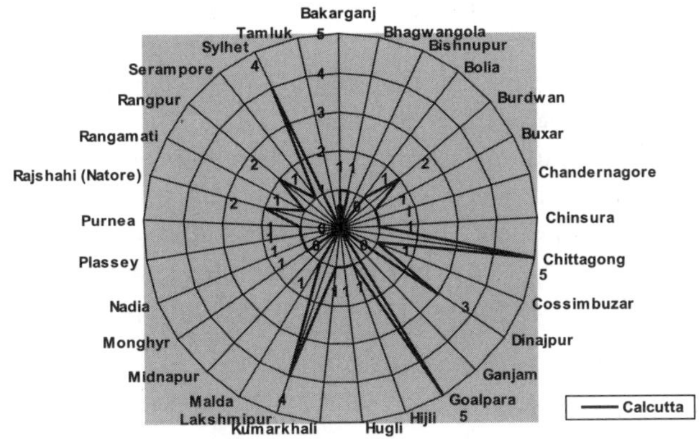

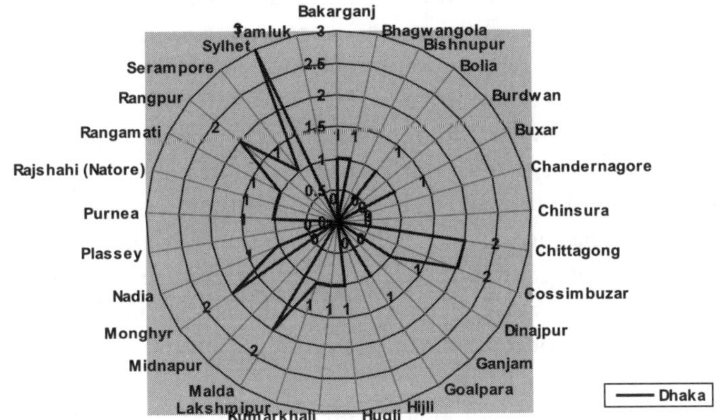

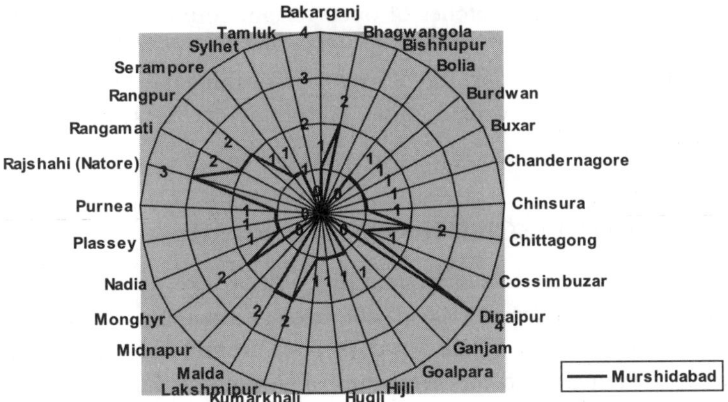

Source: Based on James Rennell, *The Bengal Atlas* (*Memoirs of a Map of Hindostan or Mughal Empire and the Bengal Atlas*, ed. B.P. Ambashthya, Patna, 1975).

River navigation had its disadvantages. The system was vulnerable to rains or rather the absence of it, which rendered many of them operational only during certain periods of the year. However, unlike land transport, it was not beset with the problems of loading and unloading in the deltaic areas. It worked out to be relatively cheaper and enabled movement of huge quantities of diverse kinds of goods to both near and far-flung places. The Company records, folk tales, ballads and paintings acknowledge their importance by their repeated portrayals. The river networks were truly the lifelines of the region supplying and connecting the peasant, the chain of intermediaries, merchants and a host of diverse array of people who moved about on their myriad tasks or for leisure in this extraordinary century of movement and enterprise.

The Surface Grid

As Deloche observes, 'the existence of road network necessarily implies the collaboration of nature and man and corresponds therefore to a geographic and human choice'.[75] The topographical layout inevitably determines the system of communications, which compelled land transit to follow the natural contours of the landscape. Notwithstanding the extremely variable quality of these roads, it was due to all these that, to use Fernand Braudel's eloquent words, the rural community was able to exist, to breathe the outside air, and learn to draw on resources other than its own.[76] Eighteenth-century communication networks showed seasonal integration as well as society's extensive use of existing modes of transport. An extensive network of roads existed in eighteenth-century Bengal, despite the constraints, which communications were faced with in the low-lying zones cut by rivers.[77]

At this time, even major roads were mere strips of presumed solid way, between unsafe verges and these trails frequently traversed rivers, marshes and other bodies of water. Stretches of open terrain could be used as roads, while pathways through dense vegetation called for intense efforts in clearing of forests and levelling of ground. There was nothing beyond pounded brick, which was generally overlaid upon a stratum of the same material, in its unbroken state.[78]

PATHWAYS AND FACILITIES ON THE ROAD

Road building was a well-established function of state systems in the medieval period. The Grand Trunk Road ran from Chittagong to Mymensing via Noakhali, Comilla, and Dhaka, and the one from Rajshahi to Rangpur via Bogra, and extending to the south probably as far as the Sunderbans by way of Malda and Murshidabad.[79] Though the region lacked durable materials needed for road construction,[80] by 1804 some had been laid in the vicinity of the European enclaves.[81] Other routes would have had to follow embankments.[82] In lower level areas like the 24 parganas, the top of the *bund*s.[83] Despite the difficulties of land transport, it was frequently used, given the seasonal nature of many rivers of the region.[84] As noted previously, at times, both land and river transit was used for reaching one's destination.[85]

Travellers and traders would not have taken to the roads in large numbers without some basic facilities in the form of resting places, drinking water and safe transit guaranteed by the ensemble of ruling authorities. All levels of local society involved themselves in the welfare of travellers, influenced by both religious injunctions, as well as other material considerations. Landed proprietors, other well-heeled groups, numerous religious establishments, house owners, and ordinary villagers—all were part of this supporting network. The provisions were distributed without monetary returns to Brahmins or other religious mendicants in the *choubaries* of Bengal.[86] *Akhara*s or convents too served similar functions to inns in a city such as Dhaka.[87] Travellers could halt for the night near wells or mosques.[88] Lodges in the premises of mosques and Sufi institutions served similar functions.[89] Ruling authorities alienated villages and lands exempt from revenue for the specific purpose of supplying travellers with provisions and other necessities.[90] *Zamindars* also contributed by feeding pilgrims and accommodating them.[91] If there were no habitations near roads, villagers came to the *chauki*s to sell grain to the travellers.[92] Although *sarai*s were present,[93] on the whole they were more common in north India[94] and in the Deccan, where travelling by land was more frequent, unlike Bengal.[95]

SAFETY OF ROADS

Travelling was not easy in the eighteenth century. Besides fear of beasts and robbers, exactions by local authorities were an additional source of distress. The first part of the century was relatively safe it appears with both the *Nazims* and the landed proprietors trying to ensure a certain level of security, and speedy redressal of thefts. It is in the later half of the century, when the hazards appear to have increased, with a great degree of political, ecological and economic flux. Despite all the inconveniences, people frequently moved about on their differing tasks with striking frequency.

The *Nizamat* and landed proprietors were much concerned about their subjects, especially merchants, and they tried to ensure that the pathways remained open and secure.[96] European travellers wrote that local people retrieved lost goods and hung them from the nearest tree and informed the nearest *chauki*. Richard Lufrano, for instance, notes the importance that early modern Chinese manuals placed on flexibility and honesty. The authors of these tracts advised their readers to return lost goods, money, etc., to avoid harming the reputation of shops and businesses.[97] Similar considerations must have influenced their contemporaries in Bengal too.

This picture of relative safety of travelling on roads[98] is contradicted in other reports. Writing in the mid-seventeenth century, Banarasidas, a Jain merchant in his autobiography mentions his frequent travels in northern India, as well as Bihar and Bengal. According to his account, normally merchants seemed to have moved about in groups.[99] Mustepha[100] recommended travellers to obtain protection from the village headman in addition to procuring guides.[101]

Contemporaries sometimes note the presence of armed travellers.[102] The Company's government decided to disarm wayfarers, which was later reconsidered and then imposed the prohibition on sanyasis. Numerous references to dacoities creep into the later records.[103] Highways running through villages could even cause the *raiyats* to desert fearing undue impositions and troubles.[104]

How widespread these incidents were it is very difficult to state with certainty. It might have been localized in certain remote areas. It could be a ploy for the Company's officials to seek greater intervention through deployment of soldiers. They petitioned the Company for sepoys to be stationed at various places, particularly on the borders to prevent dacoits escaping from one district to another.[105] Proposals were sent for *thanas* to be set up in each pargana.[106]

INVESTMENT IN SURFACE TRANSPORT

The Nawabs and particularly the landed proprietors took an active part in maintaining roads and inns. Occasional, as well as established contributions from the local inhabitants were required for works of general utility, such as dams, dikes, reservoirs, bridges and roads.[107] The charges for building a road were by no means very negligible for the cost of laying 7500 yards of pathway could amount to Rs. 600. Once laid, this would not incur charges for repairs for some years.[108] Within the cities, streets were laid out 'without the least design'.[109] Rennell goes on to say that they were narrow, confined and crooked.[110]

Land transport was heavily dependent on animal haulage, with a great variety of carts[111] and cattle in use. They had differing carrying capacities and were used for different kinds of commodities and distances. The highly specialized carrying industry was pre-requisite for the extensive commerce of the region. Raw material and an infinite range of finished goods had to be moved within the locality, the landed proprietors' estate, to other districts, ports, and indeed other parts of the subcontinent regularly, safely and without any damage in transit. It had to be a highly organized system, as the evidence amply testifies.

The bullock carts and the pack animals, principally the ox were the chief means of land transport.[112] Sometimes *qafilas* or caravans of as many as 10,000 and 20,000 animals were observed moving from Bengal to Agra and to Surat, led by the Banjaras.[113] A caravan of 100,000 oxen loaded with broad cloth, 'perpets', tin, pepper, spices and other commodities started from Radanogore in the month of March and arrived in less than two months at Delhi just to cite one example.[114] In the eastern parts of Bengal,[115] buffaloes were employed both for draft and for carriage, though it was slower than the ox and was unable to carry a much greater burden.[116] Cattle could be hired for Rs.5 a month. The cost arising from land carriage did not exceed one rupee a maund for a hundred miles. Ownership was eclectic, with specialized contractors and sometimes merchants and peasants investing in transport facilities.

Numbers are hard to quantify for the eighteenth century, but evidence from Rangpur suggests a fairly large number of cattle were engaged in carrying commodities to markets.

FIGURE 2. NUMBER OF LIVESTOCK USED FOR CARRIAGE IN THE DISTRICT OF RANGPUR

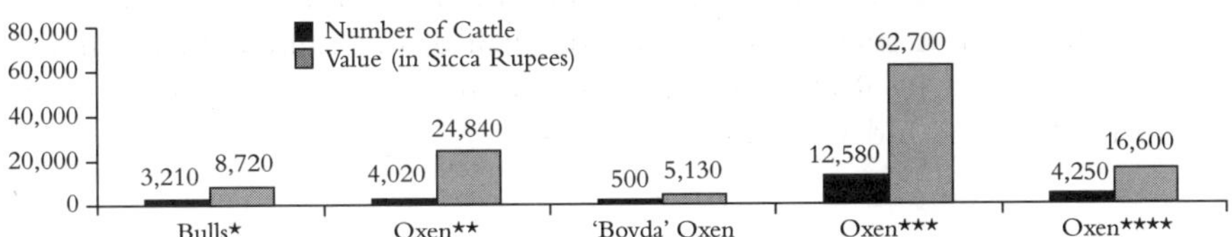

Note: *Reserved for trading in villages, **for carriage by traders, ***carriage by farmers, ****employed both in carriage and plough. Traders reserved 'Boyda' oxen
Source: CSSSC, Mss. Eur. G 12, mf, Table No. 33.

If we compare these figures with those of Germany and north India, Rangpur's[117] 4,464,037.34 ton miles per year,[118] or 5,135,211.104[119] are not too modest. The working days for Bengal might in fact have been higher than 115, which have been used for calculating the carrying capacity for north India. The working days for weavers, according to a contemporary report stood at 298 days,[120] but then it is a sedentary occupation not dependent on the seasons as the transport sector would have been. Besides the high probability of errors in the Rangpur figures, the comparatively smaller number could be because for Bengal, land transport was very much of a secondary importance, and river navigation accounted for the major portion of the movement of commodities. Though, the traffic on the roads was increasing towards the end of the century.[121]

PORTERS EMPLOYED IN THE TRANSPORT SECTOR

Human labour was often employed for carrying goods and people. Hiring rates were cheap,[122] and consequently economical to use. Landed proprietors controlled the supply of such labour in their respective estates, though some must have been outside their purview and control. Some labourers immigrated to the bigger towns and cities to earn a living. The landed proprietors and their agents generally gave lands at low rates to these groups, and called on them when required; they were paid only for the day's labour.[123]

Bearers and coolies were organized bodies of men. The hiring rates appear to have been regulated and fixed, rather than being ad hoc. The rate of duty too varied sometimes, depending on the mode or weight.[124] A few colonies of immigrants could be found in Dhaka for instance. An observer writing in the 1840s noted that the street coolies—numbering about 300—were from Purnea and Bhagalpur and had settled there for above a century and a half.[125] They lived in groups, each consisting of about twenty people, under the supervision of a *sardar*, who regulated their work, and divided their earnings among them at the end of the month.[126] A large number of palanquins were present in major cities, which could be jointly owned.[127] Neither the *Nizamat* nor the Company state appeared to have been able to regulate the activities of coolies and boatmen. The kind of argument put forth by scholars that the colonial state's labour interventions were consistent in objectives and practice and attained institutional forms in the latter half of the eighteenth century[128] is difficult to detect, ironically in the first province to come under the domination of the English East India Company.

ROADS IN LATE EIGHTEENTH CENTURY BENGAL

An enabling cause, as well as the consequence of commercialization, the transport network was of crucial importance. Certain areas, even though able to produce a particular crop, would not do so for lack of proper systems of communications in that specific area, as a 1775 report from Dinajpur makes clear: 'as there are no navigable Rivers in that district it is little worth the attention of the inhabitants to cultivate and the impossibility of exporting it'.[129]

There were four levels of networks in operation, as it appears from the sources. At the primary plane were the routes connecting the villages to their markets; then the linkages between towns; between the towns and the big cities; and finally the connection from one city to another. There are very few population centres of any importance, which appear to be poorly connected if Rennell's surveys, as well as the numerous surveys that were incorporated in the Orme collections,[130] are taken into account. At the lowest point the interconnections between the village and the *hats* and *ganjs* has been noted by historians in recent years while delineating the commercialized environment, though of course it is difficult to map out each of these roads. The communication links existing between towns are documented. The connections between the centres, other than the big cities also manifest a high level of connectivity.

Again, as in the case of river navigation, from the number of direct roads for each of the given towns, it appears even among these, there was some kind of ranked gradation where places such as Chittagong, Midnapur, Monghyr, Purnea, Rangpur and Sylhet occupied a higher rung in comparison. At the highest level were the cities of Calcutta, Dhaka, Murshidabad and Patna and the linkages among them are well recorded.

The difficulties of land transport were many as pointed out at the outset. Inclement weather would have played a role in making travel slow and unpredictable. All year mobility was not possible though it was less vulnerable

FIGURE 3. ROAD CONNECTIVITY OF CALCUTTA, DHAKA AND MURSHIDABAD

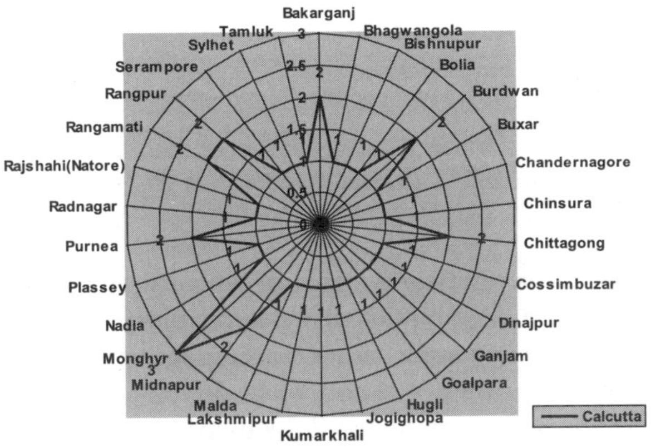

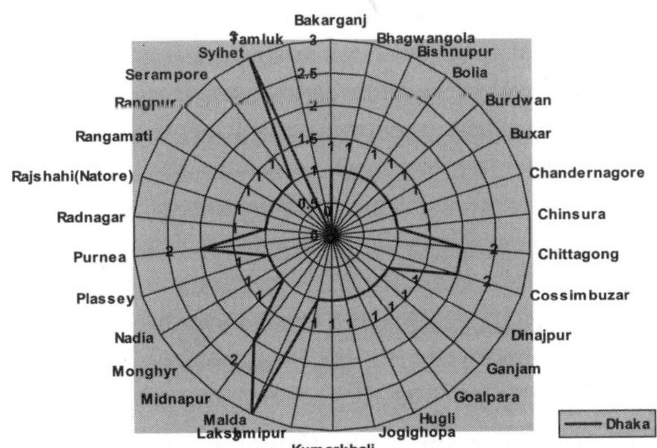

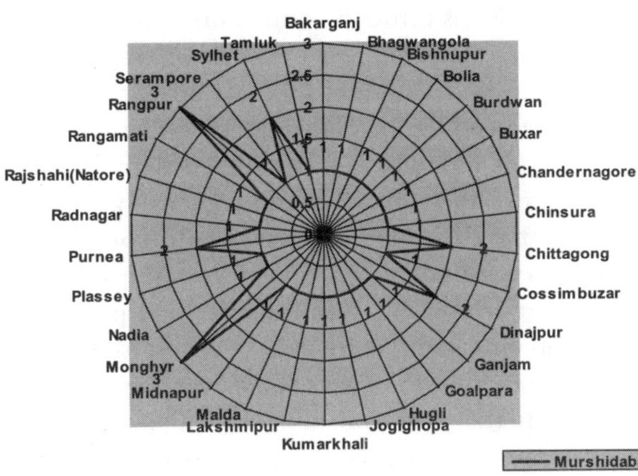

Source: Based on James Rennell, *A Description of Roads in Bengal and Bahar, etc.*, London, 1778, pp. 3–263.

TABLE 2: DIRECT ROADS BETWEEN THE FOUR MAIN CITIES

City	Calcutta	Dhaka	Murshidabad	Patna
Calcutta		2	2	3
Dhaka	2		2	3
Murshidabad	2	2		4
Patna	3	3	3	

Source: Based on James Rennell, *A Description of Roads in Bengal and Bihar, etc.*, London, 1778.

in this respect to the river networks, many of which were of a seasonal nature. The problem posed by large bodies of water on the pathways, routes cut at numerous junctures necessitating either fording the streams and the upper courses of large rivers in the dry season, or by ferrying rivers swollen by rain and non-fordable streams which required the unloading of cattle. Although the inconveniences of transit on land were many, yet they do not appear to have been viewed as major obstacles and the routes continued to be plied, especially in the dry season. Even if they were taken to be a hindrance, their continued use only indicates the level of commercialization of the region. The servicing of markets was essential, as the livelihood and survival of a sizeable section of society depended on it.

TRANSPORT AND THE ECONOMY

An extensive network of rivers and roads therefore existed in eighteenth-century Bengal. Direct routes linked the cities, towns and villages. The means of conveyance was varied, ranging from oxen, horses and camels to palanquins and carts. Portage was common as the employment of coolies for carrying goods was noted. Our evidence also indicated the organized system of hiring of bullocks and other animals through contractors as also an ordered structure for engaging boats. The finely meshed network of rivers offered a mode of transport that was cheap, with different kinds of boats plying at different rates subject to change influenced by the demand factor. The theoretical availability of transport services is of little use if it is coupled with prohibitive costs that do not appear to be from their frequent and regular usage. Their regularity is gauged from the fact that the Company was able to complete its investments and send the various commodities on time for its ships to embark on their way home, season after season. All year mobility was possible with the land and waterways complementing each other. When the roadways became impassable as during the rains, the rivers took over the traffic, and conversely during the dry periods the carts and bullocks supplanted waterways. Often, however, both were utilized, climatic factors permitting. Sometimes, the conveyance of a particular commodity necessitated the use of both the modes in order to reach the final destination. Boats provided a means of livelihood to many, and for the bigger merchants acted as a source to supplement and augment their resources. The keeping of a large number of boats and bullocks indicate their regular employment in eighteenth-century Bengal. Access to these means of conveyance did determine, to an extent, the degree of manoeuvrability in an economy where though various sectors were organized, there was no central body—whether the state or some other economic organization—which was controlling the whole system.

Jan de Vries[131] observed that the Dutch *trekvaart* network stood out in the context of pre-industrial economy because of its large requirement of fixed capital investment, its needs for an immense, relatively specialized labour force, its need to supervise and co-ordinate activities which were spread over a large geographical area, and its production and sale of a service that depended on a mass market. Despite the differences between the two cases, we can see some similarities with Bengal, though fixed capital investment as regards construction was not usually made for the roads, given the high cost of materials which precluded their use, and the rivers did not need any manmade improvements except the repairing of banks. Charges of *poolbundy*[132] (maintenance of embankments) were a constant yearly expense, for unless the bridges and banks of rivers were annually repaired before the commencement of the rains, the 'town and part of the country' were inevitably flooded, resulting in considerable loss of revenue.[133] The means of carriage would of course require capital, but even poorer sections of society could own them (especially smaller boats). The large and specialized labour force was also not lacking, as the sources reveal time and again, the existence of the ubiquitous *majhi, dandi, ghat majhi*s and cart drivers. The geographical ambit within which they functioned and the latter point about the existence of a ready market appears to be applicable in the case of eighteenth-century Bengal.

Transport costs on land were estimated to be twenty-eight times more expensive than on the rivers.[134] Since W.H. Moreland, it has been repeatedly asserted that the frailty of the routes of communications had a major bearing on the difficulties of Indian commerce. Transport costs in India obstructed the development of long distance trade in many items.[135] However, the evidence from the records suggests a different picture.

Despite these developed features, there was a degree of fragility in the system, dependent as it was on climatic and ecological factors. River courses altered, and with it the economic fate of the towns and markets situated on its banks. Even if a particular area was suited to the growth of certain crops and production of certain commodities, unavailability of a suitable mode of transport, cheaper and less circuitous routes, hampered the marketing of the produce and hence offered little inducement for its development. Although we have spoken about the presence of numerous boats and roads, the mere presence of these did not mean a perfect transport network. Instances of loaded boats stranded in the rivers due to insufficient water are also numerous. Sometimes there were not enough boats to be procured.[136] It appears from the records that the whole process of manufacturing a particular product was synchronized with the seasons when the rivers were navigable.[137]

Functional, regular and quick communication systems not only have direct co-relationship with distance, but as well as with time. There was a pervasive engagement with time also in Bengali society in the eighteenth century. The financial aspects of inter-regional trade were transacted by means of speedy exchange of letters, bills and payments, while merchants also travelled more widely. The quickening of this commercial impulse represented a further example of an increased volume of communication between people based in distant parts. An important feature that seems to have developed in the eighteenth century was the greater regard for the importance of time. Was it then that the concept of time and money was emerging in this period as reflected in the quickening of the commercial transactions? The increasing number and frequency of markets, and fairs, specialization of merchants and other intermediaries, regional specialization of production, the state's (both the *Nizamat* and Company state) attempts at manipulating the economy, routes and information flow—might reflect this desire to save time. The time taken for the goods to reach the markets determined the prices of commodities in the markets and ports. The weavers in the early years used this as a bargaining power to pass off inferior quality goods when the sailing season was at hand, and the Company was desperate to procure the required quantities. It might be that the commercialization of the economy and polity spoken of in the historiography over the past twenty years was actually a manifestation of that importance of time discipline.

The conception of time itself does not originate in this century. Personal names often had, and indeed still have, temporal origins, linking one to the time of the day, a season of the year, or an historical event. The daily life followed a pattern with appropriate hours of the day apportioned for different activities. The conception of the week was, of course, important for traders and buyers, for the periodic markets were often scheduled once, twice, a week. Similarly, seasonal cycles determined when to sow or reap, fulfil contracts with merchants and revenue obligations, fairs, festivals, marriages and kinds of food. Life cycle stages rather than age in years appear to have been the prevalent means of categorizing individuals. Ritual time varied according to gender, with premium put on authority and age. Both linear and cyclic conceptions of time co-existed. Time was measured not with European clocks, but with water ones, as a contemporary traveller described.[138] Similarly, Sieur Luillier wrote that instead of clocks, 'garis', which were small copper vessels, with a hole in the bottom, were used. This was put into a larger one full of water, the smaller ones filled slowly through the hole, and hence sunk gradually; the time taken to be completely submerged was called a 'gari'.[139] Until the end of the eighteenth century, a single-handed clock indicated time adequately enough for most people even in England, as it shows the smallest unit of time commonly referred to the 'half-quarter'. The choice between models was governed by what people thought they could afford rather than a felt need for greater precision.[140]

Households allocated more of their labour to production or offering services for the market while saving time for that labour by purchasing some things that once were produced in the household itself. Households in the South Asian variant of the 'industrious revolution'[141] appeared to have worked more hours although it is difficult to know whether the working days increased since say the seventeenth century as was happening in Europe,[142] but certainly the numbers were comparable with 298 working days. A contemporary French observer noted the 'severe regularity' of the manners of the local inhabitants and occupations and castes, and that there was little time in the day that people employed for rest and moderated their wants,[143] which of course would have had economic implications.

The labour market in Bengal's cities was primarily characterized by considerable occupational diversity. Professional groups in big towns made up a sizeable proportion of the population. The intensification of work was also associated with greater exploitation of family members' labour—wives and children—and greater recourse to 'binge drinking and binge leisure'.[144] Similar instances can be located in Bengal in the eighteenth century, especially in the textile weaving sector, with large number of women employed, and the latter perhaps in every town of Bengal, with Calcutta alone having 396 arrack shops.[145] Although 'binge leisure' would not be a characteristic of the society at large, but perhaps only for the upper echelons.

CONCLUSION

The links between trade, transport, urban networks and a commercialized economy are quite direct. The denser the network, the greater the accessibility of the village or city or *zamindari* to the hinterland and to each other, and wider the network was, larger the area which supplied the needs of the city or town or rural area. The various routes indicate a hierarchy of centres where some were more accessible in terms of the number of routes servicing them. It was not that Calcutta alone reigned supreme. Other cities like Dhaka, Murshidabad and Patna were important too. This diffusion of focus meant that the other cities had access to the whole region directly, and not just via Calcutta, which in turn meant not only physical accessibility, but also the equalization of accessibility to information (which might be the market price or the news of scarcities). These other towns were forced, perhaps by the nature of their transport system, to be either more autonomous or more dependent on other larger cities. Below the pulsating quartet of Dhaka, Murshidabad, Calcutta and Patna, lay a string of towns dispersed all over the region.

The expansion of urban centres was due to a variety of factors and among these geographical ones was particularly significant. Trade and communications were among some of the most important factors that promoted and retarded such growth. There was a whole range of small and intermediate markets (the *hats* and *bazaars*), which were linked together by an extensive system of rivers and roads. *Ganjs* and *hats* were usually established at points where the transport of goods and the assemblage of buyers and sellers were facilitated by the existence of roads and waterways. This explains the existence of innumerable *hats* on the banks of the rivers of eastern India. The inland transit and customs dues accruing to the various political and other powers were substantial if the mere number of such *chauki*s is taken into account. The cost of bribes, presents, taxes and even thefts had in the long run to be borne by the consumer. However, despite these spiralling amounts, such levies did not seriously hamper trade and commerce along either the vertical or the horizontal planes of the hierarchy. The increasing frequency of commercial transactions at these various levels only attests to the robust nature of the economy. Given its importance to the economy, the state could not and did not remain aloof from such a critical sector and hence it attempted, at various points of time, to regulate the diverse systems of trading and transport networks.

The range of the Company's interests and ambitions was undeniably broad, as it involved itself in the construction of roads and embankments, the pioneering of some new routes and new forms of vehicle, the surveying of existing networks and the gathering of other commercially valuable information.[146] The Company intensified the state's involvement in certain areas and introduced a regime of oversight in others. Before the Company took over the reins of power, the *Nazim*s did have an interest in roads, *sarais* and bridge construction and in restricting the use of status-marking modes of transport to certain religious communities or strata, in ensuring the safety of travellers and in gathering information. The Company stepped up the degree of specificity and complexity of existing regulation, while also bringing in surveys establishing the form of capillary trade networks prior to their possible simplification. The Company was inclined to pass further resolutions regarding the transport of goods in all of its aspects, but these intentions tended to founder, given the decentralized nature of Bengal's economy, with the Company being forced in most instances to fall back on the local structures of intermediaries. The vast array of agents tended in practice to prosecute their private interest, turning the system of Company jurisdiction into a set of uneven or heterogeneous dispensations, in which trade followed other pathways than those foreseen by the Company's government.

NOTES

1. K.K. Datta, *Economic Condition of the Bengal Subah in years of transition 1740-1772*, repr. Calcutta, 1984.
2. Sukumar Bhattacharya, *The East India Company and the Economy of Bengal from 1704 to 1740*, Calcutta, 1969.
3. Sabyasachi Bhattacharya, 'Regional Economy—Eastern India', in Dharma Kumar and Meghnad Desai, eds., *Cambridge Economic History of India*, vol. II, Cambridge, 1983, pp. 270-95.
4. Datta, *Economic Condition of the Bengal Subah*.
5. Bhattacharya, *The East India Company and the Economy of Bengal*.

6. Sunil Kumar Munsi, *Geography of Transportation in Eastern India under the British Raj*, Calcutta, 1980. Idem., A brief survey in 'Transportation and Communication', in Sirajul Islam, ed., *History of Bangladesh 1704-1971*, vol. II, 2nd edn. Dhaka, 1997, pp. 606-27.

7. C.E.A.W. Oldham, 'Routes, Old and New, from Lower Bengal "up the Country"', part I, 'Old Highways and Byways', *Bengal Past and Present*, vol. XXVII, nos. 55-56, 1924, pp. 21-36.

8. Bejoy Kumar Sarkar, *Inland Transport and Communication in Mediaeval India*, Calcutta, 1925. Also for a wide survey of the Mughal period see A.K.M. Farooque, *Roads and Communications in Mughal India*, Delhi, 1977.

9. Sarkar, *Inland Transport and Communication*, pp. 53-4, 71.

10. Jean Deloche, *Transport and Communications in India prior to Steam Locomotion*, vols. I and II, trans. James Walker, Delhi, 1993-4.

11. H.R. Ghosal, *Economic Transition in the Bengal Presidency, 1793-1833*, Calcutta, 1966.

12. Works on transport history are indeed very few for the pre-railway period even for other regions in India. Nothing comparable to studies in Europe exists. See for instance Jan de Vries, *Barges and Capitalism: Passenger Transportation in the Dutch Economy, 1632-1839*, Utrecht, 1981.

13. Sanjay Subrahmanyam, 'Towards a Conclusion', idem., *Penumbral Visions—Making Politics in Early Modern South India*, Ann Arbor, 2001, p. 254.

14. David Arnold and Ramachandra Guha, eds., *Nature, Culture, Imperialism: Essays on the Environmental History of South Asia*, Delhi, 1995, p. 11.

15. Richard M. Eaton, *The Rise of Islam and the Bengal Frontier, 1204-1760*, Berkeley and London, 1993, p. 3.

16. O.H.K. Spate and A.T.A. Learmonth, *India and Pakistan—A General and Regional Geography*, London, 1967, pp. 572-5.

17. Ghulam Hussain Salim, *Riyazu-s-Salatin*, trans. A. Salam, repr. Delhi, 1975, p. 20.

18. Ibid., p. 3.

19. Deloche, *Transport and Communications in India*, vol. II, p. 5.

20. James Rennell, *Memoirs of a Map of Hindostan with an introduction illustrative of the geography and present condition of that country; and a map of the countries situated between the head of the Indus and the Caspian Sea. To which is added an appendix, containing an account of the Ganges and Burrampooter rivers*, 3rd edn. London, 1793, p. 335.

21. Ibid.

22. 'Mustepha's Journey', in A. Dalrymple, *Oriental Repertory*, vol. II, London, 1808, p. 232.

23. West Bengal State Archives (henceforth WBSA), Board of Trade, Commercial (henceforth BOT-Comm), vol. 71, 23 September 1788, pp. 373-4.

24. Rennell noted that 'after leaving Barraset, we seldom found the roads good, they being excessive narrow, rough and crooked and very frequently running across paddy fields so that when the ground is ploughed there are no traces of road to be found'. LaTouche, ed., 'The Journals of Major James Rennell, First Surveyor-General of India', *Memoirs of the Asiatic Society of Bengal*, vol. 3, 1910, p. 87. Also see, Datta, *Economic Condition of the Bengal Subah*, p. 186, fn.

25. Willem van Schendel, ed., *Francis Buchanan in Southeast Bengal (1798): his journey to Chittagong, the Chittagong Hill Tracts, Noakhali and Comilla*, New Delhi, 1992, p. 3.

26. H.T. Colebrooke, *Remarks on the Husbandry and Internal Commerce of Bengal*, repr. Calcutta, 1884 (1st publ. 1795), p. 98.

27. Thomas Twining, *Travels in India a Hundred Years Ago with a Visit to the US*, London, 1893, p. 469.

28. For a higher estimate, see, Walter Hamilton, *The East India Gazetteer; containing particular descriptions of . . . Hindostan, and the adjacent countries, India beyond the Ganges, and the Eastern Archipelago: with sketches of the manners . . . of their various inhabitants,* 1st edn. 1815, London, 1828, p. 186.

29. Twining, *Travels in India*, pp. 72-3.

30. See Oriental and India Office Collections, British Library (henceforth OIOC), Prints, Drawings and Photographs, Oriental and India Office Collections (henceforth PDP), Add. Or. 2733. It might have created problems near cities where masts of boats lining the shores impeded tracking. However, stationary boats also helped in the operation; Twining, *Travels in India*, pp. 72-3. See also G. Bouchon and L.F. Thomaz, eds. and trans., *Voyage dans les Deltas du Gange et de l' Irraouaddy* (1521), Paris, 1988, p. 317. See OIOC, European Manuscripts (henceforth Mss.Eur.) F 95/ II, ff. 39(a)–39(b) for a description of the preparation of cords used in navigation.

31. National Archives of India (henceforth NAI), Home Department Public Branch, Proceedings (henceforth Home Public), vol. 30, 6 October 1766, p. 1003.

32. NAI, Home Public, vol. 20, 3 March 1763, p. 368.

33. NAI, Home Public, vol. 18B, 20 December 1762, p. 752. NAI, Bengal Public Consultations (henceforth BPC), microfilm (henceforth mf), 11 May 1743, p. 250.

34. James Taylor, *A Sketch of the Topography and Statistics of Dacca*, Calcutta, 1840, p. 294.

35. NAI, Dacca Factory Records (henceforth DFR), mf, 25 August 1743, p. 271 (b). See also NAI, BPC, mf, 29 April 1743, p. 227. WBSA, Calcutta Committee of Revenue (henceforth CCR), 19 January 1776, p. 148.

36. NAI, Home Public, vol. 11, 7 June 1759, p. 497. NAI, Home Public, vol. 20, 14 March 1763, p. 433. NAI, Home Public, vol. 20, 19 February 1763, pp. 216-24.

37. NAI, Home Public, vol. 11, 7 June 1759, pp. 496-7. NAI, Home Public, vol. 11, 14 June 1759, p. 512. Also see OIOC, Board's Collections, F/4/35, Extract of Bengal Public Cons, 27 March 1797, about experiments regarding marine pyramids.

38. S.C. Majumdar, *Rivers of the Bengal Delta*, Calcutta, 1942, pp. 45-9.

39. NAI, Home Public, vol. 20, 7 February 1763, p. 150. Colebrooke, *Remarks on the Husbandry*, p. 86.

40. *Proceedings of the Controlling Council of Revenue at Murshidabad* (henceforth *CCRM*), vol. VII, 21 October 1771, p. 80.

41. Ibid., p. 81.

42. OIOC, Bengal Revenue Consultations (henceforth BRC), P/52/28, 1 March 1791. Also cited in Rajat Datta, *Society, Economy and the Market. Commercialization in Rural Bengal c. 1760-1800*, Delhi, 2000, p. 214.

43. Twining, *Travels in India*, p. 469.

44. Ibid., p. 470.

45. Ibid., p. 83. Also William Tennant, *Indian Recreations, consisting chiefly of strictures on the domestic and rural economy of the Mahommedans and Hindoos*, vol. II, Edinburgh, 1803, pp. 59-60; William Hodges, *Travel in India During the Years 1780, 1781, 1782 & 1783,* London, 1793, p. 39. See OIOC, PDP, Add. Or. 3240, Add. Or. 3237. Buchanan Hamilton, *An Account of the Districts of Bihar and Patna 1811-12*, vol. II, repr. Patna, 1986, p. 704.

46. Also see Walter Hamilton, *The East India Gazetteer*, p. 186.

47. NAI, BPC, mf, 25 March 1744, p. 31.

48. See Buchanan Hamilton, *An Account of the District of Purnea in 1809-10*, repr. Delhi, 1986, p. 589.

49. Before 1787, Dhaka was entirely dependent on the neighbouring marts for their daily supply of grain. The famine of that year reduced a number of families to poverty and they took up their residence in the town itself and supplied the bazars with grain. The wealthiest of this class traded in grain and kept boats, which they let out for transporting of these and other bulky articles. Taylor, *A Sketch of the Topography and Statistics of Dacca*, p. 310.

50. NAI, BPC, mf, 25 August 1745, pp. 378-9.

51. NAI, Home Public, vol. 20, 4 March 1763, p. 388.

52. Buchanan Hamilton, *An Account of the District of Purnea*, pp. 591-2.

53. WBSA, Board of Revenue-Sayer (henceforth BOR-Sayer), vol. 3, 27 May 1791, p. 395.

54. Ibid., p. 396.

55. Centre for Studies in Social Sciences, Calcutta (henceforth CSSSC), Mss. Eur. D 75, mf, bk V, fol. 98.

56. Ibid., pp. 98-9.

57. NAI, Home Public, vol. 18A, 21 January 1762, p. 32.

58. WBSA, BOR-Sayer, vol. 3, 15 June 1791, p. 484.

59. Ibid.

60. Taylor, *A Sketch of the Topography and Statistics of Dacca*, p. 182.

61. NAI, Home Public, vol. 30, 11 August 1766, pp. 806-7.

62. There were categories of boat workers, with differing skills and pay. NAI, Home Public, vol. 20, 10 March 1763, p. 412. Colebrooke, *Remarks on the Husbandry*, p. 99. Also see Indrani Ray, 'Journey to Kasimbazar and Murshidabad', *Bengal Past and Present*, vol. C, pt. 2, no. 191, 1981, pp. 49-72. Taylor, *A Sketch of the Topography and Statistics of Dacca*, pp. 234-8. Thomas Bowrey, *A Geographical Account of Countries Round the Bay of Bengal, 1669-1679*, Cambridge, 1905, p. 149.

63. NAI, Home Public, vol. 18 A, 11 February 1762, p. 81.

64. NAI, BPC, mf, 19 July 1745, pp. 321-2. WBSA, Committee of Revenue, vol. 15, 24 June 1782, p. 2761. Reference to insurance by the bankers, merchants and traders of Calcutta is also found in the records, WBSA, Board of Revenue—Judicial Criminal, vol. 23, 30 July 1795, pp. 23-4.

65. The boat employers, the Company, as in this particular reference, advanced money to the *majhis* for their food during the journey. Adverse conditions could detain them on the way, leaving nothing of the advance payment. In such situations, a further sum was also given which was deducted from the remaining dues for freight. NAI, BPC, mf, 7 November 1743, p. 629.

66. Contemporaries noted the small sizes of these farms; see Willem van Schendel, ed., *Francis Buchanan in Southeast Bengal*, p. 8. Also see CSSSC, Mss. Eur. D 75, mf, bk V, fol. 107 for assignments of lands to *majhis*.

67. Rajat Datta, 'Rural Bengal: Social Structure and Agrarian Economy in Late Eighteenth Century', Ph.D. thesis, University of London, 1990, p. 99.

68. Ibid., p. 102.

69. NAI, Home Public, vol. 18A, 11 February 1762, p. 81. WBSA, BOR-Sayer, vol. 2, pt. 1, 29 September 1790, p. 154.

70. Taylor, *A Sketch of the Topography and Statistics of Dacca*, p. 315.

71. Sara Pennell, 'Consumption and Consumerism in Early Modern England', *The Historical Journal*, vol. 42, no. 2, 1999, pp. 549-64.

72. Taylor, *A Sketch of the Topography and Statistics of Dacca*, p. 315.

73. Ibid.

74. Though a number of travellers have recorded their observations about the various routes (for instance Hamilton, Twining, and others), Rennell's *Bengal Atlas* (*Memoirs of a Map of Hindostan or Mughal Empire and the Bengal Atlas*, ed. B.P. Ambashthya, Patna, 1975) and his journals enumerate the distances between the stages as also the distance of the several stages from Calcutta, Dhaka and Murshidabad, the location of the place and whether the route was along the current of the river. The figures are based on these sources.

75. Deloche, *Transport and Communications in India*, vol. I, p. 5.

76. Fernand Braudel, *Identity of France*, trans. Siân Reynolds, vol. II, repr. London, 1991, p. 465.

77. Deloche, *Transport and Communication in India,* vol. I, p. 7.

78. Tennant, *Indian Recreations*, vol. II, p. 68.

79. Sarkar, *Inland Transport and Communication*, p. 32.

80. Colebrooke, *Remarks on the Husbandry*, p. 101.

81. Ibid., p. 101

82. 'Route from Poonah to Ballisore as travelled by Lt Colonel Upton, on his return to Bengal', in A. Dalrymple, *Oriental Repertory*, vol. I, p. 494.

83. WBSA, CCR, vol. 13, 12 August 1776, pp. 118-19. Also see CSSSC, Mss. Eur. D 74, mf, fol. 220.

84. *CCRM*, vol. X, 14 April 1772, p. 193.

85. WBSA, CCR, vol. II, pt I, 13 May 1776, p. 126.

86. Francis Buchanan, *Journey from Madras etc. Through the Countries of Mysore, Canara and Malabar*, vol. I, London, 1807, p. 15.

87. For instance, Krishna devotees resided there during their short visits to the city. Travellers by land found board and lodging at these *akharas*, or at shops of *moodies* or dealers in grain. Many Hindu weavers too resided in these *akharas*. They paid the *bairagis* a certain sum of money daily or monthly for their board—the normal rates being one *anna* per day, for which they had two meals. Taylor, *A Sketch of the Topography and Statistics of Dacca*, pp. 294-5.

88. 'Route from Poonah', in A. Dalrymple, *Oriental Repertory*, vol. I, p. 483.

89. WBSA, Committee of Revenue, vol. 24 (typed copy), 7 April 1783, pp. 292-4. Also see WBSA, Committee of Revenue, vol. 31, pt. 1 (typed copy), 7 August 1783, pp. 28-35.

90. WBSA, CCR, 27 September 1774, pp. 85-6.

91. CSSSC, Mss. Eur. 74, mf, bk III, fol. 182.

92. 'Route from Poonah', in A. Dalrymple, *Oriental Repertory*, vol. I, p. 494. A French account recounted that though there were no 'hoteleries' in Bengal for lodging travellers, one could find some kind of covered shelter founded by some pious and charitable persons for perpetuating their memory. Travellers found bread or other food items with great difficulty on the routes. Gautier Schouten, *Voiage de Gautier Schouten aux Indes Orientales*, Tome Premier, Amsterdam, 1707, pp. 248-9.

93. OIOC, Orme Manuscripts (henceforth Orme Mss.), OV. 41, 'The Road from Cassimbazar to Rajemehal', ff. 117, 121, lists places like Dewan Sarai, Rani Sarai, 'Ninghen Sarai', Balakishen Sarai, etc., on the road between Kasimbazar and Rajmahal.

94. For north India, Manrique observed that *sarais* were staffed with attendants; Fray Sebastien Manrique, *Travels of Fray Sebastian Manrique, 1629-1643: a translation of the Itinerario de las missiones orientales*, introdn by, C.E. Luard and H. Hosten, vol. II, Oxford, 1927, pp. 109-15. Forster too spoke of 'the stationary tenants' of the *sarai*, with all possible facilities, *A Journey from Bengal to England: through the northern part of India, Kashmire, Afghanistan, and Persia and into Russia, by the Caspian Sea*, vol. 1, repr. Patiala, 1970, pp. 86-7, 92. Tennant, however, noted that 'caravanserais' consisted merely of an unfurnished empty house; Tennant, *Indian Recreations*, vol. II, p. 69. For similar observations, see Joannes de Laet, *The Empire of the Great Mogol, A Translation of De Laet's 'Description of India and Fragment of Indian History'*, trans. J.S. Hoyland and annotated by S.N. Banerjee, Bombay, 1928, pp. 81-2. Also see Iqtidar Alam Khan, 'The Karwansarays of Mughal India: a Study of Surviving Structures', *Indian Historical Review*, vol. XIV, no. 1-2, 1987-8, pp. 111-37; and Ravindra Kumar, 'Administration of the Serais in Mughal India', *Proceedings of Indian History Congress*, 39th session, 1978, pp. 464-72.

95. Colebrooke, *Remarks on the Husbandry*, p. 55.

96. See Tilottama Mukherjee, 'The Co-ordinating State and the Economy: The *Nizamat* in Eighteenth-century Bengal', *Modern Asian Studies*, vol. 43, no. 2, 2009, pp. 389-436.

97. Richard John Lufrano, *Honourable Merchants. Commerce and Self Cultivation in Late Imperial China*, Honolulu, 1997, pp. 120-1.

98. Also see NAI, Home Public, vol. 13, 17 December 1759, p. 1143.

99. Banarasidas, *The Ardhakathanaka*, trans. Mukund Lath, Jaipur, 1981, p. 43.

100. OIOC, Orme Mss. OV. 6, 'Mustapha's Journey, from Calcutta over the Western Mountains to Rambhur (1761)', fol. 14. Also in A. Dalrymple, *Oriental Repertory*, vol. II, pp. 236-7.

101. OIOC, Orme Mss. OV. 6, fol. 16. Also in A. Dalrymple, *Oriental Repertory*, vol. II, p. 241.

102. A. Dalrymple, *Oriental Repertory*, vol. II, pp. 253, 262.

103. About incidents of *zamindari* collusion with bandits, see J. Long, *Selections*, pp. 481, 514-15, 574; *Calendar of Persian Correspondence*, vol. I, Delhi, 1970, p. 355.

104. *CCRM*, vol. VII, 18 November 1771, p. 178. *Proceedings of the Committee of Circuit at Rajmahal*, vol. VIII, 15 February 1773, p. 177. *NAI*, Provincial Council of Revenue Dinajpur (henceforth PCRD), mf, 23 July 1774, p. 293. OIOC, Mss. Eur. B 165, ff. 3, 4. Ahuja similarly notes that in Orissa the local inhabitants resisted the construction of the mail road. Ravi Ahuja, '"Captain Kittoe's road". Early colonialism and the politics of road construction in peripheral Orissa', paper presented to the World History Workshop, Cambridge, 3 December 2001.

105. *CCRM*, vol. V, 17 June 1771, p. 286.

106. *CCRM*, vol. X, 5 March 1772, pp. 8-9.

107. Colebrooke, *Remarks on the Husbandry*, p. 46, CSSSC, Mss. Eur. D 75, mf, bk V, fol. 144.

108. NAI, Home Public, vol. 9, 1 May 1758, p. 69.

109. La Touche, ed., 'The Journals of Major James Rennell, First Surveyor-General of India', *Memoirs of the Asiatic Society of Bengal*, vol. 3, 1910, p. 78. *Du Bengale, et des Autres Possessions Anglaises Dans l'Inde*, Paris, 1797, p. 49 fn. 'Il n'ya pas de grands chemins au Bengale...'.

110. James Rennell, *Memoirs of a Map of Hindoostan*, 3rd edn., London, 1793, p. 58.

111. See OIOC, PDP, WD 1864-65. Tennant, *Indian Recreations*, vol. II, p. 75.

112. For visual representations, see OIOC, PDP, WD 1860-1867, Add. Or. 1161.

113. Sarkar, *Inland Transport and Communication*, p. 48. See also Irfan Habib, *The Agrarian System of Mughal India, 1556-1707*, London, 1963.

114. OIOC, Orme Mss. OV. 134, fol. 112.

115. CSSSC, Mss. Eur. D 75, mf, bk IV, fol. 90.

116. Colebrooke, *Remarks on the Husbandry*, p. 100. NAI, Foreign Secret, 28 April 1770, p. 287. NAI, Home Public, vol. 18A, 11 February 1762, pp. 82-3.

117. The figures for north India's banjaras are 821,000,000 metric ton miles per year and for Germany it is 325,000,000 ton miles per year. Kenneth Pomeranz, *The Great Divergence. China, Europe, and the Making of the Modern World Economy*, Princeton, 2000, pp. 34, 301-2.

118. 21,350 (no. of oxen) ×227.27 kg × 8 miles × 115 days.

119. If the total number of both bulls and oxen are used.

120. WBSA, BOT-Comm., vol. 71, 2 September 1788, p. 29.

121. CSSSC, Mss. Eur. D 75, mf, bk V, fol. 114.

122. Except at the capital, carts, oxen, or porters could be procured for hire in Dinajpur and according to Buchanan Hamilton no person made the carriage of goods a profession. Francis Buchanan Hamilton, *A Geographical, Statistical, and Historical Description of the District, or Zila, of Dinajpur, in the Province, or Soubah, of Bengal*, Calcutta, 1833, p. 328. However, evidence to the contrary exists too.

123. CSSSC, Mss. Eur. D 77, mf, bk III, fol. 38.

124. WBSA, BOR–Sayer, vol. V, 30 August 1790, p. 580.

125. Taylor, *A Sketch of the Topography and Statistics of Dacca*, p. 241.

126. Ibid., p. 310.

127. A Dalrymple, *Oriental Repertory*, vol. II, p. 233.

128. Ravi Ahuja, 'The Origins of Colonial Labour Policy in Late Eighteenth-Century Madras', *International Review of Social History*, vol. 44, no. 2, 1999, pp. 159-95. Also see idem., 'Labour Relations in an Early Colonial Context: Madras, c.1750–1800', *Modern Asian Studies*, vol. 36, no. 4, 2002, pp. 793-826. Also see Prasannan Parthasarathi, *The Transition to a Colonial Economy: Weavers, Merchants and Kings in South India, 1720-1800*, Cambridge, 2001.

129. NAI, PCRD, mf, 26 June 1775, p. 247.

130. See also OIOC, Verelst Papers Mss. Eur. F 218/103 for his correspondence regarding surveys ff. 1-14. Also OIOC, Mss. Eur. B 165 (March from 'Birdwan to Surseram') ff. 2-45, Orme Mss. OV. 65 ('The Particulars of the Road between Calcutta and Massulipatam') ff. 95-97. Orme Mss. OV. 65, 'Route from Calcutta to Dilly', ff. 179-187, Orme Mss. OV. 41, 'The Road from Cassimbazar to Rajemehal', ff. 117-134, OIOC, Orme INDIA XVII, 'Distances from Patna to Burdwan through the Hills', ff. 4668-4702, Orme INDIA XVII, 'Distances from Burdwan by way of Mungal Coot through the pass of Chuckeyah' etc., ff. 4703-4861, OIOC, Orme INDIA XVII, 'Captain de Gloss's Surveys in Behar 1766', ff. 4862-4884, Orme INDIA, XVII, 'Journal of an Expedition for the discovery of the nearest navigable river or branch of a river leading from the great river Ganges to Rang Cutta Creek', ff. 4905-4910, 4885-4930, 'Geography of Bengal from Mr Ironside—'The Road from Cassimbazar to Rajemehal', ff. 4938-4954. Orme Mss. OV. 4, ff. 33-54, Orme Mss. OV. 9, 'Journals', ff. 1-34. OIOC, Bengal Military Cons. P/18/41, 5 September 1771, pp. 335-6. Orme Mss. OV. 134, ff. 185-188. Vansittart Papers— Mss. Eur. F 331/36, 'Journal of a Survey of the Passes and Western Boundary of

Bengal taken by Thomas Carter', 1767, ff. 1-20, etc., Mss. Eur. F 331/35, 'Journal of a Circuit of the Midnapore and Jallesore Provinces'. Also see, for a later period, John Bourne, *Indian River Navigation—A Report Addressed to the Committee of Gentlemen Formed for the Establishment of Improved Steam Navigation Upon the Rivers of India*, London, 1849, pp. 3-4, 7-8, 11, 39-52.

131. Jan de Vries, *Barges and Capitalism*, p. 97.

132. See OIOC, Revenue Letters from Bengal, L/E/3/14, 1803-05, Misc. Cons. of 28 January, 24 March, 21 April, 5 May, pp. 12-20, 295-301, 441-5 for the expenses associated with *poolbundy*.

133. NAI, PCRD, mf, 13 February 1775, p. 51.

134. John Crawford cited in Douglas Peers, *Between Mars and Mammon: Colonial Armies and the Garrison State in India 1819-1835*, London, 1995, p. 113.

135. W.H. Moreland, *India at the Death of Akbar, An Economic Study*, London, 1920. See also recent works such as Peers, *Between Mars and Mammon*, p. 114. Also Anand Yang, *Bazaar India—Markets, Society, and the Colonial State in Gangetic Bihar*, Berkeley and London, 1998, pp. 26-7.

136. WBSA, BOT-Comm, 8 January 1779, pp. 35-6; WBSA, BOT-Comm, 2 February 1779, pp. 217-18.

137. WBSA, BOT-Comm, 2 February 1779, pp. 227-8. WBSA, BOT-Comm, 12 February 1779, pp. 280-82.

138. W. Bruton, 'Newes from the East Indies: or a Voyage to Bengalla....With the State and Magnificence of the Court of Malcandy', (1638) in *A Collection of Voyages and Travels, consisting of Authentic Writers in our own Tongue*, vol. II, London, MDCCXLV, p. 272.

139. Sieur Luillier, 'A Voyage to the East-Indies', in W. Symson, *A New Voyage to the East Indies. . . . To which is added, a particular account of the French factories in those parts, and of the general Trade throughout all India. With many excellent remarks by the Sieur Luillier*, London, 1715, p. 289. Epstein notes that an impulse towards greater precision in marking the day preceded the arrival of the clock. S.R. Epstein, 'Business Cycles and the Sense of Time in Medieval Genoa', *Business History Review*, vol. 62, no. 2, 1988, pp. 238-60.

140. Alun C. Davies, 'Rural Clockmaking in Eighteenth-Century Wales: Samuel Roberts of Llanfair Caereinion, 1755-74', *The Business History Review*, vol. 59, no.1, 1985, pp. 49-75, esp. pp. 52, 60.

141. Jan de Vries, 'The Industrial Revolution and the Industrious Revolution', *The Journal of Economic History*, vol. 54, no. 2, 1994, pp. 249-70.

142. Jan de Vries, 'Between Purchasing Power and the World of Goods: Understanding the Household Economy in Early Modern Europe', in John Brewer and Roy Porter, eds., *Consumption and the World of Goods*, London, 1993, pp. 85-132, esp. p. 110. His thesis is that 'by the mid-seventeenth century the 307 day maximal work year was definitely in place'. Also see Hans-Joachim Voth, 'Seasonality of Conceptions as a Source for Historical Time-Budget Analysis: Tracing the Disappearance of Holy Days in Early Modern England', *Historical Methods*, vol. 27, no. 3, 1994, pp. 127-32.

143. A. Legoux de Flaix, *Essai Historique, Geographique et Politique Sur l'Indoustan, avec le Tableau de son Commerce*, Paris, Tome Premier, 1807, pp. 402-3.

144. Jan de Vries, 'The Industrial Revolution and the Industrious Revolution', p. 260.

145. WBSA, BOR–Sayer, vol. 9, 30 July 1793, pp. 92–5. See also WBSA, BOR–Sayer, vol. 3, 10 January 1791, p. 31. WBSA, BOR–Sayer, vol. 2, pt II, 29 October 1790, p. 317, etc. See also OIOC, Bengal Law Council, P/166/88, 13 June 1794, WBSA, BOR–Sayer, vol. 1, 7 May 1790, p. 52.
146. See Tilottama Mukherjee, 'Markets, Transport and the State in the Bengal Economy, c. 1750–1800', Ph.D. thesis, University of Cambridge, 2004, pp. 274–87.

The Politics and Culture of Early Modern Warfare on the Konkan Coast of India during the Seventeenth and Eighteenth Centuries

Anirudh Deshpande

Whoever invents false revolutionary legends for the people, amuses them with lyrical tales, is no less guilty than a geographer who draws up misleading maps for navigators.

LISSAGARAY, *Histoire de la Commune* in Jean Chesneaux,
Pasts and Futures or What is History For? London, 1978, p. 30

THIS ESSAY IS divided into three sections. The first examines the concept of 'military revolution' which is central to the great majority of narratives regarding the evolution of early modern warfare. This examination is carried out in the light of qualifying examples from Indian history. The second section looks at the dynamics of war in a relatively unexplored though historically rich coastal region of western India. The third section comprises the conclusion.

THE EARLY MODERN MILITARY CONTEXT: SALIENT ISSUES

In former European colonies notions of military defeats and feelings of cultural inferiority continue to colour the historical perspective of a significant number of people even decades after decolonization. Therefore analysing historical military victories and defeats assumes a particular political and cultural

*A draft of this paper was presented at a workshop on coastal histories in India held at the Jawaharlal Nehru Institute of Advanced Studies, JNU, New Delhi in February 2007. I thank all those present at the workshop for their constructive comments.

significance in the context of this 'colonial complex'. Given this state of affairs it is easy to comprehend why the defeat and surrender of various Indian powers to British imperialism during the eighteenth and nineteenth centuries comprises a subject of an inconclusive debate. This debate is often located in the context of the political economy of the eighteenth century in India which remains contested between historians of differing persuasions.[1] Generally speaking, various issues of the early modern period in Indian history have remained unsettled since the colonial period. This chapter engages with this debate and re-examines the dynamics of British colonial expansion in a coastal region of India with reference to the concepts which have dominated colonial and post-colonial Indian military history, if such a subject exists at all, till date. Undoubtedly the most important concept which has dominated world military history since the end of the Second World War is the 'military revolution'. The concept evolved during the Cold War during which Western military doctrines were being effectively challenged by revolutionary people in many parts of the world like Vietnam. It was also located in the context of widespread decolonization following World War II. On the whole the narrative of 'military revolution' is based on an ideologically constructed difference between modern and pre-modern military practices. In the theorizing of global military experiences the 'military revolution', quite like the industrial revolution in connection with world economic history, has played a considerable role since the 1950s. However the 'military revolution'—according to which the West owes its dominance over non-Western regions to historically specific military developments and superiority—has serious political and prescriptive implications. Historians cannot ignore this because their questions are always raised with relevance to the present. The past is invariably researched in response to the demands of the present. Since the 'past has value only through what it means to us' and is 'the product of our collective memory and constitutes its essential tissue' this essay is justified in raising some of the questions mentioned later.[2]

The first concerns the historical context of the concept. Was the development of the 'military revolution' soon after the Second World War, and largely during the Cold War, a sheer coincidence or does the concept justify a cultural imperialism that characterizes much of Western writing on non-European history, culture and politics? Does the concept and the reading of history underscoring it imply that this military superiority is necessary for renewed Western domination of the post-colonial world and should, therefore, be occasionally demonstrated in case of people who refuse to conform to the West? These questions are legitimate in the light of new perspectives on culture and modernity as well as the submissions of those historians who, following the leads provided by Edward Said, describe much of Western writing on Asia and its history as 'orientalism'. These questions are also necessary if we must make sense of the paths chosen by direct and indirect imperialism since 1945. My findings suggest that it may be convenient but not always easy to argue in favour of European technological superiority while researching early modern

Indian military history. The 'military revolution', a term which is by no means free of internal contradictions, cannot be reduced to a teleological narrative of historical determinism and European supremacy. European dominance in the civilized areas of Asia was also made possible by well timed military alliances, political intrigue and collaboration of the indigenous elites. The issues arising from political sovereignty hence cannot be ignored in this debate. Nor can they be laid to rest in a discussion on the contemporary, though increasingly challenged, supremacy of the West. The coherence between political imperialism and the narratives which justify it must be addressed by everyone, including historians, in search of alternative histories of themselves.

Hitherto ignored local histories divert our attention to unexamined sources and the people who generated them. The potential of such histories to challenge the establishment narratives on the subject simply cannot be taken lightly. Contrary to elite opinion, based on textbooks as it mostly is, British imperialism did not triumph easily in South Asia. Closer inspection confirms that the smaller battles, were as important as the decisive ones. The period between the late seventeenth and mid-nineteenth centuries was marked by fluctuating fortunes and intensive warfare on the Indian subcontinent only to be followed by the upheaval of 1857. The people were of utmost importance to this process. While assessing the 'military revolution' in the light of Indian history it should be remembered that unlike the Native American civilizations, which were devastated by the Europeans between 1500 and 1550, India was very much a part of the gunpowder empires which arose in Asia from the mid-fifteenth century. Two recent contributions to the subject have managed to rescue many aspects of India's experiment with gunpowder from the sixteenth to the nineteenth centuries from the condescension of posterity.[3] Randolf Cooper has recently examined Maratha military methods specifically in the context of a discourse dominated by European cultural imperialism since the colonial times. His meticulously researched monograph reveals how conventional history has coloured our vision of Maratha warfare which is based, in most of writing on the subject, on ill-defined notions like the 'light horse'. Cooper has directed our attention to many crucial aspects of the political economy of warfare in eighteenth century India. He has also fruitfully re-investigated the role of artillery and infantry in the battles fought by the Marathas in the early eighteenth century. The results of this research are worthy of notice and historically productive. Cooper's viewpoint, based on a healthy respect for the well trained and brave Maratha troops abandoned by their European officers in adversity, reduces the difference between British and Indian methods of warfare in the early modern period. Exaggerating these differences has played a major role in the ideological propaganda of colonial modernity in much of history writing in India.

The 'military revolution' has been applied by scholars like Geoffrey Parker to the early modern period of history, i.e. AD 1500 to 1800. The main question addressed by 'military revolution' theorists is quite simple. How, they ask, did the European powers gain their colonial empires given the smallness of Europe and its geographical location during these three hundred years? The process is attributed to a combination of technological and organizational developments called the 'military revolution'.[4] This essay examines whether histories of localities hitherto ignored by mainstream military history tend to substantiate or qualify the military revolution thesis. According to the direction of research taken by this intervention focusing our attention only on defeats and victories diminishes the overall historical significance of the rich, multifarious and popular military process. Cooper's recent analysis of the Anglo-Maratha conflict and William Dalrymple's account of the final Anglo-Mysore war in the much acclaimed *White Mughals* illustrates this beautifully. While commenting on the 'military revolution', and its relevance to Indian history, I argue later that above all it is important to derive military conclusions from the social, political and economic aspects of an age. I contend that military activities symbolize culture; these activities are neither merely contingent upon technology nor only informed by them. This essay utilizes the military aspect of history as a stimulant to deepen our understanding of an under-researched locality of India.[5] Its conclusions critique the role of military technology in deciding crucial issues in the military-naval history of the western coast of India specifically in the eighteenth century. The eighteenth century in India was marked by the decline of the Mughal Empire and the rise of several big and small regional powers like the Marathas, Sikhs, Jats, Rohilkhand, Awadh, Bengal, Hyderabad and Mysore. All these powers were gunpowder states. By the late eighteenth century many of them were also producing fairly good firearms.[6] The enthusiasm shown by Indians for firearms and gunpowder in the seventeenth century paved the way for the rise of a system called 'military fiscalism' based on the use of infantry and gunpowder in South Asia during the eighteenth century.

The 'military revolution' is an addition to the various 'revolutions', like the scientific and industrial revolutions, which were crucial to the rise of some European countries since the Renaissance period (1350-1550).[7] The concept was christened by Michael Roberts in 1955. Soon thereafter it was praised by Sir George Clark in his Wiles lectures at Belfast in 1956. Having gained salience in European history writing by the 1970s it assumed fresh relevance in international military history after Parker published his attractive book in 1988. According to Parker between 1560 and 1660 four important changes occurred in the European way of warfare which ultimately underwrote European military dominance over non-European regions. First came a revolutionary change in tactics with the emergence of massed archers and organized musketry. Second, the size of most European armies grew appreciably. Third, more complex and ambitious strategies began to evolve to make these larger armies more effective in battle. Finally these three factors in combination impacted on European society and polity changing them into a network of institutions supportive of the new technologies. The concept owes its success to the ability of many historians to stress the dialectics of the modern state and the evolution of military technology. For instance it is doubtful whether the 'military revolution' would have occurred in the absence of the agricultural and industrial revolutions of Europe. In short the concept increasingly appears allied to certain significant developments in European history such as the rise of the nation-state, the transition from feudalism to capitalism, the age of overseas expansion and the institutionalization of European, West European in particular, military and society in general. Although some of these developments are themselves problematic military history in Europe is increasingly conceived in the 'war and society' paradigm.[8]

The 'military revolution' refers to a period before 1800 despite several limitations experienced by warfare in it. Early modern European warfare generated many improvements in guns. Khan also confirms the advantage some Europeans developed in gunpowder technology during this period.[9] The development of artillery spurred changes in fort design—there is evidence of this in medieval India as well—while the increasing reliance on hand guns increased the importance of infantry. However these developments often caused a stalemate by cancelling each other. Since the development of artillery caused improvements in fortifications many wars were reduced to attrition. While the infantry was conspicuous in open battles it might have been ineffective against strong forts. The consequences were clear: Most wars fought in Europe before the French Revolution were not brought to an end by a strategy of extermination, but through a strategy of attrition, via the patient accumulation of minor victories and the slow erosion of the enemy's economic base.[10] In India the native powers were subjugated in the late eighteenth century and the first half of the nineteenth century. The war against Tipu Sultan (1798–9), which ended with Tipu's death after a prolonged and frustrating siege of his island fortress Seringapatam by the combined forces of the East India Company and the Nizam, is a good example of how the economic base of an enemy was first destroyed before the final enagagement took place. Here victory was obtained less due to a 'military revolution' mastered by Arthur Wellesley and more because of the Nizam's assistance and the neutrality of the Marathas. In Tipu the English met a ruler who too had learnt the lessons of the military revolution rather well. But there was more to learn from the military experiences in the Indian subcontinent. Indeed, as Cooper shows, the future Duke of Wellington drew important lessons from the Marathas during the Deccan campaign following the defeat of Mysore in 1802-3. In the north, after winning the hard fought battles of Patparganj (Delhi) and Laswari (Agra) against the Marathas, the English were forced to retire after a frustrating siege of Bharatpur. The visible strength of the Jat forts at Bharatpur and Deeg and the variety of firearms deployed by the Jats in battle shows that like the sultan of Mysore the Jats too had mastered the technologies usually associated with the military revolution.[11] Before the late eighteenth century attrition also proved more important than outright victories over defiant Indians, like the Angre family, who were apparently less well armed but heavily fortified. The 'military revolution' obviously ran into problems in localities with which the Europeans were unfamiliar. The Europeans often borrowed from indigenous military practices to gain eventual success in a dialectical and long-drawn process. The historical richness of this process is lost to a paradigm which elevates what appears to be modern at the cost of what seems to be traditional in the early modern period of Indian history.

The European 'military revolution' raised the costs of war to unprecedented levels. Expanding and strengthening forts called for higher taxation. Hiring professional gunners and infantrymen was neither easy nor inexpensive. In sum, therefore, financing wars became a preoccupation of the early modern state. Most commanders preferred attrition because long range heavy cannons were capable of enormous destruction both to men and material. This meant fielding armies with baggage for many months at a stretch with the attendant problems of supplies and recruitment to overcome wastage. The matters of war, consequently, remained decentralized for a long time. The standardization in uniforms and armament desired by generals was rarely achieved in practice. Brigandage and the rabble, not unlike the Pindari bands attached to all armies in early modern India, were integral to military campaigns. Many European countries still had a long way to go before attaining standing armies and an all round professionalism in military affairs. All countries were not Prussia and even Prussian victories were not guaranteed. In fact the 'military revolution' made the state and economy preponderant in history.[12] This militates against expanding the idea into a universalist narrative.

When land war produced stalemates, states turned to naval power. Parker contends that land failures deflected strategy in Europe towards the sea. Whether this was significant for the future of colonial warfare or whether naval warfare was a logical outcome of maritime rivalry is a question difficult to answer. Here we are concerned with major developments and their implications for international history. Maritime rivalry, spurred by the discovery of the new world and sea routes to Asia (1450-1550), sharpened in the sixteenth century. This undoubtedly highlighted the importance of naval power. The profits of colonial trade underlined the need to secure sea lanes and overseas trading posts. Trading expeditions sailed from Europe at enormous cost and risk. This expensive trade was protected by gunpowder. The idea of naval power and supremacy at sea on the basis of ships armed with cannons emanated from this context. In a period characterized by fragile treaties, piracy and mercantilism naval power emerged from the economic need and geographical placement of the Atlantic powers namely Portugal, Spain, Holland, Britain and France. In the second half of the sixteenth century, as Tudor shipbuilding efforts and the defeat of the Spanish Armada show, it became clear that countries like England and Holland could defeat Spain by following an aggressive naval strategy.

In early modern Europe, when only the outlines of national policy were visible, no one equated sea-power with a mastery over the oceans stretching from Australia to South America. Early merchant companies were formed to trade with, and not conquer, the tropical regions although they began to wage war against regional powers quite early. Company officials appreciated naval support but without conceiving of acquiring a naval superiority over opponents. In any case most European states were economically incapable of exercising an overwhelming naval supremacy on the high seas. The capital of maritime trade was still not large enough to dictate policy at home. Evidence in support of prescribing, though not easily achieving, naval superiority was growing though not decisive at this stage. Nonetheless the era between the great discoveries of the 1490s and the 1840s when the Royal Navy was at its peak of dominance is called the golden age of sea-power.[13] Despite this the American revolt of the eighteenth century proved successful and England failed to dictate terms in North America. The master narrative of British sea-power, expressed in the works of Alfred Mahan, is dated to the late nineteenth century. These qualifications do not deny the fact that the navy was crucial to the rise of British imperialism. They only bring back into focus various other causes which complemented British naval power in the historical process which eventually produced the Empire.

Several writers criticize the notion of sea-power. For instance, G.S. Graham asserts that in the 'eighteenth and nineteenth centuries sea power was probably most influential when it was least conspicuous.... Even in time of peace it functioned as a powerful instrument of diplomatic action and compulsion.'[14] This is a shift from the 'salt water narratives' of naval power. Maritime history, a much broader subject compared with naval history, contains new perspectives on a complex historical process which resulted in the domination of the East by the West. The maritime achievements of Arabs, Chinese and Indians have been recognized and accorded a place in history. It is a well-known fact that maritime trade in the East was cosmopolitan much before the Europeans appeared on the scene with their armed ships.[15] However, it has become clear that naval power or hegemony could only be effective in the long-term in regions where land power complemented it. In crisis situations, like the Carnatic War in India, naval power could tilt the balance in favour of the British but on its own it could not assure victory. It could not, and never did, substitute institutional failures on land. Had that been the case Britain would not have lost America. If armed ships were enough the Portuguese and Dutch possessions in India would have been more substantial.

Between 1500 and 1800 the method of European naval warfare on high seas was marked by two important developments. First, due to the mounting of a row of cannons in the centre of the ship, naval tactics began to change in favour of larger ships commanding greater firepower. The second development was a consequence of the first. Due to improved ships armed with long range cannons European powers began to dominate the seas. This was most evident in regions where gunpowder technology was weak and not sufficiently harnessed to seafaring. For some unexplained reasons the Chinese, who had pioneered the use of sea artillery, were stagnating by about 1550 onwards. Earlier, in 1522, the Ming Navy had used artillery to defeat the Portuguese near Tunmen delivering the captured Europeans to execution.[16] In the same period Indian ships rarely carried guns and their design, due to historical needs, militated against their conversion into ships of the line carrying rows of cannons.[17] In contrast the Europeans began to develop naval firepower from the fourteenth century. The development of firepower on sea, like all new technology during pre-industrial times, was slow and in the fifteenth century ships generally carried light pieces. The feudal galley, like the Indian ships, only permitted guns in the stern or stem. The most important change occurred in Venice. The modified Venetian *galeass*, a sail and oar ship fifty metres long by nine, could carry heavy guns at both ends and anti-personnel pieces along the sides. A broadside could now be fired at enemy ships by manoeuvring the ship into position. Thus the Turks were defeated at Lepanto in 1571.[18] It is worth speculating why the Ottoman Turks, who had mastered the use of guns in massed formations, proved deficient in naval power.[19] This was perhaps a price they paid for being primarily a land power. Hence they first lost control over the Mediterranean Sea and were eventually bypassed by the Europeans who, beginning with the sea voyages of the late fifteenth century, started dominating the Indian Ocean from the early sixteenth century.

The evolution of the well armed man-of-war transformed naval warfare. Gradually designs affording crew reductions, to lower the costs, gathered momentum. First, as the galleys proved, experiments were made with light guns like the four pounders. In the sixteenth century heavier muzzle loading artillery pieces were mounted on the sides of the ships in specially created gun ports and thus England came to possess a strong navy. In Tudor shipbuilding drives ships weighing over five hundred tonnes and armed with more than fifty guns were built to defeat the Spanish navy. These ships were made of several layers of reinforced planks and could withstand rough weather and heavy punishment. But this 'capital' gunship was slow and required large crews. Improvements in gun carriage design—specially made to handle the recoil of heavy cannons—aided this process and demonstrated the resourcefulness of the Atlantic powers. Mastering the use of cannons on sea, specially compared with their use on land, was very difficult. Firing stable cannons from moving and lurching ships with a certain degree of accuracy required great skill and technological innovation. Stocking ammunition and keeping a supply of dry powder ready to fire the cannons at a steady pace during running gun battles required great proficiency. This explains the demand for European gunners in cosmopolitan regions like south Asia in the early modern period. By the eighteenth century these experts were increasingly replacing the Rumi or Ottoman experts on whom the Mughals had relied since the sixteenth century to counteract the rapid spread of firearms and their popular consequences in India.

The 'capital' ships lacked speed and their large crews and victualling needs made them unprofitable for long voyages. The regions where European commercial interests were growing were far-flung and naval conflicts often occurred in the tropics. These developments created the need for new ships appropriate for commercial and naval purposes in the seventeenth century. Finally, the forty gun, three hundred ton galleon frigate was perfected by the Dutch. This was an ace for waging war on the high seas. As a general purpose ship designed for piracy, as the Mauritius based French squadrons demonstrated later in the Indian Ocean, the frigate was ideal. By combining speed, range, firepower and mobility these frigates became the 'ships of the line' and played a crucial role in the 'savage and prolonged naval rivalry' among the Atlantic states in the seventeenth century.[20]

It is common knowledge that guns were crucial to the Spanish conquest of America simply because the American Indians were helpless against these new weapons. With modern weapons, and diseases imported from Europe, the conquistadores waged a genocidal war against native Americans. In contrast European dominance in other parts of the world did not arise so easily despite the alluring submissions of the 'military revolution' thesis. In South Asia the early modern Europeans came across kingdoms which were integral parts of the gunpowder empires. Many of these states, small and large, had to be liquidated before the Portuguese, French and the British could exercise hegemony in the southern seas. All non-state actors including peasant communities or bands of mercenaries in Asia were armed with guns by the time the Thirty Years War broke out in Europe. Local histories confirm that during the conquest of India, achieved at great human and economic cost in the eighteenth and nineteenth centuries, the 'military revolution' was challenged and modified. On the whole, since most of India was free of direct British rule before 1800, the applicability of concepts like the 'military revolution' to Indian history appear problematic. Graham, for instance, contends that even during the colonial domination of the eastern seas by Britain the Royal Navy was rarely strong enough to ensure victory in any major battle without substantial support from outside.[21] Indeed, as this essay shows, exaggerating the differences between the European and Indian way of warfare in the early modern period of history could easily produce a preconceived view of European naval supremacy in Indian waters. Matters were not made easier for the Europeans because their gunners were quite eager to desert and serve Indian employers often ready to pay higher wages. The fluidity of the Indian military labour market, and uncertain political loyalties, during the seventeenth and eighteenth centuries easily undermined the military edge desired by anyone.

THE CASE OF INDIA

While it is true that technological superiority, especially in the production and use of firearms, gave the fighting edge to Europeans in the long run, in the short-term it was still possible for them to lose battles despite this superiority.

That the Europeans could be, and indeed were, defeated in battles by non-Europeans was demonstrated time and again. Early modern technology seemed to have rubbed off on certain Asian powers in the period concerning this chapter. In the seventeenth century the Ottoman ruled Egyptians put up stiff naval resistance to the European presence in the Red Sea and other regions. In the latter half of that century the Oman Arabs became powerful enough to challenge European dominance. Finally, in 1698, they wrested the strategic port of Mombassa from the Portuguese. While the Arabs fought back, on the western coast of India the Maratha navy was founded by Shivaji towards the middle of the seventeenth century. At the same time the 'Malabar pirates' continued to harass the Europeans. Between 1659 and 1679 the Marathas conquered most of the important Konkan forts and this set up a tripartite naval conflict involving the English and the Sidis of Janjeera.[22] Sources indicate that all the warships in the region were using firearms. However the greatest challenge to the rising European hegemony in the Eastern seas was posed by the Chinese. The famous Chinese captain/admiral Cheng Cheng Kung, better known to his rivals as Coxinga, cut deep inroads into the Dutch naval monopoly from Fukien. It is reported that by 1655 he commanded two thousand warships and a hundred thousand troops. Fortunately for the Europeans he died at thirty-seven in 1662 leaving behind a son who continued resistance for another twenty years. This family of admirals was ultimately defeated by an alliance between the Dutch and its Ching dynasty of China. This defeat proved momentous for European colonialism in Asia because it established a lasting naval equilibrium in South East Asian seas.[23] This equilibrium broke with the rise of Japan in the late nineteenth century.

Before examining some naval events on the western coast of India in the seventeenth and eighteenth centuries it would be fruitful to mention the stage of gunpowder technology in India at which they occurred. Locating these events properly will make us do better justice to the actors involved in them.[24] In general, the history of gunpowder and firearms in medieval India can be divided roughly into the following phases. In phase one, gunpowder was introduced to India towards the middle of the thirteenth century and its use became well-known by the middle of the fourteenth century—the period from when *barud karkhanas* (gunpowder factories) became a regular feature of medieval Indian states. In the fourteenth and fifteenth centuries gunpowder was used for military purposes in India and even artillery in its rudimentary form was known. It is a different matter that Indian warriors, like the Rajputs, failed to pay adequate attention to these revolutionary weapons despite being aware of them.[25] The second phase was influenced by Ottoman technology. Rumi experts, matchlocks, artillery and casting methods were imported into India by the Mughals who continued to cultivate their central Asian contacts well up to the eighteenth century. Babur's dependence on artillery and massed matchlockmen in the battles of Panipat (1526), Khanwa (1527), Chausa (1539) and Kanauj (1540) are well-known. The second phase, beginning with Babur's invasion and ending with Akbar's death in 1605, was underlined by innovation, development and proliferation of gunpowder technology in all parts of India. During Jahangir's reign the Mughal Empire was part of the 'gunpowder empires'.

The third phase spanned the crisis and decline of the Mughal Empire. It was probably marked by technological stagnation. From the reign of Jahangir to the mid-eighteenth century battles of Plassey (1757), Panipat III (1761) and Buxar (1764) Indian gunpowder technology, according to the problematic view of Iqtidar Alam Khan, had stagnated for almost a hundred and fifty years at levels achieved during Akbar's reign. However, the battles of Plassey, Panipat and Buxar came as eye-openers to the Indian powers of the eighteemth century for reasons associated with the organization and deployment of military manpower. Consequently during the fourth phase, spanning the latter half of the eighteenth and first few decades of the nineteenth century, the Sikhs, Marathas, Nizam and Mysore produced the latest firearms and deployed them in battle. This phase ended around 1857 with the complete establishment of British rule and military monopoly in India. After that colonial India imported British guns for military purposes. Since 1947 India and Pakistan have remained large importers of expensive firearms at great cost to their forex reserves and general development. Whether we agree with Alam or not it is certain that between 1500 and 1800 Indian gunpowder technology underwent several changes. In contrast to Alam's submissions connecting the stagnation of medieval Indian gunpowder technology with the decline of the Mughal Empire, Cooper argues in favour of south Asian military technology. According to Cooper, military technology in South Asia evolved in response to contextual needs and could have been superior to European technology in some respects. Since most of Indian military history has been written from a Eurocentric perspective it is easy to see that the topic remains open to new research.

When the English arrived in India the Portuguese had preceded them by a century. In that century the Portuguese had gradually built a coastal empire in India comprising strategic islands, land bases and forts.[26] They did not command a great land mass because the Mughals and Marathas were too strong for them but they were successful in controlling and milking a large part of India's sea trade by using systematic force which was hitherto unusual in the Arabian Sea.[27] Following the example set by the Portuguese the English East India Company (the Company) began its career in India from small factories (trading posts) on the Indian coasts. The Company also created a small navy comprising Indian warships and local sailors to protect its trade. By 1615 this navy had become a permanent feature of its presence on the Indian coast. It was a small force of ten *ghurabs* and *galbats*—called gallivats by the English record keepers of those days. These vessels, mounting light cannons, were manned by lascars (professional low-caste or Muslim seamen from the Konkan) and proved convenient for coastal navigation and skirmishes. Whether the lascars carried guns at this stage is difficult to say but it would be reasonable to assume that they were expected to chip in with musketry during combat. These warships, commanded by commissioned English officers, also carried on board a contingent of lower class, armed European sailors more than willing to desert as and when opportunities arose. After 1650 the Company began to register a greater presence in the Arabian Sea. Among the factors responsible for this was the acquisition of Bombay by the Company from the English Crown in 1665.[28] But this did not prevent the Marathas from inflicting regular defeats on the Company forces from the mid-seventeenth to the mid-eighteenth century.[29]

Records show that the Company directors had thought of the strategic importance of the Bombay islands from 1654. This was expected because by then the most important ports, which also doubled as naval bases on the western coast, were either under Portuguese or Maratha control. The Konkan was the only stretch of the Indian littoral not controlled directly by the Portuguese. It had a number of redoubtable coastal forts owned by native powers like the Angres of Colaba and Gheria, and the Sawants of Sawantwadi. It is difficult to say why the Portuguese, who had established shipyards at nearby Bassein, neglected the possibility of developing the excellent harbour of Bombay for commercial and naval interests. The Company, for its part, wasted little time in utilizing the services of Bombay for the purpose of which technical experts were called from England. In 1669, to secure trade, construction was begun on three small armed ships at Bombay. The atmosphere in those days was charged with numerous insecurities. Surat had been sacked by Shivaji in 1664 and the Maratha navy was active near Bombay and Surat. So strong was the association of Shivaji with rising Maratha power in the region that we find English sources referring to Maratha or other local warships as *Shivajees*. Too close for comfort near Bombay was the island fortress of Janjira. It was the base of the well armed Sidis, who often acted as the protégés of the Mughals. In this context Warwick Pett of the celebrated Pett shipbuilder family of Deptford was appointed the Company's shipbuilder in Bombay and orders were placed for necessary equipment in England. In 1670 orders were issued for the building two brigantines, i.e. smaller warships, to equip the Company's Marine with fast sailing ships mounting light cannons capable of defending English trade against aggressive local powers.[30] The years were also overshadowed by Maratha naval expansion in the region, and the ongoing Maratha-Sidi rivalry.

These measures taken during the late seventeenth century bore fruit and English confidence in the Arabian sea began to grow. The Company was not only building or acquiring *gurabs* and *galbats* but these Indian ships underwent some amount of innovation which allowed them to carry more guns than they usually did. In 1715 the Bombay Marine, as the Company's naval arm was then called, comprised a ship of presumably European design sporting thirty-two guns, four *gurabs* mounting twenty to twenty-eight guns and twenty smaller *gurabs* and *galbats* carrying between five to twelve pieces each. This 'grab service', as the Marine was pejoratively called by some contemporaries, gave the Company enough confidence to prematurely attack some Angre strongholds in 1717. At this stage we can view the technologically and socially hybrid Bombay Marine as an excellent example of how the 'military revolution' progressed in the Arabian sea. It is certain that the innovative Marine gave the Company increased firepower and mobility near Bombay and Surat but whether this was enough to guarantee supremacy against the Indian powers of the day remained to be seen.[31] In the same period the local powers of the region were also upgrading their military technology. The stalemate of technology thus achieved underwrote much of military conflict in the region for many decades. For instance in 1722, to subdue the Angre strongholds, from where the *Shivajees* sailed forth to plunder English ships, the Company forged an alliance with the Portuguese. Alibag, an important Angre fort, was besieged by a combined Anglo-Portuguese force but the Europeans, momentarily united by their fear of a common enemy, failed.[32]

The Company's 'military revolution' was propelled by Indian manpower, expertise and resources. In the early eighteenth century the Company employed Parsi shipwrights to supplement English expertise. Some of these, like the famous Wadias of Bombay, became reputed shipbuilders and later prominent capitalists of Bombay. The Company, following the practices of the Indian powers, also realized that the Western Ghats were an important source of quality teak, while the coastal communities could provide competent shipbuilders and sailors. The Bombay shipyards thus established remained productive well into the nineteenth century. They went into decline only after the emergence of iron steamers manufatured in Britain later. Dominated by the famous Wadia families, Bombay produced long lasting and large ships some of which were gunships of the Royal Navy. Later the professional expertise present in the Bombay dockyard was acknowledged by the Parsi Chief Engineer of the dockyard, Cursetjee, before the Parliamentary Select Committee in 1853.[33] Ultimately the development of India as a dependent colony of industrial Britain killed the promising shipyards of Bombay.

However, these developments did not mean that the Marine had become a professional navy. Its officers were usually opportunist factory clerks engaged in part-time military work and private trade. They also remained perpetually in search of prize. The crews comprised a motley crowd of the European proletariat generally found in an inebriated state thanks to copious supplies of cheap spirits sold to them at a profit often by their officers. Disease and infections were rampant, life expectancy was abysmal and the tropical weather did not suit European constitutions in general. This made a constant supply of European proletarians, called dregs in history often enough, necessary to bolster European presence in India. The European crews were supported by the hardy lascars who, the sources candidly admit, were forced to work long hours under the tropical sun. The lascars had their own hierarchy and were commanded by the native petty officers called *sukhanis* and *tindals*. Sailors of mixed Indian and

European parentage, the *topasses*, were also conspicuous on the Company ships and their social position made them extremely vulnerable to exploitation. Drunkenness was so common and detrimental to the health of men and navy alike that in the 1730s express orders were issued restraining the officers from selling liquor to the men. However, it is difficult to ascertain whether such orders were obeyed by the greedy English officers. To lessen the subaltern wretchedness in the Marine several incentives were announced for the European sailors. These included pension to widows (an important incentive, given the high probability of European seamen losing their lives quickly in the tropics) and compensation for disability. But pay remained low and desertion was common. Attempts were regularly made to capture and regain expensive European drafts in a highly competitive military labour market simply because these men were often artillerymen and were necessary to 'stiffen' native sailors and troops. Such measures usually failed because deserters, who had every right to be as opportunist as their officers, were easily lured away by the high wages offered by some of the Indian powers. Difficult terrain, thick forests and the pursuers' ignorance of local topography helped desertion which remained high till fairly late in the eighteenth century. Desertion rates finally came down in the late eighteenth century: as the Indian powers declined, the British grip on the coast tightened and marine surveying began in earnest.[34]

It is often understood that regular payment in cash and paying attention to the greater welfare of men constitute important steps in military modernization. However, in the eighteenth century the Company was not the only one trying to modernize its forces as the earlier mentioned measures tend to indicate. Maratha records kept at the Peshwa Daftar in Pune for the period 1764–80 and the abstracts of budget records and naval administration records for the *subah* of Vijayadurg under the command of *Sarkhail* Anandarao Dhulap mention many instances of pensions being granted to widows of men killed in action or accidents. Although these sources are hardly enough to write a history of that period from the popular perspective, they provide important information including names of people affected by warfare. The following instances have been chosen to highlight this point from the budget records of the years concerned.[35] In 1767, Dalji Dharmoji Shelatkar received a gun shot in his leg in an engagement at Halemadi when the factory of Shaba Ali was attacked by the Marathas. Since the wound left Shelatkar permanently disabled he was granted a disability pension of seven rupees (per month). Injured subalterns could also petition the Maratha state for improvements in compensation thus making a case for investigating popular volition in early modern Indian history. For example, in 1771-2, within a week of naval skirmishes with the Portuguese, during which the Marathas captured an enemy ship, about twenty Maratha sailors were killed and more than a hundred injured. The state distributed Rs. 5000 among the crews as special rewards and for presents offered to the coastal deities who were the protectors of the sailors in popular imagination. This sum, not inconsiderable in those times, failed to satisfy the men who had fought so well. The recipients petitioned the government and asked for at least Rs. 10,000. Since these men were important to the state, the government conceded the demand on the condition that during the current year the sailors would bring home loot worth many lakhs of rupees.

Widows were also granted pension in case their husbands died on official duty. For instance, Mahadaji Shelatkar was wounded in a naval engagement in 1785 after which he worked for a while as a sentry before succumbing to the infection. Since he had worked as a government re-employee, his widow was granted a pension of three rupees in cash per month. When the Maratha warship (*galbat*) Raghunath engaged a *ghurab* Haidari, i.e. belonging to Haider Ali of Mysore, five men were killed in action. Thirty-seven rupees were sanctioned to their widows as pension per month. Lakshmanrao Pote received a gunshot wound in the left leg in an engagement with the English during 1782-3 but died as a result of complications arising from the injury sometime during 1784-5. His widow was sanctioned a pension of seventy-five rupees (most probably per annum) including payment in cash and kind. Pote must have been a man of some stature because the pensions appear considerably less in the case of the labouring classes or in cases when men on duty died or were injured not fighting. For example, Kanoji Mayekar employed aboard the *Naranag* got his femur broken by a log while the ship was being launched in the year 1784-5. Some time later he succumbed to the wound upon which an annual pension of twenty rupees was sanctioned for his widow and children.[36] Sanctioned amounts as pensions were carefully marked in the annual budgets as the complete annual budget for 1770-1 for the naval *subah* of Vijaydurg in charge of Anandrao Dhulap shows.[37]

The Company was only one of the several powers in the Arabian Sea region during the seventeenth and eighteenth centuries. Indeed there is enough evidence to suggest that from the twelfth century trade was accompanied by force in the region. Independent rulers, both Muslim and Hindu, of coastal areas often maintained fleets of warships to raid merchant ships.[38] In addition there were seafaring adventurers of all hues—the corsairs—whose primary source of wealth was piracy. The period also witnessed large scale organized piracy which not only exposed the limitations of contemporary military technology but left many rulers helpless. The Mughal state and the various European companies were often powerless in the face of high sea piracy now undertaken by the rovers with the aid of large and well armed ships. The only thing the Mughals could do when the pirates in question turned out to be Europeans was to exert pressure on the trading companies in a bid to recover losses. The Malabar, Cambay, Persian Gulf and various island groups in the Indian Ocean provided good bases and logistical support to several bands of pirates targetting unarmed or even armed merchantmen and Indian trading and pilgrim ships. In the middle of the seventeenth century the elusive Sindanians, who could well have been an efficient community of pirates based on the Makran coast, raided the relatively lesser known stretches of the sea lanes between Surat and the Persian Gulf. The Sidi posed a constant threat to the Company around Bombay itself.[39] Strung on the Konkan coast, from the 1680s onwards, were the formidable Angre forts commanded first by Kanhoji Angre and later his sons Sekhoji, Manajee and Tulaji. Kanhoji was the son of Tukoji Angre a comrade in arms of Shahaji Bhonsale, Shivaji's father. To begin with, the Angres were Maratha *sarkhails* assured of political and military support from the hinterland. Later, in the first half of the eighteenth century, as the squabbling between Pune and the Maratha sardars grew, some of the Angre forts emerged as autonomous power centres disliked both by the Peshwa and the Company. Piracy also operated in the spaces created by constant political rivalry which distracted precious naval forces from the task of going after the pirates. Hence, when the tripartite rivalry between the Sidis, the Company and the Marathas escalated on the Konkan coast-based pirates became bold enough to attack and loot large ships near the Malabar coast.[40] In 1683 a large ship *President* was attacked by Arab freebooters commanding two ships and four *gurabs*. In the engagement that followed the *gurabs* were destroyed by the *President*'s guns and the Muscat Arabs were taken prisoner. The Arabs claimed to have been mercenaries hired by Sambhajee, Shivaji's son and heir, to harass the English but when approached the Maratha Raja denied any involvement in the affair.[41] During most of the eighteenth century the Konkan remained a site of contest between several political powers. This absence of political monopoly, which arose with the establishment of British colonialism in India, left the commercial field open to anyone who could muster mercenaries, armed ships and was willing to fight. While such a state of affairs was obviously disliked by those who were trying their best to politically dominate the region it certainly assisted the non-state actors including several communities based on the coast.

The balance of power was also underwritten by geography and its insufficient knowledge in the hands of the European powers. On the western coast of India the relatively calm sailing season began after the monsoon, but local navies with superior knowledge of coastal navigation could easily register a marked presence on the familiar coast even during the stormy monsoon months. Numerous creeks and rivers flowing into the sea from the hilly hinterland offered excellent opportunities to hide and strike when the opportunity arose. The popular practice of hauling up smaller ships on shore or moving them inland through estuaries and deeper rivers before oiling them against the rains helped conserve native naval strength. The larger European ships could bombard forts from a distance but failed to pursue native craft into the shallow creeks. At many places ships straying close to the coast risked being wrecked on the yet unknown shoals. These factors checked European efficacy of firepower in Indian waters. On the other hand the mercantilist tendency of cutting costs to maximize profits occasionally proved self-defeating. For example, for many years the Company's recruitment policy actually ended up helping its rivals not to speak of the recruits themselves. Till the 1750s, lascars were discharged as soon as the monsoon began. This was done to save money. Contrary to prevalent perceptions regarding the predictability of 'seasonal' labour, it was found that many of these trained, well informed, demobilized and enterprising men joined rival navies during their period of discharge to the disadvantage of the Company. It took quite a while for the Company to reconsider the policy of lascar recruitment in the light of these facts. Later these lascars were given permanent employment. This translated into additional expense on pay and some housing allowance during the lean season.[42] Throughout the eighteenth century steps were continuously taken to overcome chronic desertion. In 1724, as a desperate measure, it was decided to hold the pay of all seamen two months in arrears.[43] We don't know whether this worked. Most probably it backfired because in the 1720s and 1730s the Angres were busy recruiting fighters. In 1754 the Mutiny Act was applied to the Marine because in 1748 a mutiny had taken place on board the *Bombay* due to the economic problems and homesickness faced by the English crew.[44] Accommodation for English sailors in the entire period remained rudimentary and wretched at best. Finally, and as late as 1779–80, the Board of Directors resolved to construct proper rooms for them. Lascars, we may safely presume, were forced to make their own arrangements. So in this period, given the instances examined earlier, it would be wrong to assert that the 'modern' Company was ahead of its 'traditional' Indian rivals like the Marathas in the matter of caring for its employees.

During the early modern period the Company did not completely control the Arabian Sea even close to Bombay. Local powers sometimes joined hands to threaten the existence of the Company's powers. In 1689, for example, Sidi warships in alliance with the Mughal Army on land successfully blockaded Bombay. This forced the desperate English into piracy to sustain themselves. But this was hardly an aberration from the norm as the English accounts of the period, despite the ingrained bias in the language used, tend to assert. The Portuguese had set a notorious example of piracy accompanied by a great brutality and bloodshed in the sixteenth century.[45] The language of the period holds important clues to the activities of the various actors involved in naval and commercial rivalries. In Company parlance the phrase 'to cruise' was in fact a euphemism for piracy. What laws could be implemented or voluntarily observed on the open Arabian Sea in those days is difficult to say. Prize money and private trade, which the Company officers were allowed to engage in, were incentives enough for greedy officers and hungry seamen to indulge in piracy on the side. The misuse of *dastaks* by Company officers who disguised private trade in Company ships often caused military conflict with the native powers. Life and conditions in those days of seafaring were violent and uncertain and opportunities for plunder were omnipresent. The Company's drawbacks against the high sea pirates were too obvious to be recounted in detail here. At the end of the seventeenth century large Arabic fleets using pack tactics freely pillaged Diu and the Gujarat coast. Almost simultaneously the heavily armed

European rovers had a free run of the Indian Ocean. By 1698 the rovers became so powerful, and their capacity to plunder expensive European trading ships was so great, that various European trading companies sank their differences temporarily in an attempt to check their activities. Particularly vulnerable to the high sea pirates were the Indian trading ships which carried few or no guns. Heavy losses in Mughal shipping prompted the Mughal state to force the Europeans into accepting joint responsibility for curbing piracy. But all these steps were insufficient to check piracy simply because the profits of piracy were great enough to lure away even those to whom the task of policing the seas was entrusted. Towards the end of the seventeenth century, for example, the crews of the *Mocha* and *Josian*, ships of the Marine, murdered their officers and turned pirates. Captain Kidd of the Royal Navy, commanding the *Adventurer* (thirty-two guns and two hundred sailors), sent from England to attack pirate strongholds in Madagascar, ended up becoming a famous pirate.[46]

The rise of high sea piracy hemmed in the Company's ships and the almost helpless Marine from one side. At the same time, on the coast, the strength of the Maratha admiral Kanhoji Angre was growing. In due course high sea piracy would be tackled by the more disciplined Royal Navy which had the technology and power to do so while the Company concentrated its attention on the Konkan. Angre power had to be broken both for naval reasons and because the Angre family controlled the adjacent hinterland of Bombay and Ratnagiri coast which was the main source of the Company's trading merchandise and shipbuilding material like teak. This is not to discount the fact that the Company had to pay tribute to the Angres in the form of buying passes from them and paying taxes at the Angre harbours. Although the Company had been formed to rake in profits for its shareholders in England at the lowest possible cost, war became an inevitable part of its activities in India. Hence till the 1750s the Company waged war against the Angres of Konkan in which early modern military technology, as the Company and the Angres both possessed, was stretched to its limits on the western coast of India. In 1700 the Company's military weaknesses also stemmed from two factors. First, the Bombay Council was almost totally dependent on England for the supply of crucial navigational equipment and cannons. Second, the Royal Navy did not always come to the Company's aid soon enough. A coordination of Crown and Company forces was desired by the Company commanders but was rarely achieved in practice. Only from the 1750s, as Anglo-French commercial and naval rivalry intensified, did the East Indies Squadron of the Royal Navy become a more or less permanent feature of British Crown presence in the Indian Ocean region.

After mentioning the growth of the Maratha navy in the 1650s Parker does not follow its history.[47] Such is the occultation of Asian history in the Eurocentric narratives. Let us complete the picture. In 1690 Kanhoji Angre was appointed deputy commander of the Maratha navy which had comprised fifty-seven *gurabs* and five thousand men at the time of Shivaji's coronation in 1674.[48] This navy carried guns supplied by the Portuguese. The gunners were a motley crowd of well paid European deserters, mercenaries, half-castes and men drawn from the low-caste coastal communities. At this stage the Portuguese were probably trying to use this navy against their rivals, including the English; not only were the Marathas their customers but could become their protectors on the Konkan. It is certain that after becoming the *surkhail* of the Maratha navy Angre began to harass the Company regularly between 1690 and 1700 from a string of well built coastal forts running down the coast from Colaba to Suvarnadurg.[49] This preoccupied the attention of the English. In the same period we find increased references to *Angria* the pirate in Company records.[50] At this stage Angre, who was beginning to emerge as an autonomous ruler of the Konkan forts, enjoyed the support of the Maratha state which controlled the hinterland of Bombay. The conflicts between the various Maratha sardars and the Pune based Peshwa which characterized the Maratha polity during the eighteenth century were not prominent during the reign of Kanhoji. In the first decade of the eighteenth century we find him issuing *dastaks* on the lines of the Portuguese *cartaz* to ships under his protection or having paid tribute to him. Ships which did not carry his *dastaks* were liable to boarding and seizure. This alarmed the Company to a great extent for reasons easily understandable. As the events of the subsequent decades were to show, the Company was averse to dealing with another sovereign authority on the Bombay coast. Hence it denied Angre legitimacy by calling him a pirate and trying to deal with him as such.

The naval skirmishes between Angre and the Company go back to the earliest years of the eighteenth century when the English were still busy consolidating their hold over Bombay. Angre, for his part, generally eluded the Marine. Since the Maratha navy comprised mainly *gurabs* whose firepower, despite mixed gun crews, was limited Angre avoided facing the heavier English ships in a line to line battle on the high seas. Instead Kanhoji resorted to guerilla tactics much like the Marathas did on land. His navy operated in the safe zone between the coast and open sea. When attacked by larger ships carrying bigger guns the Maratha *gurabs* slipped away into shallower waters under the protection of his fort artillery. The Angre forts mounted heavy long-range cannons manned by international gun crews which, we are told, were handsomely and regularly paid in cash.[51] When fired accurately these guns were capable of damaging ships in the bay seriously. As the eighteenth century progressed more Indians entered the Maratha service as expert gunners—the increasing references to the surname *golandaaz* used by some professional Muslim families of the coastal region proves this beyond doubt. Most of the stone built Angre forts had extremely high and thick walls which were impervious to naval artillery and, as two maps and layout plans of Vijaydurg (better known as Gheria) present in the Peshwa Daftar records demonstrate, comprised well appointed and thoroughly organized military bases. Many of these forts, built on hills alongside rivers and creeks, could be accessed only from land. They could fall, as eventually they did, only upon being encircled, besieged and starved from land and sea.

The map of Gheria shows how well provisioned it was to conduct offensive and defensive operations. The walls were mounted with heavy guns for which copious supplies of ammunition were stored carefully in specially constructed stone *kothis*. Adequate arrangements for accommodation, water and food were also evident, thus highlighting the role of native military planning in the early modern period.

But Angre is said to have gone beyond these tactics. Like Tipu Sultan and Madhav Rao Shinde in the latter half of the eighteenth century, Angre understood the need to develop a strong navy to defend his sovereignty. He obviously understood the connection between naval power and commercial success. As the political differences between him and mainland Marathas began to grow it became even more important for him to develop a strong navy. Towards this end he encouraged shipbuilding and teak was even sown in some parts of the Western Ghats to assure a supply of timber to the Angre forces. Although the Portuguese initially helped him in these endeavours, later their support became uncertain as the growth of Angre power ultimately threatened them as much as it did the English. We don't know whether the Angre centres produced artillery and most probably Kanhoji remained dependent upon the Europeans in this regard. Whether guns came from the hinterland is also not known but events show that Kanhoji and his successors knew how to use them effectively against their European and native rivals.[52] During the first half of the eighteenth century gunpowder and firearms had assumed a central role in Indian coastal warfare. Maratha sources are replete with references to the symbolic and real power of guns on the Konkan coast. All Maratha warships from the *galbats* and *ghurabs* to the large *pals* carried artillery manned by professional groups of *golandaaz*. All coastal and fort deities, both Hindu gods and Muslim *pirs*, were offered gun salutes by squadrons which sailed past them. Measures were taken to ensure that a certain number of shots were fired from cannons during the humid months to protect the precious barrels from oxidization. Steps were also taken to keep the powder dry and ammunition was stored carefully—records were also kept of the number of times cannons were discharged during military encounters and the ammunition used.[53]

The growth of Angre power changed the power equation on the Konkan coast. To redress the emerging imbalance the Company launched its first serious attack on Angre selecting Gheria, the strongest of his forts, in 1718. It is said that a fair amount of intelligence about this fort and its defences was given to the Company by a Portuguese defector. Despite the confidence underlying the attack the Company forces failed. Very soon, in November 1718, the English attacked Khanderi, another Angre fort, with a large force comprising fifty ships of all sizes, two thousand sailors and hundreds of Indian sepoys stiffened with some detachments of English soldiers. The English tried to land near Khanderi several times in an attempt to storm the fort but the comparatively much smaller Angre garrison first held out and later drove the attackers back. The entire offensive was carried out with a degree of inefficiency we would normally not associate with the 'military revolution'. The Company officers were exposed as amateurs and the English failed to besiege Khanderi either with men or artillery. During this battle for Khanderi the bulk of the Angre navy avoided battle by slipping away into shallow waters while the thick stone walls of the fort could not be breached even by the heavier of English guns.

Following this failure a desperate assault was launched on Gheria in 1720. Evidence suggests that in the early eighteenth century European artillery was not powerful enough to seriously damage the Angre forts which had thick stone walls. This militates against assuming that Indian fort design was inferior to European designs which developed in response to developments in artillery.[54] Probably because of the limitations of their field artillery the Company thought of using a secret weapon against Angre in 1720. This weapon, the sources tell us, was called the *phram*. Actually this *phram* was a floating platform carrying a number of forty-eight pounders. The idea was to take the *phram*, which was supposed to float on shallow water because of its low weight, close to the fort in an effort to bombard it more effectively. The landings of 1718 had obviously proved too expensive in material and manpower; the Directors in London would not have sanctioned another Khanderi. In the event a considerable amount of preparation went into the invasion of 1720. This included, we presume, a copious supply of rum to the troops, many of whom would have shrunk from combat upon remembering the failures of the recent past. What exactly went wrong with this attack is difficult to say but the *phram* sank even in the shallow waters of the low tide during which Gheria could be approached. Either the contraption was too heavy to float or probably it received a direct hit from one of the several Angre guns trained on the assaulting forces. This proved to be an ill omen for soon widespread drunkenness prevailed in the demoralized Company ranks and this finally put paid to the expedition. The following year another desperate assault was launched against Colaba with the aid of the Portuguese and some ships of the Royal Navy both from the sea and on land. At low tide, six thousand infantry and two hundred cavalrymen with sixteen major artillery pieces were deployed against Angre in battle, whereas from the sea three hundred big guns from ten major ships incessantly rained fire on the fort. The Anglo–Portuguese alliance deployed Europeans, African mercenaries, *topasses*, sepoys from Bombay and even some Pathan adventurers. The coalition plan to storm the fort after the heavy shelling had softened it from the sea failed because a large Maratha force commanded by Pilaji Jadhav arrived on the battlefield. This proved too much for the attackers who dispersed soon thereafter. It is said that later Peshwa Baji Rao, a friend of Angre, who had realized the strategic importance of Colaba appeared on the scene with another strong Maratha army. These failures to breach the Angre fortifications led to a great amount of acrimony between the English and Portuguese who were not the best of friends in any case. Amidst bickering and blaming, the European alliance broke and in December 1721 the defeated Company forces retired to Bombay to recover in safer conditions.[55]

Between 1718 and 1721 three major battles at great cost had been fought against Kanhoji Angre. In the last attack the Royal Navy had also been involved and this meant that the Crown was also beginning to take Angre seriously. But, in the event, the large cannons of the Royal Navy ships proved insufficient to settle the issue. The extent of destruction visited upon the Colaba fort by these guns is difficult to assess in the light of English sources but once again, like in 1718, Angre held out successfully. The timely appearance of the Maratha army on the battlefield tends to suggest that Angre was well prepared for the attack and possessed admirable powers of military coordination. For once the unity of the Marathas stood out in stark contrast to the disunity which prevailed in the Anglo-Portuguese coalition; the arrival of Pilaji Jhadav and Baji Rao was not incidental but a purposeful tactic to outflank the Company's land forces. Angre obviously knew how to play his cards well. The failure of 1721 made it clear to the Company that neither naval superiority nor an invasion from land would result in a quick and decisive victory over Angre. At this stage the Company leadership could not match Kanhoji's personal leadership of his forces and better local intelligence which can hardly be overemphasised as a decisive and sufficient factor in battle. The Portuguese had proved themselves fickle allies on both sides and this did not help the Company much. Ultimately the failures of 1721 led to a stalemate till 1729 when Kanhoji Angre died. In these years, rather emboldened by the Company failures, he remained defiant and victorious on the coast. However, soon after his death the Company, in 1730, tried to enlist the support of the Bhonsales of Sawantwadi against his successors in an attempt to encircle the Angre forts. This move to outflank the Angres from land and sea seems to have failed, probably because the rulers of Sawantwadi were more afraid of the Angres than the Company.

During the 1730s, after Kanhoji's demise, various steps were taken to enhance the efficacy of the Marine and more ships were added to it.[56] The Company also began to pursue alliance diplomacy more seriously. In this regard, its overtures to Sawantwadi have been noticed above. In the early 1730s the Company first aligned with the Sidis against Manajee one of Angre's sons and successors. This alliance having proved unproductive from the viewpoint of enhancing English power substantially, the Company, in 1738, sent Captain Inchbird of the Marine with money and stores to assist Manajee against his brother Tulajee. This is perhaps the first instance of a Maratha chief demanding and receiving English assistance. These diplomatic and military steps were accompanied by the usual cruising operations against the opposing Angre faction. It is easy to see that the growing disunity between the successors of Kanhoji must have made this easier from this point onwards. However, even as late as the 1730s the Marine commanders had insufficient knowledge of local navigation and this allowed the native navies to escape into shallower waters. Thus on 22 December 1738 a certain Commodore Bagwell failed to destroy a hostile Angre fleet which slipped away into the Rajapur river near Gheria. They, we are told, 'made off under his heavy broadsides, until he found himself with only four fathoms of water locked in by the rocks. Ignorant of the navigation he was compelled to give the signal for returning.'[57] There could have been other reasons for this retreat for I may add that had he continued the fleet would have come within range of Gheria's reputed defensive batteries.

Alliance diplomacy, more than the methods of the 'military revolution', finally paid dividends to the Company. After Kanhoji's death and the brief admiralship of the elder sons Sekhoji and Sambhaji, the Angre kingdom and navy were split between his sons Manajee based in Colaba and Tulajee based in Gheria. The former, as the Company records assert, came over to the English by the 1730s against his brother. This initiated the fall of Angre power. From the 1730s the Angres were divided into the friendly and hostile Angres as far as the English were concerned. On the other hand, after Kanhoji's death, differences began to grow between the Brahmanical power centre of the Peshwas based in Pune and the various Maratha sardars who refused to act as mere military agents of the Peshwa.[58] As Maratha power expanded, the uneasy relationship between the largely non-Brahmin sardars and the Peshwa began to affect coastal politics and warfare. This discord between the Maratha field commanders and the Peshwa ultimately proved disastrous for the Maratha polity on the whole. Right from the beginning Tulajee abandoned the shrewdness exhibited by his father in dealing with the Peshwa and made it clear that he would not become a pawn of Pune. In fact Tulajee carried out widespread corsair activities in the Arabian sea capturing several English ships and holding their crews to ransom. Leaving important coastal forts in the hands of autonomous rulers like Tulajee also meant a substantial loss of revenue to the Peshwa. These factors disturbed the balance of power in the area and made Tulajee a source of discomfort for Bombay and Pune both. By the middle of the eighteenth century the Marathas and the Company began to draw closer in an effort to contain Tulajee's defiance and growing power. In 1755 the Peshwa, angered by Tulajee's rebellion, allied with the Company to inflict the first serious defeat on some Angre strongholds in Tulajee's possession. This completely outflanked Tulajee from the land side but he refused to submit. However, with the Pune forces against him the rebel did not stand a chance. This time artillery was brought close to the fort of Suvarnadurg and probably played some part in demoralising the garrison but once again the ten thousand Marathas who stormed the fort successfully proved decisive. The fall of this important fort weakened Tulajee considerably and prepared the stage for the final assault on Gheria. In February 1756 an allied force of the Marine, the Marathas and a well equipped squadron of the Royal Navy under Admiral Watson blockaded and bombarded Vijayadurg which surrendered after a five day siege. On this occasion military intelligence seems to have been better because of Maratha support and the fall of Suvarnadurg earlier. The Marathas proved invaluable in the field once again. Following the capitulation of Gheria, the Angre fleet was burnt in the harbour and Tulajee was made a prisoner for life. Indeed during the eighteenth and nineteenth centuries extensive ship burning was a feature of all successful naval operations of the Company. This was done to destroy military hardware built over a long time and to deny recovery to the Indian powers in question. Manajee did not live long enough to relish this victory and died in 1758. After the Battle of Vijayadurg the majority of the Angre forts were taken over by the Marathas and annexed to their naval *subah* which was formed after the conquest of Bassein from the Portuguese in 1739. Gheria was first systematically plundered by the British then handed over to the Peshwa as part of the Anglo-Maratha deal. In the 1760s and 1780s the Maratha navy was commanded first by Rudraji Dhulap and later (his son?) Anandarao Dhulap. In the latter half of the eighteenth century, as and when the Marathas were at war with either the English or Hyder Ali of Mysore, the Maratha navy took part in offensive operations against enemy ships. In 1818, upon the conclusion of the Third Anglo-Maratha War, the Peshwa was exiled to Bithur and the Angre family became a feudatory of the British. The small Angre state was finally annexed to British India in 1840 when suitable male heirs among the Angres could not be found.

Our story of the Company's rise in the eighteenth century ends in 1756 which proved to be a turning point in the history of Bombay. After the liquidation of Angre power local powers posed a minor threat to the Company's growing hegemony in the region. The process of attrition brought about by the balance of power, geography and political economy in the late seventeenth century on the Konkan finally resulted in an Anglo-Maratha victory over Tulajee Angre, the more ambitious of Kanhoji's sons. In the ultimate analysis the Angre forts were politically isolated and Tulajee's navy, bereft of allies, was burnt in the harbour.[59] The fall of the Angres paved the way for the rise of the Company as the most important political player on the west coast. After the victory of 1756 the Marathas turned their attention to the Nizam, and to the north where Ahmad Shah Abdali, the founder of the Durrani Kingdom, was threatening their recently acquired dominance. This left the Company supreme in Bombay, a collection of islands which ultimately grew into the most important British commercial centre in South Asia. Three decades after the fall of Gheria, the so-called Gibraltar of Konkan, the British considered Bombay indispensable to the British naval presence in the Indian Ocean region.[60] At the beginning of the nineteenth century the Company's efforts of the previous century had secured Bombay for the British. The French posed little danger now. Tipu Sultan, the tiger of Mysore, was defeated and killed in 1799 and with him died all hope of an indigenous Indian power trying to undermine British naval supremacy in the Indian seas. Bombay could now be safely used as a launching pad for subduing, invading and conquering regions which lay to the north of the Arabian Sea. Towards this end the Company had an excellent para-military organization dedicated both to the collection of varied regional information and mounting expeditions in the form of the Bombay Marine.[61]

CONCLUSION

This essay has examined the limitations of the 'military revolution' thesis in the context of coastal warfare on the western coast of India during the seventeenth and eighteenth centuries. To begin with it was observed that the 'military revolution' refers to the emergence of modern warfare as a typical European way of conducting war in the early modern period. Its theoretical attraction lies in the widespread acceptance of historical difference between pre-modern and modern societies—a difference marked by the reference of social scientists to chronology, science and technology as well as the emergence of modern civil and military bureaucracies and the establishment of colonialism in several parts of the world. Thus victories and defeats are constant companions in the inevitable march of modernity over historiography; victories remaining on the side of modernity and defeats being credited to traditional societies. However closer inspection reveals that the theoretical framework of the 'military revolution' does not escape significant internal contradictions. Within it a stalemate of important military developments is clearly visible; artillery versus development in fort design and attrition, cavalry versus infantry, mobile field artillery versus infantry and set piece battles versus irregular warfare. There are three other problems with the 'military revolution'. First, its application to the new world, where gunpowder was unknown, compared with the early modern experiences of the gunpowder states of Asia, yields much quicker results. Second the 'military revolution' fails to answer certain important questions which may be raised in the context of military history read differently. It cannot be overlooked, for example, that throughout early modern history the technologies associated with the 'military revolution' were something two could easily play at—perhaps this was first demonstrated during the American war of independence. The concept of conventional military superiority, which may logically follow a belief in the 'military revolution', or even the superiority of modernity might prove inadequate to the writing of a better informed military history of modern Russia, Vietnam, Afghanistan and more recently, Iraq. Quite obviously warfare is a complex phenomenon in which the uncertainty of events, not to speak of popular participation, has the uncanny ability to throw theory and planning off track. Third, master narratives of modernization, to which the 'military revolution' is no exception, carry the tendency of overlooking local histories because of their ideological bias and hence their generalizations suit all those historical viewpoints according to which there existed a widespread and identifiable difference between early modern European and Asian people. These perspectives can be colonial, nationalist or even so-called Marxist. This difference may appear increasingly exaggerated in viewpoints supporting modernization and modernity in comparison with an equally biased reconstruction of traditional or so-called pre-modern societies. This paper in its exposition of local history has suggested that this techno-sociological difference can often result from reading the nineteenth century backwards into the largely pre-colonial periods of the seventeenth and eighteenth centuries in Indian history. While British victories in the late eighteenth and early nineteenth centuries were difficult to obtain in south Asia, they were even more difficult in the period prior to 1757.

This essay has shown why and how the Angres could hold out against the Company from the late seventeenth century till 1756. Local history suggests that much more can be accomplished in the field by not remaining constrained by the fashionable, attractive and convenient paradigms provided by our contemporary times. Directing our gaze at several aspects of coastal warfare in the mentioned period reveals a richly textured local canvass. The question why it took the Company more than half a century to overcome a powerful family entrenched on the Konkan has been answered in this essay in relation to the various factors conditioning military conflict and collapsing the difference between strictly modern and native methods of warfare in the early modern period in India.

These factors were:

1. The role of geography—rivers, creeks, hills, rocks and knowledge of navigation.
2. The strength of Maratha forts and resourcefulness of the coastal fighters.
3. The symbolic and real importance of artillery.
4. The hybridization of all major coastal military forces in the period.
5. The competitive military labour market on the western coast.
6. The role of Anglo-Maratha politics and family feuds in ensuring the allied victory over Tulajee Angre.

The formation of the Maratha Navy by Shivaji in the 1650s and Angre resistance (1690-1756) as well as the Maratha sources of the eighteenth century lead us to question several assumptions underlying the emergence of modern warfare in the region. What, for instance, does the endemic desertion of European seamen from the Company's Marine and other ships indicate but the capability of the Marathas to seduce these men with better terms of service? Throughout the period, the Angres seemed to have been successful in exploiting the local military labour market to their advantage. Their 'mixed crews' comprised an effective combination of lower class Europeans, men of the low-caste seafaring communities, Muslim *golandaaz* and *topasses*. Since no power was in a position to monopolize demand in the military labour market between the middle of the seventeenth century to the beginning of the nineteenth century the demand for trained professional fighters remained high. The pay must have been good enough to encourage desertion, national loyalties among the men were absent and subaltern opportunities for the utilization of military skills were plentiful. The survey of Maratha sources of the period also indicates that the Peshwa admiralty, probably following the trend set by Shivaji and *sarkhail* Kanhoji Angre, took care of soldiers and seamen in ways usually associated with modern methods of manpower management. The fact that sailor-soldiers could successfully petition the authorities asking for a raise in their compensation speaks volumes about the importance accorded to professional soldiering by the native powers. Other welfare measures mentioned in the essay suggest that the difference between 'modern' and 'native' powers was not as great as the votaries of modernization would have us believe; the cause of defeat on either side was often more political than military.

NOTES

1. For a résumé of the debate on the eighteenth century see Sekhar Bandyopadhyay, *From Plassey to Partition*, Orient Longman, 2004, chap. 1.
2. The words of Chesneaux, *Pasts and Futures*, p. 11.
3. Iqtidar Alam Khan, *Gunpowder and Firearms: Warfare in Medieval India*, New Delhi, 2004; Randolf G.S. Cooper, *The Anglo-Maratha Campaigns and the Contest for India: The Struggle for Control of the South Asian Military Economy*, Cambridge University Press (published in South Asia by Foundation Books, New Delhi), Cambridge, 2005. Also see my reviews of both books: 'Firearms in Medieval India'—Review of Iqtidar Alam Khan, *Gunpowder and Firearms: Warfare in Medieval India*, 2004 in *Economic and Political Weekly*, 15 January 2005 and 'War and the Military Economy'—Review of Randolf G.S. Cooper, *The Anglo-Maratha Campaigns and the Contest for India: The Struggle for control of the South Asian Military Economy*, Cambridge University Press, 2005 in *Economic and Political Weekly*, January 28, 2006.
4. Michael Roberts was the first to coin the concept of 'military revolution' in 1956 in *The Military Revolution 1550-1650* (Belfast, 1956). The concept was developed by Geoffrey Parker in the well researched and well known *The Military Revolution: Military Innovation and the Rise of the West, 1500-1800*, Cambridge, 1988. Since then it has been elaborated in a longer term perspective by Jeremy Black in three books on the subject dated 1991, 1994 and 1998. For more details of the progress of the 'military revolution' concept see Jos J.L. Gommans and Dirk H.A. Kolff (eds.), *Warfare and Weaponry in South Asia 1000-1800*, New Delhi, Introdn. Shelford Bidwell, *Swords for Hire: European Mercenaries in Eighteenth Century India*, John Murray, London, 1971, intro., displays a good grasp of the 'European method' of warfare and its contrast with a so-called Indian way of warfare.
5. For the debate on 'military revolution' see Clifford J. Rogers (ed.), *The Military Revolution Debate: Readings on the Military Transformation of Early Modern Europe*, Colorado. Much material on the concept is available on the internet. For example Kurt Kuhlman's review of Parker's 'Military Revolution', 2/25/92, Warhorse Simulations, webmaster@warhorsesim.com does not find it Eurocentric. Also see a review of Parker's book by Danny Yee (1997) on http://dannyreviews.com. This review lavishes praise on Parker while calling his work 'complementary in many ways to [Braudel's] Civilization and Capitalism.'
6. Khan, *Gunpowder and Firearms: Warfare in Medieval India*; following the assertions made by Khan it can be said that these regional powers comprised minor 'gunpowder' states. They were made in opposition to the Mughals with the help of artillery and matchlocks and later relied on European expertise to upgrade their arsenals.

7. For the 'military revolution' see the following: Geoffrey Parker, *The Military Revolution: Military Innovation and the Rise of the West, 1500–1800*, Cambridge, 1988; Paul Kennedy, *The Rise and Fall of the Great Powers: Economic Change and Military Conflict from 1500 to 2000*, New York, 1987; Eric Wolf, *Europe and the people without history*, Berkley, 1982 (for the sociological implications of the rise of the West). A survey of main issues in Indian military history in the light of the master narratives of the subject is provided in Anirudh Deshpande, NMML Monograph Number 3, *The Stigma of Defeat: Indian Military History in Comparative Perspective*, NMML, Teen Murti Bhavan, New Delhi, 2003.

8. The Fontana 'War and Society' series was a shift in this direction. G. Best, *War and Society in Revolutionary Europe 1770–1870*, Fontana, 1982 offers a comparative analysis of several early modern European conuntries and warns against over-emphasising technological factors. For a later period a critical survey of warfare is provided by V.G. Kiernan, *European Empires from Conquest to Collapse 1815–1960*, Fontana, 1982. The emergence of modern armies, arms races, military professionalism and impact of total war on European society is analysed by Brian Bond, *War and Society in Europe 1870–1970*, Fontana, 1984.

9. Khan, *Gunpowder and Firearms: Warfare in Medieval India, passim*, op. cit.

10. Parker, *The Military Revolution*, pp. 24–43.

11. Although both the massive forts mentioned have been encroached upon by modern habitation this author could not help being impressed by them during a visit in November 2007. Both the forts rise from the plains on promontories and are surrounded by deep and broad moats which were almost impossible to cross during the early nineteenth century. The breadth of the moat and the extraordinary height and thickness of the stone masonry walls, bastions and ramparts gave these strongholds sufficient protection against the field artillery of the period. Special canals and aqueducts fed the crocodile infested moats which, local people claim, never dry up. I saw no evidence of crocodiles in 2007 but it is likely that during the eighteenth and nineteenth centuries, to deter enemies bold enough to swim across the moats to try an escalade, the highly adaptable Indian *magar* was bred in the moats.

12. The works of Best and Bond cited above inform us that professional standing armies with academies and a trained officer corps were finally created in some European countries not before the late eighteenth century. In parts of east Europe, which experienced nationality and nation-state formation in the nineteenth century, the process was further delayed.

13. For details see Parker, *The Military Revolution*, pp. 82-114 for an assessment of European victories at sea. Statistically detailed information pertinent to naval developments in Europe is present in Kennedy, *The rise and fall of the Great Powers*, pp. 100-39.

14. G.S. Graham, *The Politics of Naval Supremacy: Studies in British Maritime Ascendancy*, Cambridge, 1965, p. 1. The influence of Mahan can be seen in old fashioned naval histories such as G. Callender and F.H. Hinsley, *The Naval Side of British History 1485-1945*, London, 1952.

15. *The Politics of Naval Supremacy*, p. 33, points out that the Gujarati pilot of da Gama's ship Ahmad Ibn-Majid was the author of thirty-five treatises on navigation. The instruments aboard Gama's ship, it is noted, did not impress him either.

16. *The Military Revolution*, p. 83.

17. For the kind of ships used on the Arabian sea by the Indian powers of the period see B.K. Apte, *A History of the Maratha Navy and Merchantships*, State Board for Literature and Culture, Bombay, 1973.

18. Ibid., pp. 86-9. For more details see Hugh Trevor-Roper (ed.), *The Age of Expansion*, London, 1968.

19. *Gunpowder and Firearms: Warfare in Medieval India*, notes the Ottoman influence on artillery and musketry in India.

20. *The Military Revolution*, p. 103.

21. *The Politics of Naval Supremacy*, p. 27.

22. *A History of the Maratha Navy*, pp. 72-3: Shivaji's naval strength has been estimated between 200 and 400 ships probably including merchant ships specially fitted out to trade with Mocha and Muscat. Daulat Khan and Mai Nayak Bhandari were the well-known admirals of Shivaji.

23. *The Military Revolution*, p. 114.

24. *Gunpowder and Firearms: Warfare in Medieval India,* op. cit.

25. I have examined this problem in the NMML, Monograph Number 3, *The Stigma of Defeat: Indian Military History in Comparative Perspective*, NMML, Teen Murti Bhavan, New Delhi, 2003.

26. The Portuguese legacy in the Arabian Sea is examined by Sanjay Subrahmanyam, *The Career and Legend of Vasco da Gama*, 1997.

27. Ibid., p. 112.

28. My account of the early Marine is based on the following: *Materials towards a Statistical Account of the Town and Island of Bombay*, vol. 3, *Administration*, Bombay, 1894, comprising consultations of the Company Directors and the Bombay Council from the 1680s to 1780s; *Gazetteer of Bombay City and Island*, vol. 2, Bombay, 1909; C.R. Low, *History of the Indian Navy*, vols. 1 & 2, London, 1877, reprint New Delhi, 1985 (extremely detailed monograph by an ex-Indian Navy officer covering the period 1600 to 1863); James Douglas, *Bombay and Western India: A Series of Stray Papers*, vol. 1, first published 1893, reprint New Delhi, 1985. Useful details of local sailing ships are to be found in the *Gazetteer of the Bombay Presidency*, vol. 2, Bombay, 1877; *Cambridge Economic History of India*, vol.1, 1984.

29. An account of such encounters is to be found in *A History of the Maratha Navy*, pp. 73-118.

30. *Materials towards a Statistical Account of the Town and Island of Bombay.*

31. I have examined the history of this 'hybrid' Bombay Marine in, 'The Bombay Marine: Aspects of Maritime History 1650-1850', *Studies in History*, 11, 2, NS (1995).

32. *History of the Indian Navy*, vol. 1, p. 90; *Gazetteer of Bombay City and Island*, vol. 2, pp. 275-6.

33. *Reports from Select Committees on the Affairs of the East India Company: First Report from the Select Committee on Indian Territories with Minutes of Evidence and Appendix, 1852-53*, Shanon, 1970, pp. 86-90.

34. *Materials towards a Statistical Account of the Town and Island of Bombay.*

35. Chosen extracts from the Maratha naval records can be found in Appendix C–1 in Apte, *A History of the Maratha Navy and Merchant Ships*.

36. Ibid.

37. Ibid., Appendix C–3.

38. *The Career and Legend of Vasco da Gama,* p. 8. Asian warships, unlike the European ships, did not use cannons on a large scale to begin with, but Maratha records show that by the early eighteenth century guns had acquired a new ritual, symbolic and military significance on the Konkan coast.

39. D.R. Banaji, *Bombay and the Sidis,* Bombay, 1932, is an example of high class Indian research and scholarship in late colonial India.

40. *History of the Indian Navy,* vol. 1, p.71.

41. Ibid.

42. Herbert Richmond, *The Navy in India,* London, 1931, p. 382.

43. *Gazetteer of Bombay City and Island,* vol. 2, p. 275.

44. Ibid.; *Materials towards a Statistical Account of the Town and Island of Bombay,* pp. 203–52, makes it abundantly clear that between the 1690s and 1780s the Company authorities in Bombay remained preoccupied with desertion among other endemic problems.

45. *The Career and Legend of Vasco da Gama,* mentions several instances of early Portuguese 'corsair activity' in the entire Arabian sea including the seizure and destruction of the hajj ship *Miri* returning to Calicut from Mecca by Gama in October 1502 (pp. 205–8). On this occasion Gama had the *Miri* bombarded and burnt with hundres of pilgrims including women and children on board. Once the Portuguese understood the value of trade between the Red Sea and the western coast of India their policy was 'to capture hostile shipping, and to make prizes of them'. The activities of Vincente Sodre, Gama's maternal uncle, and his brother Bras Sodre highlight the Portuguese moves to undermine Muslim shipping by extremely violent means (pp. 229–31). In general the policy was to capture trading ships, confiscate the cargo, murder the crew and later destroy the ships.

46. For details of piracy see History of the Indian Navy, vol. 1, pp. 80-1; *Gazetteer of Bombay City and Island,* vol. 2, p. 274. There is nothing in A.C. Roy, *A History of the Mughal Navy and Naval Warfares,* Calcutta, 1972, to even remotely suggest that Indian ships could stand up to the well armed and ruthless high sea pirates.

47. *The Military Revolution,* p. 112.

48. Manohar Malgonkar, *Kanhoji Angre Maratha Admiral: An Account of His Life and Battles with the English,* Bombay, 1959. According to Malgonkar, Kanhoji was an enterprising local Maratha sardar based in Colaba.

49. Ibid.

50. *Materials towards a Statistical Account,* p. 206; *Gazetteer of Bombay City and Island,* vol. 2, pp. 278–80. A brief history of the Angre family is contained in *Maharashtra State Gazetteers—Kolaba District,* 1st edn, 1883, 2nd edn, 1964, pp. 57-129.

51. *Kanhoji Angre,* p. 100.

52. Indian military weaknesses are analysed by Satish Chandra, *The Eighteenth Century in India: Its Economy and the Role of the Marathas, Jats, Sikhs and the Afghans,* Calcutta, 1986. Also see *The Military Revolution,* chap. 4. I have also examined several Indian military drawbacks in NMML, Monograph No. 3, *The Stigma of Defeat: Indian Military History in Comparative Perspective,* NMML, Teen Murti Bhavan, New Delhi, 2003.

53. Apte, *A History of the Maratha Navy and Merchantships,* pp. 245–81.

54. For more details of Indian fort plans, etc., see J.N. Kamalapur, *The Deccan Forts,* Bombay, 1961 and Ramachandra Murthy, *Forts of Andhra Pradesh,* Delhi, 1996.

55. My description of these battles is based, among other works, on a reading of *Kanhoji Angre.*

56. *Materials towards a Statistical Account*, pp. 206–7.

57. *History of the Indian Navy*, vol. 1, p. 107.

58. Indeed *Peshwa* Baji Rao fought an important battle against the young *Senapati* Trimbakrao Dhabade who refused to hand over control of revenue rich Gujarat, won by his father for Shahu Maharaj, to the Peshwa. For details of this revealing episode see N.K. Wagle and R.G.S. Cooper, 'The Battle of Dabhoi' (1 April 1731) in A.R. Kulkarni et al. (eds.), *Medieval Deccan History*, Bombay, 1996.

59. My account of the Angre defeat is based on the following sources: *Materials towards a Statistical Account*, p. 221; *History of the Indian Navy*, vol. 1, pp. 133-4; *Gazetteer of the Bombay City and Island*, pp. 279-80 and Douglas, *Bombay and Western India*, pp. 110-25. In praise of Kanhoji, Douglas wrote: 'Our readers would not thank us for a history of Kanhoji Angria's exploits. Each of the three great European nations in Indian waters had a shy at him and his family and each came off second best.'

60. Admiral Hughes, Commander of the East Indies Squadron, Royal Navy, considered Bombay crucial to British successes against the French—'Admiral Hughes to the Governor and Select Committee of Bombay', 13 May 1784, India Office Records, Home-Misc., p. 174, cited in Appendix viii, in Richmond, *The Navy in India.*

61. I have examined the history of this service including the colonial conquests achieved by its officers and men in 'The Bombay Marine: Aspects of Maritime History 1650-1850', *Studies in History*, 11, 2, NS 1995.

Coastal Polity and the Changing
Port-Hierarchy of Kerala

Pius Malekandathil

THE DIVERSE PORTS of medieval Kerala used to operate on the basis of a remarkable degree of hierarchy, in which one port in its evolution as the central and pivotal exchange centre used to attract the bulk of overseas trade, while other ports were made to depend on it by converting them into its feeding satellite units, or as its minor distribution centres. Consequently foreign trade used to get concentrated in this pivotal port making wealth from foreign trade increasingly accumulated in this exchange centre, and inviting the politically ambitious rulers to convert it into their chief base for exercising power. These rulers used to resort to diverse means and methods for minimizing and deterring the volume of foreign trade with other ports, and to maximize exchange activities in their port, a mechanism that ultimately ensured the maintenance of a port-hierarchy that furthered their political ambitions. However this hierarchical structure of ports did not remain static all through history. The hierarchical structuring of the ports used to change considerably over time, with one leading principal port giving way to another and relegating itself to the background as a feeding port for the emerging one, against the backdrop of changing geo-physical environment and coastal politics. The diverse rulers of coastal Malabar, who wanted to bag wealth coming from sea-borne trade, were very much instrumental in the shaping and re-shaping of port-hierarchies in Kerala. This was done either by shifting their headquarters to the commercially prominent port-area or by way of conquering other highly potential maritime exchange centres and making them remain subservient to the principal port. However, their political decision was conditioned very much by the geo-physical location of the leading port that they had chosen for its potentiality and ability for resource mobilization.[1]

The central purpose of this essay is to see how far the changing geo-physical environment of the coastal regions was utilized by the rulers of coastal regions for developing a structure of port-hierarchy in Kerala for the purpose of generating wealth for setting up power-exercising devices and state structures in their kingdoms at different points of time. This was done, on the one hand, by analysing the geo-physical changes of the coastal regions of Kerala, and, on the other, by examining the consequent attempts of the coastal rulers to make use of the changed situation for controlling the maritime trade traffic in differing degrees. The study looks into the long-term trends from the ninth century to the eighteenth century, highlighting the links of continuity and the multiple levels of changes over time that were instrumental in shaping the polity and economy of coastal Kerala.

HISTORICAL SETTING

During the first centuries of the Christian era, when vessels from Bernice or Myos Hormos from Roman Egypt were coming in large numbers to India,[2] about 120 per year as Strabo writes,[3] almost 50 per cent of them used to frequent the ports of Limyrike (the present-day Kerala) and south India, while the rest used to sail to Barygaza (Braukaccha or Broach) after having crossed the straits of Bab-el-Mandeb.[4] As *Periplus of Erythraen Sea* mentions, there were several ports in Kerala during the early centuries of Christian era, whose economic importance varied so significantly that there eventually evolved among them a hierarchical order keeping the highly activated port of Muziris (Cranganore)[5] at the pivotal position and other ports of Naura (Cannanore), Tyndis (Ponnani), Bakare (Purakkad), Nelkynda (Niranam) and Comorin (Cape Comorin) as its satellite feeding ports.[6] Muziris, to which commodities were brought from other feeding ports of Kerala with the help of small *tonis* and barges for their further transshipment to overseas markets,[7] eventually turned out to be the principal trading centre of the Indian Ocean region during this period because of its better geo-physical location and its access to the spice hinterland of Kottanarike spread between Muziris (Cranganore) and Bakare (Purakkad).[8] Over and above this, the political interventions of the Chera chieftains also seem to have played a more vital role in determining the hierarchical order among the ports. What the Roman vessels brought were gold and silver coins, which the Chera chieftain might not have wanted to reach other ports and share with others, without getting his share from it. The creation of a principal port with several feeding ports was conceived of as a part of the political strategy to get the wealth coming from maritime trade concentrated in the pivotal port of Muziris controlled politically by the Chera chieftains, however crude this wealth was in form, content and manifestation. The other side of the same politico-economic development was the attempts of the Chera chieftains to keep the various minor ports dependent on the pivotal port for the purpose of their trade, which in fact was developed as a mechanism to keep the regional economies depend upon the principal maritime exchange centre, capacitating the ruler to have more power concentration. The accumulation of sizeable wealth in the hands of the Chera chieftains from the trade of Muziris and the consequent process of conversion of a share of that wealth for minting coins for the purpose of activating maritime trade is attested to by the discovery of Chera gold coins from the excavations of Pattanam near Parur in Kerala.[9] In fact the process of converting Muziris into a pivotal port fed by a chain of feeding satellite ports for the purpose of wealth accumulation and power concentration took place at a time when the tribal Chera chieftain was increasingly engaged in conflicts with the Pandyas and the Cholas, in the process of eventual transformation from being the leader of cattle-raiders to the position of a ruler with sizeable land-territory under his control.

PARTICIPATION OF QUILON IN THE LONG-DISTANCE TRADE AND THE STRUCTURING OF A NEW PORT HIERARCHY IN KERALA

From the ninth century onwards, a new port hierarchy began to take shape in Kerala with the increasing visits to its coast by the Arab *dhows* for trade on their onward voyages to China. In the newly evolving port hierarchy, Quilon, was made to emerge as the most important port in Kerala and other ports including the old port of Muziris (Muyirikode or Cranganore), Pantalayani Kollam, Ezhimala (Mount Eli), etc., were made to function as its feeding satellite ports. Ezhimala, which was the original seat of the Kolathiris, was an important centre of maritime trade from an earlier period,[10] around which the minor ports of Ilangopatanam,[11] Acalapatanam,[12] Kachilpatanam[13] and Maday[14] evolved as other important ports in northern Kerala.

Quilon (Koulam Mali) emerged as the principal port of Kerala during the period between the ninth and fourteenth centuries, because of its selection as the principal halting centre by the merchants involved in the long-distance movement of commodities between al–Basrah or Muscat or Sohar (Oman) in the Persian Gulf and Canton in China.[15] Kurakkeni Kollam, which was known differently as Koulam Male or Koulam Mali in the Jewish Genizza papers[16] and in Arabic sources[17] as well as Gu–lin (in the Song period)/Ju–lan (in the Yuan period) in Chinese documents,[18] does not appear in any source prior to AD 823, which also suggests that this port emerged as a principal centre of call only in the ninth century. The *dhows* that started from Persian Gulf used to take about 30 days to reach Koulam Mali (Quilon) and another 30 days to reach Qalah (Kedah in the Malaya Peninsula) and yet another 30 days to reach Canton. Thus the one-way trip from Persian Gulf to China took 90 days. On the way, these *dhows* invariably took shelter in Quilon for want of favourable monsoon wind and for taking water and food provisions.[19] With the increasing concentration of Abbassid merchants in Quilon, the old Muziris (by this time known as Muyirikode[20]) started losing its importance of being the prime pepper-exporting port, and this position was eventually taken over by Koulam Mali, whereby the old Muziris-centred port hierarchy started giving way to a new port hierarchy with a chain of satellite ports like Pantalayani Kollam, Mount Eli and Cranganore revolving around Quilon.

The Chera rulers[21] attempted to utilize the changes in the port hierarchy to their political advantage and to promote the trading activities of Quilon for bagging wealth to counter initially the commercial and political expansionist moves of the Pandyas and later of the Cholas. The first step in this direction is seen in the conferring of a set of commercial privileges by Ayyanadikal Thiruvadikal, a feudatory of the Chera rulers in AD 849, on the Persian Christian merchants, who came to Quilon in 823 under the leadership of Maruvan Sapor Iso and on the church of Tharisapally, set up by the latter.[22] In fact the local ruler Ayyanadikal Thiruvadikal granted the diverse privileges on seeing the instrumentality of Maruvan Sapor Iso in activating the overseas trading activities of the port of Quilon. This set of privileges legitimized the claim of the church of Tharisapally on overseas trade as well as control over the weights and measures of the city, besides gaining several tax-collecting rights over the diverse settlers coming under this church, by way of the grant.[23] These diverse privileges were given to strengthen and empower the mercantile Christian community that in turn would emerge as a sufficiently capable agent for mobilizing resources for trade and for bringing wealth to empower the state.

At a time when the Cholas attempted to monopolize the trading activities of South-East Asia by controlling the exchange centres of Ceylon, the Coromandel coast and the Sailendras,[24] the Chera rulers and their feudatories made increasing use of the church of Tharisapally, *Manigramam* and *Anjuvannam* merchant guilds of Kerala[25] that had links with the ports of Persian Gulf and Red Sea, for organizing and controlling the commercial affairs of Quilon and the Arabian Sea.[26] In fact the merchant guilds, viz., *Anjuvannam*, *Manigramam* and *Arunnoottuvar* were entrusted with the right to protect the church and its privileges newly granted by the ruler.[27] *Anjuvannam*, which is considered generally to be a Jewish merchant guild and *Manigramam* of Kollam which is generally considered to be a Christian guild,[28] were given the power of *karalars* of the city,[29] a fact that would evidently suggest that they had considerable authority and forces of power with them, with which the protection of the church and its privileges could be ensured.

These foreign merchants were favoured and encouraged as a part of the Chera strategy to attract more overseas trade to Quilon and thereby to ensure more wealth-concentration in the city for the purpose of meeting the political challenges from the Pandyas in the south. The Pandyas, having invaded the Ay-Vel territory and captured the ruler along with his relatives, besides a huge volume of treasures, posed a serious challenge to the Cheras. The latter managed to regain only the Ay territory and its political headquarters of Quilon (Kurakkeni Kollam).[30] By keeping Quilon as the central and pivotal port and making others dependent upon it, wealth from trade was made to get concentrated on it strengthening not only the Chera ruler but also his feudatory at Quilon, on whom a major share of the responsibility of defending the southern territories of Chera kingdom actually fell.

Later, we find several attempts being made by the Chera rulers and his feudatories to encourage trade of the foreign and local merchants operating at different places of Kerala, particularly after the defeat of Chera naval power at Vizhinjam and the loss of Quilon to the forces of Raja Raja Chola (985-1014). In the war council that was convened, Joseph Rabban, who was the head of the *Anjuvannam* merchant guild linked with the Jews of Muyirikode (Cranganore), placed at the disposal of the Chera ruler his ships, men and materials for the conduct of the war with the Cholas. In return he was conferred with seventy-two privileges and prerogatives of aristocracy about the year AD 1000 by the Chera ruler Bhaskara Ravi Varma (962-1020).[31] The conferring of privileges upon the leading Semitic traders is to be seen as a strategy to ensure the mercantile support of the Jewish merchants operating from different parts of Kerala for realizing his plan of regaining Quilon from the Cholas and maintaining it as a mechanism for resource mobilization.[32]

It was from Koulam Mali (Quilon) that the principal navigational lines of the merchants belonging to the guilds of *Anjuvannam* and *Manigramam* moved towards the exchange centres of Abbassid Persia and Fatimid Egypt.[33] Consequently Quilon turned out to be the principal destination for the Jewish merchants, travelling from Aden to Munaybar (Malabar or Kerala), via Sindabur (Chandrapura in Goa) and selling storax and coral collected from Mediterranean ports, as is mentioned by the Jewish merchant Allan b. Hassun (AD 1116-17).[34] Concomitantly, Quilon was also the principal destination for the Chinese junks, as well, both during the Yuan (Mongol) and early Ming periods and it participated in the Chinese tributary trading system by sending a commercial mission to the court of Kublai Khan. The commercial mission from Quilon was reciprocated by an official Mongol mission dispatched to Quilon in 1283 carrying with it a golden badge for Wa-ni, the king of Koulam Mali, on whom the Chinese emperor conferred the title of 'Fu-ma' or imperial son-in-law.[35]

With the acceleration of trade in Quilon, the amount collected as customs over there also augmented. Each cart (*vayinam*) that used to take merchandise by land into the market of Quilon was to pay eight *kasu* and each boat (*vediyilum*) that was used to carry cargo to the port had to pay four *kasu* as customs duty to the Tharisapally that turned out to be the principal mobilizing institution of trade in the port, as is mentioned in the Tharisapally copper plate of AD 849.[36] However, much greater was the share of trade surplus that went into the coffers of the rulers as *ko pathavaram* (share of wealth from trade to the Chera ruler Sthanu Ravi Varma) and *pathi pathavaram* (share of wealth from trade to the local ruler Ayyanadikal).[37] This is evidently suggestive of the reason why the Chera rulers and his feudatories tried their level best to maintain a port hierarchy with Quilon to be the dominant exchange centre and other ports as secondary ones.

EMERGENCE OF CALICUT AND THE POLITICS OF SUBORDINATION OF SEA PORTS

A new type of port hierarchy started appearing in Kerala from mid-fourteenth century onwards, with Calicut emerging as the leading port, while other maritime centres of exchange were reduced to the status of satellite feeding centres to it, or as minor centres of sea-borne trade. It coincided with the fragmentation of the central authority of the Cheras (the Kulasekharas) of Mahodayapuram in AD 1124, when there evolved different petty power structures under nomenclatures like *svarūpams, nātuvāḷis, dēsavāḷis, kaimals, karthas,* etc.[38] Many of these local rulers competitively tried to keep various seaports of Kerala under their control, with a view to gaining a share of maritime trade for their political assertions. It was a time when al-Karimi merchants linked with Mamluk Egypt were increasingly visiting the northern ports of Kerala, following the slackening of trade in Quilon, which started with the decline of international trade emanating from the Persian Gulf after the Mongol conquest of Abbassid Baghdad (1258). The al-Karimi traders started concentrating more and more on the port of Calicut, because of the predominant presence of the Jewish and Christian traders in the southern port of Quilon. It was against this background of intensified trade that the chief of Nediyirappu Svarupam shifted his royal residence from the inland agrarian pocket of Nediyirappu in Ernad (Malappuram District) to the maritime trade centre of Calicut, which he had meanwhile captured from the ruler of Polanadu, an action which indicates his desire to bank upon profits from its sea borne trade.[39] This in fact gave rise to a new type of port hierarchy in the fourteenth century with Calicut in the pivotal position and other ports to revolve around it with the functional roles to feed it commercially. The chief of Nediyirappu Svarupam, who is popularly known as Zamorin, conquered and occupied the port of Mount Eli and Cannanore, from the rulers of Kolathunadu in the fifteenth century, and later towards the end of the fifteenth century, captured Cochin from the chief of Perumpadappu Svarupam and converted them into feeding exchange centres for the port of Calicut. On capturing Cochin, Zamorin insisted that the traditional trading community of the Nazarenes (or the St. Thomas Christians) should be ousted from the port of Cochin and in their stead Cochin's commerce should be transferred into the hands of the Muslim merchants of Calicut.[40] By installing his Muslim commercial allies as the sole merchant group of Cochin, the Zamorin ensured the process of subordinating Cochin's trading activities to the greater interests of Calicut.

In fact, attempts to take away the independent trading nature of the neighbouring seaports and subjugate them to Calicut as its commercial feeding units went hand in hand with Zamorin's efforts to carve out sizeable spice-producing hinterland from the rulers of interior Kerala for the commerce of Calicut. With the money and personnel supplied by the al-Karimi and other Muslim merchants, the Zamorin conquered the neighbouring principalities of Kolathunadu, Nilambur, Manjeri, Malapuram, Kottakkal, Cranganore and even Cochin and made them units of hinterland for the commerce of the port of Calicut, besides amalgamating them for the process of making a territorial state centred around Calicut.[41] The process of transformation of a large tract of spice-producing areas from the newly occupied territories as hinterland for his port and the subordination of important seaports of central and north Kerala to the commercial designs of Calicut, enabled the Zamorin to emerge as a political figure with a great amount of wealth, authority and influence in South Asia. Meanwhile the accumulation of wealth and the consequent wielding of authority by Zamorin went hand in hand with a process of legitimization resorted to by him through the mechanism of the pan-Kerala festival called Mamankam. As the right to preside over *Mamankam* was a prerogative of the most powerful ruler in Kerala, the Zamorin, who acquired this privilege from the Kolathiris as a part of his political conquests, could easily thrust his authority on the other rulers of Kerala through cultural articulations. The entire historical process, besides ensuring the legitimacy and official sanction required for the expansion of his state, conferred upon the Zamorin a relative pan-Kerala identity, as well. [42] The political assertion of Zamorin went hand in hand with the economic process of the restructuring of the port-hierarchy, in which the exchange activities of all other ports of Kerala were made subservient to those of Calicut.

LOCUS OF COCHIN IN THE CHANGING PORT-HIERARCHY

Cochin, which emerged as a port in 1341 following the great flood in Periyar, became an important maritime trade centre following the shifting of the headquarters of the chief of Perumpadappu Svarupam from the inland agrarian region of Vanneri to the port area around 1405. [43] His movement from the land-locked area towards the maritime centre of exchange is evidently suggestive of his deeper desire to appropriate wealth coming from the maritime trade of Cochin. In order to compete with Calicut both economically and politically, the chief of the Perumpadappu Svarupam encouraged a wide variety of merchant groups to engage in the trade of Cochin, as is evidenced from the account of Ma Huan (1409). The groups included besides the Muslim merchants linked with West Asian trade, Chetties associated with the coastal trade of Coromandel and Canara, Kelings or Klings linked with South-East Asian commerce[44] and the Nazarani Thomas Christians who were once connected with the Manigramam merchant guild. Out of the seven voyages that the Chinese Admiral Cheng-Ho conducted between 1405 and 1433, as emissary of the Ming ruler Yung Lo, it is believed that he had visited Cochin at least five times. [45] This is suggestive of the commercial importance that Cochin had grabbed in the first part of the fifteenth century. However the Zamorin, who wanted to have only one major port in Kerala and that too in Calicut for the purpose of getting the wealth from foreign trade concentrated in his hands, attacked Cochin and reduced it to the status of a feeding port to Calicut by the end of the fifteenth century. Consequently, the kingdom of Cochin was narrowed down to the island of Vaipin and Cochin. [46] Moreover, elements of sovereignty were also taken away from the king of Cochin, who was in turn made to become a vassal to the Zamorin. The perpetual subordination of the king of Cochin to the Zamorin was realized by depriving the former of the rights to mint coins and by preventing him from roofing his palace with tiles.[47] But the ultimate subjugation of Cochin port to Calicut happened with the ousting of the traditional merchant groups (the Nazaranis or St. Thomas Christians) from its port and the consequent coercion to transfer its commerce to the Muslim traders of Calicut.[48]

The beginnings of the transition of Cochin from the status of a feeding port to that of a pivotal port started with the advent of the Portuguese in this small city in 1500.[49] Within no time, the presence of the Portuguese power in Cochin began to manifest itself in the forms of artilleries, guns and huge ships creating a tilt towards Cochin's politico-economic hegemony, whose substance actually rested in its wealth accrued through maritime trade. The fortified settlement of the Portuguese in Cochin was made the seat of the Portuguese Government in India by Francisco da Almeida in 1505 and it was supported by a chain of armed fortresses in Cannanore, Anjedive and Kilwa.[50] The system of fortresses, monitoring the commodity traffic through rivers and the regular patrolling along the coast, weakened considerably the trade-potentials of non-Portuguese trade centres in Kerala.[51]

Cochin was made the political capital of the Estado da India (till 1530) [52] and this eventually became the focal point of Indian Ocean trade. Different commodity streams of the Indian Ocean trade were made to intersect at Cochin for purposes of making available cargo for Lisbon-bound vessels. On the one hand, the Lusitanians made the ports of Quilon, Cranganore, Purakkad, Parur and Kayamkulam as subsidiary feeding units for Cochin by making the cargo from these ports reach Cochin regularly for further transshipment to Lisbon.[53] *Tonis* were regularly plying between these ports and Cochin, distributing the wares brought to the latter from Portugal. By frequent movement of commodities from Cochin to these ports and vice versa they were integrated into the new economic order evolving in Cochin, making it as the pivotal port around which other ports of Kerala were made to revolve as satellites.[54] Though Calicut continued to stay out separately and feed the traditional caravan route,[55] the Portuguese tried to keep the major spice-production centres of Kerala integrated with the circuits of the Cochin-Lisbon commercial axis by erecting minor forts and trade centres at Chaliyum, Ponnani and Palayur, besides the fortress and factory at Cannanore.[56] Concomitantly, the Lusitanian mechanism of coastal patrolling forced the movements of commodities to the port of Cochin, making it the apex port of Kerala in the evolving port hierarchy.

The different commodity streams that used to intersect at Cochin through the medium of the *casado* traders included silk, copper and porcelain from China, sandalwood, nutmeg and mace from South-East Asia, silk, rice, long pepper (known as *thippili* is a spice-variety obtained chiefly from Bengal and Assam), wheat and salt-petre from Bengal in exchange for pepper from Kerala, rice, cereals and textiles from Coromandel, cotton clothes, indigo and opium from Gujarat, cowries and coir from Maldives, cinnamon from Ceylon, horses from Hormuz.[57] The pattern of movement of commodities revolved around Cochin because of it being the principal port from where copper and silver brought from Portugal as pepper money was made available for further transactions in other ports.[58] The Portuguese crown wanted to keep Cochin as the principal port for the simple reason that everything that was needed for Lisbon–bound vessels was made available there by the Portuguese *casado* traders and their commercial allies, like the banias and the local Muslims, in the process of their visit to it for collecting cargo coming from Europe as well as for taking pepper for intra–Asian trade with the help of callous officials.[59] The king of Cochin in whose territory the port was located, was extremely supportive of the Portuguese ventures to promote the port hierarchy structure with Cochin at the apex and he was most eager to keep its import-export indexes over and above those of Calicut, as its trade fetched for him an yearly income of 60,000 *pardaos* in 1605,[60] which rose to 80,000 *pardaos* by 1612.[61] The king of Cochin tried to attract more trade to his port by reducing the customs duty to 3.5 per cent for all Portuguese *casado* traders, a step that ultimately helped to prevent trade from being scattered to other Portuguese ports, where the rate of customs duty was 6 per cent.[62]

Thus the diverse politico-economic strategies and the mechanisms developed both by the Portuguese and the local ruler for maximizing profit created an atmosphere that favoured the emergence and sustenance of the new port hierarchy in the sixteenth and seventeenth centuries with Cochin at the apex and a chain of other ports as feeding units to it.

TERRITORIAL CONQUESTS AND RESHUFFLING OF PORT-HIERARCHY

The leading role that the port of Cochin enjoyed, thanks to the intense trade of the Portuguese and the Dutch, started receiving a set back by the middle of the eighteenth century, whereby Cochin was eventually relegated to the background by Alleppey, which by this time had emerged as a prominent maritime exchange centre in Kerala as a result of complex coastal politics.[63] Marthanda Varma of Travancore, who defeated the Dutch at Colachel (1742), realized that the power of the Dutch was based on their commerce that had flourished at Cochin. Consequently, he chalked out a policy of conquering the major spice-producing areas supplying cargo for the Dutch trade in Cochin and of grabing its hinterland evidently with a view to depriving the Dutch any access to spices and other commodities. Thus, most of the spice producing territories like Quilon, Kayamkulam, Thekkumkur, Vadakkumkur and Porcad were conquered one after aother and annexed to Travancore between 1743 and 1752, following which the flow of spices and other commodities from these principalities to the port of Cochin got blocked, inflicting a severe blow on the commerce of the Dutch.[64] With the position of major spice-producing centres of central and south Kerala, trade in pepper was declared a state monopoly in Travancore in 1743, which not only severely reduced the amount of pepper available to the Dutch, but also created serious difficulties for the local merchants in its purchase and sale according to the needs of the market.[65] Later in 1763, during the reign of Dharmaraja, a new port was established at Alleppey by his chief minister Raja Kesava Das with a view to re-routing the spices from the newly conquered territories of the north through a Travancorean door to international markets.[66] New trading groups represented by Thachil Mathu Tharakan (but linked with the trade of the English) were made to take over the trade of Alleppey, which in course of time paved the way for another hierarchical ordering of ports in Kerala with Alleppey at its apex.[67] *Pandikasalas* (warehouses) were constructed in most of the important spice-producing regions of the kingdom for purchasing and storing spices after the harvest. New roads and water channels were constructed all over the kingdom linking the different *pandikasalas* with the port of Alleppey, ensuring the regular movement of commodities. The facilities for checking and monitoring the entire process of purchase as well as transportation of commodities and the prevention of pepper diversion were very much guaranteed by the military camps, which were often located closer to the *pandikasala* structures.[68] In the evolving process, the Travancorean ports of Anjengo, Quilon and Vizhinjam were made feeding units to Alleppey, while Cochin, Calicut and Cannanore that were facing vicissitudes of a different nature were eventually losing their prominent positions.[69] In the changed situation, where Cochin was relegated to the background by Alleppey, the king of Cochin Raja Rama Varma (1790–1805) alias Saktan Thampuran began to develop Trichur as the seat of his power and the core area of his politico-economic activities, undermining the base of Dutch commerce in Cochin.[70]

Thus we find that the different ports of Kerala had different levels and degrees of exchange activities at different phases of time. Muziris, that was the leading port up to eighth century, was pushed to the background by Quilon (Koulam Mali) that came to the apex position by ninth century, thanks to the politico-economic policies of the imperial Cheras and their feudatories. By the fourteenth century, relegating Quilon to background, Calicut emerged as the leading port, which position went to Cochin by mid-sixteenth century, mainly because of the commercial and political interventions of the Portuguese. However, by mid-eighteenth century the lead position in the port-hierarchy was occupied by Alleppey, with the entry of the Travancoreans in the coastal politics. On the basis of the scale and volume of trade carried out therein, one can evidently speak of a port hierarchy, in which one leading port dominated the entire exchange scenario of Kerala at one specific point of time by keeping other ports as its feeding units or as insignificant exchange centres. However the prominent position of the port changed over time on the basis of the coastal politics, by which the ports of the weaker rulers were made to depend upon the port of the paramount ruler. This was a strategy to reduce chances for the inferior rulers to get access to wealth and to keep them always subjugated to the paramount ruler. Besides strengthening the hands of the ruler through wealth of trade, the types of cargo exchanged, the people with whom trade was conducted and the volume of trade carried out in the port were in those days considered as visual devices to manifest the extent of the power that the ruler wielded. It was against the background of the desire of the coastal rulers to materialize their political ambitions and to display their strength through the medium of ports that they tried to define and structure the port hierarchy of Kerala differently at different time phases.

NOTES

1. In fact, Ashin Das Gupta had earlier referred to the existence of a port-hierarchy in Gujarat, where one principal port dominated the neighbouring maritime ports at different time periods, causing the latter to evolve as feeding and supplementing ports for the central one. (Ashin Das Gupta, *Indian Ocean Merchants and the Decline of Surat, c.1700-1750* , Wiesbaden, 1979, p. 1). However, its inner dynamics or the manner in which the port-hierarchy functioned over time in any coastal region were not much elaborated upon till date. A sketchy formulation of this theme was done earlier by the author in two of his works. Pius Malekandathil, 'Urban Growth of Cochin in the Sixteenth and Seventeenth Centuries', M.Phil. thesis submitted to Pondicherry Central University, 1992, pp. 24–45 and Pius Malekandathil, 'Maritime Malabar and a Mercantile State: Polity and State Formation in Malabar under the Portuguese, 1498-1663', in K.S. Mathew (ed.), *Maritime Malabar and the Europeans, 1500-1962*, Gurgaon/London, 2003, pp.198, 216.
2. For details see Vimala Begley and Richard Daniel de Puma (eds.), *Rome and India: The Ancient Sea Trade*, Delhi, 1992; Rosa Maria Cimino, *Ancient Rome and India: Commercial and Cultural Contacts between the Roman World and India,* New Delhi, 1994.
3. Strabo, *Geographia*, 2, 118.
4. Lionel Casson, *The Periplus Maris Erythraei,* Princeton, 1989, p. 225.
5. There are differences of opinion regarding the identification of Muziris. Till recently historians used to equate Muziris with the present day Cranganore. However, the archaeological excavations conducted at Pattanam in 2004 and the discovery of several artifacts showing cultural contacts with Rome make many historians believe that Muziris is the present day Pattanam located near Parur. See K.P. Shajan, 'Discovery of Muziris: Geological and Archaeological Evidences', paper presented at the seminar on *Muziris* organized by Kerala Historical Research Society at Kerala Sahitya Academy, Trichur, 11 July 2004. On seeing the site and the artifacts, I too am convinced of the antiquity of the place and of it having been a trade centre during the early centuries of the Christian era. However, so far nothing has been found at the site to substantiate the claim that it was Muziris. For want of a link to connect Pattanam with Muziris, I maintain here the old view among scholars that Muziris must have been Cranganore. It must also be said that the newly unearthed site must have been highly contemporaneous with Muziris and probably acted as a feeding port for Muziris. See also See Pius Malekandathil, 'Ships, Navigation and Maritime Exchange Centres of Kerala: Channels of Civilizational Contacts, 100 BC to 1500 AD', *Summer Hill* (IIAS Review, Shimla), vol. X, nos.1&2, 2004, pp. 42-4.

6. L. Casson, *The Periplus Maris Erythraei*, Princeton, 1989, pp. 83-5; Ranabir Chakravarti, *Trade and Traders in Early Indian Society*, New Delhi, 2007, pp. 54-5.

7. Ibid., pp. 83-7.

8. For the geographical identification of Kottanarike, see A. and M. Levi (eds.), *Itineraria picta: Contributoallo studio della Tabula Peutingeriana*, Rome, 1967; R. Champakalakshmi, *Trade, Ideology and Urbanization: South India 300 BC to AD 1300*, Delhi, 1996, p. 120; Pliny, *Natural History*, 6.105; Claudius Ptolemy, *Geografia*, 7.I.9; P.J. Thomas, 'Roman Trade Centres in Malabar', *Kerala Society Papers*, vol. II, Trivandrum, 1999, p. 267.

9. So far two gold coins belonging to the early Cheras have been unearthed from the site. The first Chera coin was discovered in 2004. For details see M.G.S. Narayanan, 'Muzirissine Thedi', paper presented at the seminar on *Muziris* organized by Kerala Historical Research Society at Kerala Sahitya Academy, Trichur, 11 July 2004, pp. 4-8. The second Chera coin was obtained in 2007.

10. M.G.S. Narayanan, 'Mushakavamsa as a Source of Kerala History', *Re-Interpretations in South Indian History*, Trivandrum, 1977, p. 65.

11. Ilangopatanam appears in the Ramantali stone inscription. See *Annual Report on South Indian Epigraphy*, Madras, 1932, pp. 86-7.

12. Acalapatanam is mentioned in the *Mushikavamsa* of Athula. Cf. K. Raghavan Pilla (ed.), *Mushikavamsam*, Trivandrum, 1983.

13. Kachilpatanam is mentioned in the *Payyannur Pattu*, which is said to have taken shape in the thirteenth or fourteenth century. P. Anthony (ed.), *Payyanur Pattu: Pathavum Pathanavum*, Kottayam, 2000, p. 3; Binu Mailaparambil John, *The Ali Rajas of Cannanore: Status and Identity at the Interface of Commercial and Political Expansion, 1663-1723*, Ph.D. thesis submitted to the Department of History, Leiden University, The Netherlands, 2007, pp. 10-11.

14. The work *Mushikavamsam* mentions that it was the Mushaka King Vallabha II who established the town of Madai (Maday), which was a royal port-city in the early days of the medieval period. Cf. Raghavan Pilla, *Mushikavamsam*, p. 301.

15. George F. Hourani, *Arab Seafaring in the Indian Ocean in Ancient and Early Medieval Times*, Princeton, 1951, pp. 70-4.

16. For details of Koulam Mali mentioned in Genizza papers, see S.D. Goitein, *Letters of Medieval Jewish Traders*, Princeton, 1972, p. 64.

17. The earliest Arab source is Suleiman's account of AD 841 entitled *Salsalat-al-Taverika*. For other Arab sources on Kollam see George Fadlo Hourani, *Arab Seafaring in the Indian Ocean in Ancient and Early Medieval Times*, pp. 70-4.

18. For details on Chinese references to Kollam, see Haraprasad Ray, 'Historical Contacts Between Quilon and China', in Pius Malekandathil and Jamal Mohammed (eds.), *The Portuguese, Indian Ocean and European Bridgeheads: Festschrift in Honour of Prof. K.S. Mathew*, Lisbon/ Tellisserry, 2001, pp. 386-8.

19. George F. Hourani, *Arab Seafaring*, pp. 60-80.

20. Elamkulam Kunjan Pillai, *Studies in Kerala History*, Kottayam, 1970; M.G.S. Narayanan, *Cultural Symbiosis in Kerala*, Trivandrum, 1972, p. 82.

21. Here the reference is to the Cheras (Second Cheras) who ruled over Kerala with political base at Mahodayapuram from AD 800 till 1124. See M.G.S. Narayanan, *Perumals of Kerala: Political and Social Conditions of Kerala under the Cera Perumals of Makotai (c.800-1124)*, Calicut, 1996.

22. For details see T.A. Gopinatha Rao, *Travancore Archaeological Series*, vol.II, Madras, 1916, pp.66-75.

23. Ibid., pp. 60-80. The most important among the privileges that the church obtained was the right to keep *parakkol* (weighing-device), *panchakandy* (a device for weighing probably with *pancha-khandas* or five measuring segments), *kappan* (device to measure liquid) of the city of Kollam under its safe custody, which this community enviously held till 1503, when these were taken away from them following the malpractices by some of its trading members. Walter de Gray Birch (ed.), *The Commentaries of the Great Afonso Dalboquerque: Second Viceroy of India,* New York, 1875, p. 15. This shows that Tharisapally was not merely a centre of worship but represented a space of corporate body of traders that ensured standardization of the weights and measures of the city with the responsibility of maintaining the integrity of trade. Moreover, Ayyanadikal Thiruvadikal made a gift of four Ezhava families, four Vellala families, one *Thachan* and one Vannan (Mannan) family to the Tharisapally and handed over to it the right to collect a wide variety of taxes from them, which Ayyanadikal used to levy earlier like *thalakkanam, enikkanam* (professional taxes from toddy tapers and tree-climbers), *mania meypan kollum ira* (housing tax), *chantan mattu meni ponnu* (tax for using the title Channan or Shanar—to show his high social status), *polipponnum (*tax given on special occasions), *iravuchorum* (*balikaram* or tax collected to feed the Brahmins, refugees and destitutes*)*, and *Kudanazhiyum* collection of a *nazhi* (a type of liquid-measurement) of toddy as tax from each pot tapped. T. A Gopinatha Rao, *Travancore Archaeological Series*, pp. 63-8. Moreover the church-men might collect eight *kasu* from each cart that used to take merchandise by land into the market of Kollam (*vayinam*) and four *kasu* from each boat that was used to carry cargo to the port (*vediyilum*). Ibid., pp. 68-71.

24. George W. Spencer, *The Cola Conquests of Sri Lanka and Srivijay,* Madras, 1983; K.A. Nilakanta Sastri, *The Cholas,* Madras 1955, pp.183; 199-200; 211-18; 252-3.

25. For details on *Manigramam* and *Anjuvannam* see Meera Abraham, *Two Medieval Merchant Guilds of South India,* New Delhi, 1988. Also see K.A. Nilakanta Sastri, *A History of South India,* New Delhi, 1999, pp.181-2.

26. This fact is evident from the Tharisapally copper plate. For details see T.A. Gopinatha Rao, *Travancore Archaeological Series,* vol. II, Madras, 1916, pp. 66-75.

27. Ibid., pp. 67, 71.

28. M.G.S. Narayanan, *Perumals of Kerala,* Calicut, 1996, p. 155. Except for Quilon, *Manigramam* seems to have consisted of Hindu traders. For a discussion on the different types and cultural composition of Manigramam guild in different parts of south India see Rajan Gurukkal, *The Kerala Temple and the Early Medieval Agrarian System,* Sukapuram, 1992, p. 92; Raghava Varier and Rajan Gurukkal, *Kerala Charithram,* Sukapuram, 1991, pp. 135-6.

29. T.A. Gopinatha Rao, *Travancore Archaeological Series,* vol. II, pp. 68, 71.

30. M.G.S. Narayanan, *The Perumals of Kerala,* p. 32.

31. Elamkulam Kunjan Pillai, *Studies in Kerala History,* Kottayam, 1970; M.G.S. Narayanan, *Cultural Symbiosis in Kerala,* Trivandrum, 1972, p. 82.

32. The most prominent Jewish trading settlements were in Palur, Pulloot, Maliankara and Madai. For details see P.M. Jussay, 'The Jews in Kerala', in *St. Thomas Christians and Nambudiris Jews and Sangam Literature: A Historical Appraisal,* ed. Bosco Puthur, Kochi, 2003, p. 128.

33. Cf. note.14.

34. S.D. Goitein, 'Portrait of a Medieval India Trader: Three Letters from the Cairo Geniza', *Bulletin of the School of Oriental and African Studies,* XLVIII, 1987, pp. 457-60.

35. Ibid., pp. 437-8.

36. See the Tharisapally Copper plate. T.A. Gopinatha Rao, *Travancore Archaeological Series*, vol. II, Madras, 1916, pp. 68-71.

37. Cf. Tharisapally Copper plate. T.A. Gopinatha Rao, *Travancore Archaeological Series*, p. 68.

38. Cf. M.G.S. Narayanan, *Foundations of South Indian Society and Culture*, Delhi, 1984, p. 22.

39. For details regarding the shifting of the royal residence of the Nediyirappu Svarupam see K.V. Krishna Ayyar, *The Zamorins of Calicut*, Calicut, 1938, pp. 1-2. By the middle of the fourteenth century or by the time Ibn Batuta visited Calicut (1343), it had become the most important port of Kerala. See Ibn Batuta, *Die Reise des Arabers Ibn Batuta durch Indien und China*, ed. Hans von Mzik, Hamburg, 1911, p. 302. Hermann Gundert (ed.), *Keralolpathiyum Mattum*, Kottayam, 1992, pp. 190-200; Pius Malekandathil, 'Winds of Change and Links of Continuity: A Study on the Merchant Groups of Kerala and the Channels of their Trade, AD 1000-1800', *Journal of the Economic and Social History of the Orient*, vol. 50, part 2,-3/2007, p. 266; Ashin Das Gupta, *Malabar in Asian Trade: 1740-1800*, Cambridge, 1967, pp. 5, 19; B.J. Schrieke, *Indonesian Sociological Studies*, vol. 1, The Hague, 1955, 7ff; Pius Malekandathil, *The Germans, the Portuguese and India*, Münster, 1999, p. 9; for details on the Al-Karimi merchants in India see Eliyahu Ashtor, 'The Venetian Supremacy in Levantine Trade: Monopoly of Pre-colonialism', *Journal of European Economic History*, Rome, 1974, vol. III, p. 27; Walter J. Fischel, 'The Spice Trade in Mamluk Egypt', *Journal of the Economic and Social History of the Orient*, Leiden, 1958, vol. I, p.165.

40. O.K. Nambiar, *The Kunjalis, Admirals of Calicut*, Delhi, 1963, p. 40; K.P Padmanabha Menon, *History of Kerala*, vol.I, p.167.

41. C. Achyuta Menon, *The Cochin State Manual*, Ernakulam, 1911, p. 43; A.Sreedhara Menon, 'A Survey of Kerala History,' Kottayam, 1970, pp. 178-9; Pius Malekandathil, *Portuguese Cochin and the Maritime Trade of India: 1500-1663* (a volume in the South Asian Study Series of Heidelberg University, Germany), Delhi, 2001, pp. 34-5. The Muslim contribution to the state-building ventures of the Zamorin was promptly rewarded, as is evident from the Muchunti mosque inscription. Cf. M.G.S. Narayanan, *Keralacharithrathinte Adisthana Silakal*, Calicut, 2000, p. 64.

42. Mamankam was held once every 12 years. For details see William Logan, *Malabar Manual*, vol.I, New Delhi, 1989, p. 278; A Sreedhara Menon, *A Survey of Kerala History*, Kottayam, 1970, pp. 178-9; Pius Malekandathil, *Portuguese Cochin and the Maritime Trade of India: 1500-1663*, pp. 50-1. In fact the project of conquering Tirunavai was proposed to the Zamorin and supported by the Calicut Koya. For details see K.V. Krishna Ayyar, *The Zamorins of Calicut*, pp. 91-2; Pius Malekandathil, 'Maritime Malabar and a Mercantile State: Polity and State Formation in Malabar under the Portuguese, 1498-1663' in K.S. Mathew (ed.), *Maritime Malabar and the Europeans*, New Delhi, 2003, p. 200.

43. W.W. Hunter, *The Imperial Gazetteer of India*, vol. IV, London, 1885, p. 11; K.Rama Varma Raja, 'The Cochin Harbour and the Puthu Vaippu Era', in *The Bulletin of the Rama Varma Research Institute*, no. 2, Cochin, 1933, pp. 49-51; C.Achyuta Menon, *The Cochin State Manual*, Ernakulam, 1911, p. 2.

44. Ma Huan, *Ying Yai Sheng lan 12*, 'Kochih', trans W.W. Rockhill, 'Notes on the Relations and Trade of China with the Eastern Archipelago and the Coast of the Indian Ocean during the Fifteenth Century', *T'oung Pao*, vol. XVI, Leiden, 1915, pp. 450-1.

45. For details on the Maritime Silk Road see Sally K. Church, 'The Colossal Ships of Zheng He: Image or Reality?', in *Zheng He: Images & Perceptions, South China and Maritime Asia,* ed. Roderich Ptak and Höllmann Thomas, vol. 15, Wiesbaden, 2005, pp. 155-76; Roderich Ptak, 'Ming Maritime Trade to South East Asia, 1368-1567: Visions of a "System" ', in Claude Guillot, Denys Lombard and Roderich Ptak (eds.), *From the Mediterranean to the China Sea: Miscellaneous Notes,* South China and Maritime Asia, vol. 7, Wiesbaden, 1998, pp. 157-91.

46. Tome Pires, *The Suma Oriental of Tomé Pires: An Account of the East, from the Red Sea to Japan written in Malacca and India in 1512-1515,* New Delhi, 1992, p. 80.

47. Duarte Barbosa, *The Book of Duarte Barbosa: An Account of the Countries Bordering on the Indian Ocean and their Inhabitants,* trans. Mansel Longworth Dames, vol. II, Nendeln, 1967, p. 95.

48. For details see O.K. Nambiar, *The Kunjalis, Admirals of Calicut,* Delhi, 1963, p. 40; K.P. Padmanabha Menon, *History of Kerala,* vol. I, p. 167.

49. F.C. Danvers, *The Portuguese in India: Being a History of the Rise and Decline of their Empire,* vol. I, New Delhi, 1988, p. 72; William Brooks Greenlee (ed.), *The Voyage of Pedro Alvares Cabral to Brazil and India from Contemporary Documents and Narratives,* London, 1938, p. 143.

50. Sanjay Subrahmanyam, *The Portuguese Empire in Asia 1500-1700: A Political and Economic History,* London, 1993, p. 72; Jaime Cortesão, *Os Descobrimentos Portugueses,* vol. VI, Lisboa, 1978, pp 141-2.

51. Pius Malekandathil, *Portuguese Cochin and the Maritime Trade of India,* pp.148-50.

52. Vitorino Magalhães Godinho, *Os descobrimentos e a economia mundial,* vol. III, Lisboa, 1982, p. 34.

53. R.A. de Bulhão Pato (ed.), *Cartas de Affonso de Albuquerque seguidas de documentos que as elucidam,* tom. II, Lisboa, 1884, p. 268, tom. III, pp. 30, 258-9; tom. VI, pp. 114, 398-9; tom. VII, p. 174; Antonio Silva Rego (ed.), *Documentação para a Historia das Missões do Padroado Portugues do Oriente,* vol. II, Lisboa, 1991, pp. 352-4; Pius Malekandathil, *Portuguese Cochin and the Maritime Trade of India,* pp. 150-3

54. We find merchants from Quilon, Kayamkulam and Cranganore bringing cargo to Cochin to be further taken to Lisbon. Cf. note 53.

55. Calicut was under the Portuguese control only for a very short period, i.e. from 1513 till 1524. For the rest of medieval period Calicut was a major port in Kerala that used to dispatch cargo regularly to caravan routes. See Pius, Malekandathil, 'From Merchant Capitalists to Corsairs: The Role of Muslim Merchants in Portuguese Maritime Trade of the Portuguese', in *Portuguese Studies Review,* 12(1), 2004, pp. 83, 87. Gaspar Correia, *Lendas da India,* tomo III, parte I, Lisboa, 1921, pp. 274-5.

56. AHU, *Caixas da India,* Caixa 2, doc. 89, dated 27-1-1613, fols. 1-11; 16-7. the fort of Chaliyum was erected in 1531.

57. Pius Malekandathil, *Portuguese Cochin and the Maritime Trade of India,* pp. 118-23, 207-18.

58. We find several *casado* traders, besides Muslim and Jewish merchants taking copper from Cochin to Gujarat for further trade transactions. Pepper money was the money sent to India for the purchase of pepper. Very often it was kept under the custody of the Franciscans and it could not be used for other purposes. See R.A. de Bulhão Pato (ed.), *Cartas de Affonso de Albuquerque seguidas de documentos que as elucidam,* tom. I, pp. 240, 265; tom. III, pp. 52, 83, 269, 396; tom. VII, pp. 41-2, 78.

59. R.O.W Goertz, 'The Portuguese in Cochin in the Mid-Sixteenth Century', *Indica*, vol. 23, nos. 1 & 2, March–September 1986, pp. 63-7; Pius Malekandathil, 'The Portuguese *Casados* and the Intra-Asian Trade:1500-1663', in *Proceedings of the Indian History Congress* (61st Session), Kolkata, 2001, pp. 384-92.

60. Francisco da Costa, in Antonio da Silva Rego (ed.), *Documentação Ultramarina Portuguesa*, vol. III (Lisboa, 1963), p. 315.

61. BNL, Cod. 11410, fol. 116v, *Orçamento de 1612* (Cochin).

62. AHU, *Caixas da India,* Caixa 2, doc. 4, dated 12.1.1612; Caixa 2, doc. 73, dated 23.12.1612; K.S. Mathew and Afzal Ahmad, *Emergence of Cochin in the Pre-Industrial Era: A Study of Portuguese Cochin,* Pondicherry, 1990, doc. 53, pp. 77-83; Sanjay Subrahmanyam, 'Cochin in Decline, 1600-1650: Myth and Manipulation in the Estado da India', in Roderich Ptak (ed.), *Portuguese Asia: Aspects in History and Economic History*, Stutgart, 1987, pp. 67-8.

63. For details about the emergence of the port of Alleppey, see S. Rajendran, 'Monopoly of Pepper trade and its abolition in Travancore', *Journal of Kerala Studies,* Vol. XVIII, Part IV, 1991, p. 245.

64. V. Nagam Aiya, *The Travancore State Manual*, vol. I, Trivandrum, 1906, pp. 343-51; Shangoonny Menon, *History of Travancore from the Earliest Times*, New Delhi, 1878, pp. 135-55; A Sreedhara Menon (ed.), *District Gazetteer of Trivandrum,* Trivandrum, 1962, p 11; Pius Malekandathil, 'Winds of Change and Links of Continuity: A Study on the Merchant Groups of Kerala and the Channels of their Trade AD 1000-1800', *Journal of the Economic and Social History of the Orient*, vol. 50, part 2, 3/2007, pp.280-2.

65. For details on the pepper monopoly of Travancore, see Ashin Das Gupta, *Malabar in Asian Trade: 1740-1800*, Cambridge, 1967, p. 4; S. Ramachandran Nair, *Social Consequences of Agrarian Change,* Jaipur, 1991, p. 11; Aswathi Thirunal, *Thulasigarland,* Trivandrum, 1998, p. 93.

66. Anila K., 'Urban Growth of Trivandrum 1760-1947', unpublished M.Phil. thesis, Sree Sankaracharya University of Sanskrit, Kalady, 2006; for details on the Travancorean trade of this period, see B. Sobhanan, 'Trade Monopoly of Travancore', *Journal of Kerala Studies* (1981), VIII, 30-1; P. Shangoonny Menon, *A History of Travancore*, pp. 135-55, 166; V. Nagam Aiya, *The Travancore State Manual*, vol. I, Trivandrum, 1906, pp. 343-51; A Sreedhara Menon (ed.), *District Gazetteer of Trivandrum,* Trivandrum, 1962, p. 11.

67. For details on Thachil Mathu Tharakan, see Ookken, *Thachil Mathu Tharakan,* Irinjalakuda, 1966; Bernard, *Marthoma Christhianikal,* Pala, 1914, pp. 640-701; Paremakkal Thoma Kathanar, *Varthamana Pusthakam*, Thevara, 1977; M. Kurian Thomas, *Niranam Grandhavari*, Kottayam, 2000, pp. 96-110.; K. Sivasankaran Nair, *Marthanda Varma Muthal Munro Vare*, Kottayam, 1996, pp. 110-20.

68. An evident case is that of Muvattupuzha, where soldiers were stationed in the fort of Sivankunnu located near the *pandikasala* (warehouse). See Pius Malekandathil, 'Kothamangalam Roopathayude Charitrapachathalavum Kraisthava Koottaymakalude Verukalum', in Pius Malekandathil (ed.), *Anpinte Anpathandu* (Malayalam), Kothamangalam, 2008, pp. 24-52; P. Shangoonny Menon, *Thiruvithamkur Charitram*, Thiruvananthapuram, 1973, pp. 128-9; Pachu Moosathu, *Thiruvithamkur Charitram*, Kochi, 1986, p. 50.

69. For details on the Travancorean trade of this period, see B. Sobhanan, 'Trade Monopoly of Travancore', *Journal of Kerala Studies*, vol. VIII, part IV, 1981, pp. 30-1; A. Sreedhara Menon, *Kerala Charitram*, Kottayam, 1967, pp. 400-20.

70. C. Achyuta Menon, *The Cochin State Manual*, pp. 174-8.

Women in Portuguese India:
Their Representation in
Sixteenth Century Sources

Joy L.K. Pachuau

THE PORTUGUESE, as is well known, established forts and town settlements all along the littoral of the Indian Ocean at places strategic to their trading interests. Such coastal entrepôts, already well known for their cosmopolitanism whether along the Malabar Coast or in the Indonesian archipelago, were made even more international with the arrival of the Portuguese. An earlier historiography had presented the Portuguese and their trading methods as causing a break from a pre-existing system of free trade,[1] whilst others like M.N. Pearson have focused on the persistence of the basic structure of Asian trade despite Portuguese intervention and interruption.[2] Whatever may be the final verdict on this matter, it is clear that the Portuguese established permanent bases along the coastal regions to facilitate their trade, which helped in the formation of unique colonial societies. Portuguese colonial society largely functioned around two imperatives, first that of trying to establish a society that was similar to the metropolises in Portugal in institutional terms as well as through the perpetuation of social mores acceptable in the metropolis. On the other hand, if such an ideal portrayed a society that was ordered and measured in its functioning, the exact opposite seemed to be the operating principle in daily negotiations, with personal aggrandizements taking the forefront over profit for state. It is in the study of such colonial societies that the place of women needs to re-considered.

The subject of women in Portuguese India in general, is a topic that has not received much attention in the historiography of Portuguese studies. The theme suffers from problems that are similar to the writing of women's histories in general, i.e. our main sources of information are usually rather evasive when it comes to affairs concerning women. Matters related to state, trade and war and the politico-social-economies of the region they encounter are given prime focus. None of the early chroniclers for Portuguese India whether Gaspar Correa, Barros or Castenheda for instance have much to say about women in their narratives.[3] However, even the information available through other sources (travelogues, letters and reports) have scarcely been looked into for histories of women as the concerns of Portuguese presence in India have mainly been the organizational/institutional aspects of trade, state and empire. Thus women have largely been silenced from speaking from the past.

The attempt in this essay will be twofold—first, to highlight or bring to focus the circumstances and contexts in which women find mention in the sixteenth and seventeenth centuries. Such an approach no doubt has the lacuna of what was often said of earlier studies on women. In other words, in such works the lack of study of women was compensated by a mere inclusion, their absence was corrected by a simple narrative of their presence as if they had no bearing within the larger frameworks of the society they lived in. Second, therefore, this essay will try and analyse the representation of women in the texts of the period. Considering that most writers were men, I try and locate their works in the context of how they conceived of gender relationships in Portuguese/European colonial society largely in the sixteenth century. What kind of women did they generally write about? What about women who received the attention of our sources or what did they find important enough to write about them? These are some of the issues I will try and deal with in this essay.

WRITING WOMEN: REFERENCES TO WOMEN IN THE SOURCES

As Donald Lach[4] has pointed out it was only in the latter half of the sixteenth century that early Portuguese works were finally produced for public consumption. Much of the early Portuguese endeavours were regarded as a national secret and not much 'concrete information' was publicized between the years 1520-50 beyond what was already known about the region. Thus, detailed knowledge about the East that emerged can be '. . . divided into three major categories: the Portuguese commercial reports and the great chronicles of conquest, the Jesuit newsletters and histories and the *fin de siècle* reports of Italian, English and Dutch commentators. . . .'[5] Interestingly, each category of writing takes up from where the other ended and thus the Portuguese material[6] narrates events in the Indies down to 1540. The Jesuit newsletters follow from that time onwards with the arrival of the first Jesuit, Francis Xavier in 1542. The travelogues of the non–Iberian writers published at the end of the century deal with India during the period from 1564 to 1591, which was a period when Jesuit letters began to deal more with the Far East and the knowledge of India declines.

The evolution of illustrations in these sources is a good way of exemplifying the concerns of the Portuguese in Asia and the manner in which they gradually come to conceive of it. The early illustrations of the Indies found in the Portuguese chronicles of conquest, similar to what was written were mostly views of bays, coastal towns, forts and other kinds of settlements, visibly demonstrating to a European audience the reality of the Portuguese presence. One of the earliest set of illustrations that we know of representing social life, including both men and women are those referred to by Georg Schurhammer in an article entitled 'Desenhos Orientais do Tempo de S. Francisco Xavier' or 'Oriental pictures of the time of S. Francisco Xavier'.[7] As the title of his essay suggests, this collection of images (Codice 1889 in the Biblioteca Casanatense of Rome) was not dated and Schurhammer, based on the artist's comments on various events attributes its completion to anywhere between 1537 to 1554. The author of this set of as many as a hundred and forty one images is not known either. However the images reveal the work of an individual who was well acquainted with the life and practices of the locals stretching from West Asia to as far as China, excluding Japan, besides also being acquainted with Portuguese ways. John Huyghen van Linschoten, whose work was first published in 1595[8] is another source to provide us with illustrations of inhabitants and ways of living. Linschoten's work was similar to other travel writing of the period. Such works were written to make known the East to the west essentially locating marvel in difference. The illustrations accompanying Linschoten's text (done by João Baptista van Doetecom), similar to those referred to by Schurhammer exemplified this concern, which was to represent life in and around Portuguese settlements. Their main interest was therefore to try and paint a picture of the different kinds of people who inhabited the colonial towns and their lifestyles. Women were amply represented in Linschoten's illustrations (as also in Codice 1889) to go along with the great emphasis he placed in their lifestyles in the text. However such illustrations of women were always part of a larger picture of men, such as those representing a Chinese man and woman (in Linschoten), or in pictures that try and illustrate life in a colonial town. This is in contrast to for instance a nineteenth century document, *Usos e Costumes da India,*[9] which had a plate entitled 'Christian lady of India' along with other discrete plates of a Hindu goldsmith, Hindu shopkeeper, etc. Perhaps one can allude that the later pictures reflected the autonomous way in which women were beginning to be seen, rather than as mere appendages to a larger picture. In other words, what the early travelogues emphasized were the context, the world in which men were making and creating space for themselves and women although present were always part of this larger background. The writings, narratives by Europeans about the East in this period reflect a similar bias when it concerns women.

While one may denounce the fact that early chroniclers did not have anything to say about the subject of women, it must, however, also be said that as far as the Portuguese/European women were concerned there were never too many of them who came to India.[10] Nonetheless, even in the early years there were a few who braved the rule against official policy of not allowing women on board ships to India. Elaine Sanceau, one of the earliest to recognize that women had not been given a voice, noted that the lack of voice did not necessarily mean a physical absence. According to her, there were all kinds of women who appeared on the shores of the East arriving from Portugal, some with legitimate passes and others who were hidden on board ships, as stowaways.[11] This was in contrast to Iberian policies as C.R. Boxer informs us, to the Americas, where women were given official permission to travel and a large number of women actually emigrated. It is significant that while the viceroys of Mexico, nearly always took their spouses, at least in the seventeenth and eighteenth centuries, no spouses of the viceroy or governor-general accompanied him to India between 1549 and 1740. When the Marquis of Tavora insisted on leaving with his wife, it caused a sensation in Lisbon, as the authorization was given with reluctance. There are a few other examples of magistrates and other functionaries as well as private individuals such as the famous doctor Garcia d'Orta, who brought along their spouses and other relatives to India legitimately, or would have asked them to follow later; but they would have constituted very rare cases in comparison especially to the Spanish Americas. Thus in the great majority of the cases, with regard to women and children accompanying their spouse and parents to India, as far as one can see of the registers of the boats to India, it is fragmentary and rare.[12] In fact Vasco da Gama in an order of 1524 ordered that no woman was to board a ship, even if she were a wife, and if found was to be flogged publicly.[13] Despite such sanctions and despite scant evidence we find that Portuguese women did reach the shores of India. A famous first was Iria Pereira, mistress of Antonio Real, the captain of Cochin who had sailed to India in 1505 along with D. Francisco Almeida's fleet. She, it is said, continued to live in India, even though Antonio Real was expelled by Albuquerque in disgrace in 1514.[14]

Although there were quite a few adventurous women, willing to take the risks of voyage and official discredit to join husbands and lovers to distant shores, it was due to this general lack of Portuguese women that non-Portuguese women begin to find mention in our sources. The earliest references to indigenous women are often in the context of sexual liaisons the Portuguese cultivated with them. References are made to female slaves (*escravas*), as in Cochin where they were kept for sexual favours. This practice became so common that it became pertinent for the state to have a regulatory policy where it was agreed that rather than promote promiscuity within the household, it was a better policy to legitimize prostitution. Thus in Cochin, all female slaves were freed and all those who wished to remain with the Portuguese were placed under a certain Beatriz, a newly converted local woman, whose house Portuguese men could now frequent. Religion was not kept outside the purview of these sexual liaisons as it was considered favourable that the Portuguese establish sexual relationships only with women (even when outside of marriage) who had converted to Christianity. It was reported that there were several conversions among the women, who made a small income through such liaisons, as a result of this ruling.[15]

Another important circumstance in which women were mentioned was in the context of the often-quoted policy of Albuquerque to promote the marriage of Portuguese men with local women.[16] It was hoped that despite the lack of

Portuguese women, marriage with local women would lead to the creation of a population loyal to the Portuguese and their interests. Albuquerque's policy thus facilitated progeny (called *mestiço/mestiça*) and the creation of settlements that were predominantly Portuguese in culture.[17] At the same time, although the Portuguese crown approved of the policy, local officials did not necessarily approve of it. The *casado* or married householder was an official rank in the Portuguese colonial hierarchy with certain privileges and it was reported that by marrying locals this privileged status was being taken advantage of by 'ordinary Portuguese soldiers and mariners, petty artisans and even exiles' who were the only ones actually resorting to this provision.[18]

It was not, however, official sanction alone that led to inter-racial alliances in the Portuguese settlements. Our sources (especially letters of priests) are replete with stories of Portuguese men forming none too holy alliances with all kinds of local women, poor and rich, low or high. A document of 1514 from Cochin, within less than twenty years of the Portuguese presence shows us the wide variety of women in terms of nationalities that were engaging and interacting with the Portuguese. This document, which essentially gives us the roll of Christians in the city provides a list of indigenous women married to Portuguese men and their progeny, both boys and girls, who were thereby of a mixed heritage. There were fifty-eight women (*mays* [*sic mães*], literally mothers) listed along with the names of their husbands and forty-eight children, and their nationalities ranged from *malavar* (Malabari) to *java* (Javanese) to *moura* (Moor) to *canrym* (Canarese) to *nayra* (Nair) to *bramena* (Brahmin) to *çacotorina* (from Socotra), and to *guzarate* (Gujarati). Besides such women married to Portuguese, another list provided for the names of women (forty of them, again of different ethnic backgrounds, with forty-five children), called *solteiras* by the Portuguese, who had had children with the Portuguese out of wedlock.[19] This was a list of all the Christians at the Cochin fort and we see from an early period that the Portuguese were intermarrying with a large number of women from Asia as a whole, and by extension the category of the *mestiça* (a woman of 'mixed' heritage) was being created.

REPRESENTING WOMEN: FACTORS OF 'RACE/BLOOD', 'CLASS' AND GEOGRAPHY

In the earlier section, I narrated the general context in which women—Portuguese/*castiça*, native and *mestiça* find mention in sixteenth century Portuguese sources. It is, however, also important to see how the authors who write about them actually represented them.

A predominant view that comes through in the sources is the attempt to essentialize women on the basis of race, class and caste. For Portuguese chroniclers there were two distinct categories of women, based, as was the norm, on 'race' or 'blood'. These were the pure races, Indian or Portuguese and the *mestiça*. Women of the pure races were almost always considered better behaved; in other words, they were seen to present themselves with 'humility' and 'sobriety'. Thus, D. Aleixo de Menezes, the famous Archbishop of Goa, who aggressively campaigned to take over the Syrian Church of Malabar on grounds of considering it heretical was all praise for the women of the Church because of their 'modesty' and sense of 'propriety'. Thus he wrote,

> . . . the women are extremely honest, both in life and in their dress, because they wear some white veils. . . which cover their entire body, and they throw it around the face, like a veil under the chin, in turns [folds?], which cover their heads, and flows down until it covers the feet, which represents high honest[y?]. . . All of them, as the[ir] Bishop enters the church, go one by one to kiss the hand with such composure, first on their knees and then placing the head on the ground and raising with great reverence they receive the blessing, so that they all seem to be very religious with great composure. . . .[20]

Good behaviour, attributed to the high caste Malabari Syrian Christian women was understood in terms of religiosity as well as submission to religious authorities.

Similarly, it was the European woman who was praised. In the heroic defence of Diu during the time of its siege in 1538 and 1545, it was said that the women distinguished themselves in treating the wounded as well as in combat along the side of their men. The most celebrated, according to Boxer was Isabel Fernandes, 'the old (lady?) of Diu' as she came to be known. In a letter sent to the Regent of Portugal in 1559, it was said that she had not received any compensation for her services wherein 17 of her 18 children had died in the service of the king. The contribution of women was also acknowledged when in periods of crisis they came (from Chaul and Goa) to help the crown, voluntarily by offering jewellery and ornaments; there were others like D. Luisa da Silva, opulent widow of Cochin and proprietor of several slaves, famous for the care with which she attended the passengers and sailors of the ships of India when they would arrive suffering from scurvy and malnutrition.[21] The *castiça*, or 'pure race' was thus seen as one who contributed to the state building exercise and also one who would withstand with courage and valour those who threatened that existence.

Thus as was the case in Portuguese colonial society in general,[22] where 'race' and 'class' were important categories of differentiation, there was a strong interplay of the same factors in the way the Portuguese understood and represented women. Whilst the representation of those who were 'racially pure' was usually favourable, this disposition was only towards those who were considered high in the social hierarchy in the case of Indian women. Thus, Albuquerque's policy of inter-marriage although taken up generally by those lower in the Portuguese hierarchy was to be with only those who were considered higher in the Indian social hierarchy—Brahmin women and only the 'fair and good-looking Muslim ladies, taken captive',[23] implying therefore that low caste native women were also not always given a good report card.

Returning to Linschoten as a source, we find in him a very important informant for the condition of women in the East. We locate in his work, an entire chapter dedicated to the customs of 'Portingales, and *Mestiços* and Indian Christian women'. Judging from the work of Linshcoten, what was of interest to the early modern man about women in India was how they dressed, customs such as the chewing of betel, their habits with regard to hygiene and how they related to men in general. All these were objects of curiosity for him. Besides, for Linschoten too, class and race components played an important role in understanding the behaviour of women. His initial remarks on Portuguese women and *mestiças* concerned the life of seclusion and their general lack of visibility stemming from what was regarded as proper behaviour for women. He thus elaborated on customs practiced which prevented women from maintaining contact with men. Thus, according to Linschoten, upon the appearance of a male visitor, 'wives and their daughters run to hide ... to leave the man to answer him....' Moreover, no man was to dwell in the same house as those of women, even if they were related. Thus women in general led very secluded lives, rarely venturing out except to church. If they did travel it was in palanquins or litters. Although Linschoten did not mention explicitly the class they belonged to presumably these were women of high social standing.

However, it was the category of the *mestiça* that was looked upon with derision. Perhaps the half-castes were seen as aberrations to the norm and, therefore, their behaviour was considered different from normal behaviour in polite society. This perhaps explains the extreme lengths other authors also go into to describe their vagrant character. According to Luillier, '. . . one finds in India *mestiças* (female half-castes), that is to say, offsprings of European blood and Indian blood'. He then went on to describe them saying, 'these women (half-castes) are luxurious to an excess and they prostitute themselves in a shameful manner; they are very ugly, to a laughable degree . . . some dress themselves in an Indian manner and others in a Portuguese manner'.[24] Pyrard de Laval also found them 'impudent'. He wrote of them thus, 'the women are so cunning and full of artifice, that they almost always succeed in their intrigues', while narrating their use of the datura plant to knock their husbands unconscious in order to have other licentious relationships.[25] Linschoten and others give the impression that there were many feuds wherein husbands were slain by their wives and lovers and vice versa. He similarly generalizes the behaviour of *mestiças*[26] by calling them, '. . . luxurious and unchaste . . . although they bee married, but they have besides their husbands one of two of those that are called souldiers, with whome they take their pleasures: which to effect, they use all the slights and practices (they can devise...).'[27]

Most authors thus talk of *mestiças* as women who could not be controlled by men, who would go to any extent to start an affair, resorting to all kinds of intrigues, the most famous example being of course the use of the datura plant referred to above, to drug their husbands in order to wander off with their lovers. Their mixed heritage, it seemed to these authors aided the *mestiças* in acquiring access to the various practices of the East, which they did not shy from using. Describing such an instance Linschoten wrote,

> ...they knoe howe to make a [certaine] poyson or venome, which shall kill the person that drinketh it, at what [time or] houre it pleaseth them; which poyson being prepared, they make it in such sort, that it will lye six yeres in [a mans] body, and never doe him hurt, and then kil him, without missing halfe an houres time. They make it also for one, two, or three yeres, monthes, or dayes, as it pleaseth them [best] as I have seene it in many, and there it is very common.

Added to race and class being important determinants of female sexuality, for a number of writers of the sixteenth and seventeenth centuries, behaviour was also linked to geographical conditions. Climatic zones were inscribed on to women's bodies making them potentially dangerous and threatening. According to Arie Pos, 'adultery, prostitution and courting women acquire a unique significance in Portuguese India'.[28] There was a general feeling that the tropics were a hotbed of scandal and torrid affairs. For instance, the island of Manaar was reported to 'accentuate sensuality'.[29] Such an attitude of fixing the behaviour of women to a geographical zone was reflected in various ways such as those which led agents of the church to institutionalize or formalize a particular kind of relationship towards them. For instance in Saint Francis' instructions to his subordinates in the Indies he asked them to never meet women in private and always to see that they did not visit them in their homes. If they were compelled to meet them it was to be in the presence of their husbands or other members of the household. Such an attitude was garnered because of the assumptions he had about women. He wrote: 'since women are generally inconstant, lacking in perseverance, and take up much time, you should deal with them in the following way....' He also believed it was more profitable to work with husbands as they were more 'constant' and it was therefore possible to provide much more fruit, in comparison to women who were 'inconstant'. It also seemed natural for Xavier to advise his followers never to 'blame a husband in public' even though it was his fault, as women, 'being indomitable' look for such opportunities to 'belittle their husbands'. And if on the other hand it was the husbands who were at fault, then the women were to be encouraged to put up with their husbands.[30]

'CHARACTERIZING' WOMEN: WOMEN IN THE CONTEXT OF SOCIO-RELIGIOUS INSTITUTIONS

Women as was shown in the preceding section were almost always documented in the context of how they behaved, acted, and related to men. If they were virtuous women they were necessarily those who were married and contributed along with their husbands in perpetuating a colonial society. The regard of the female as a potential wife was deeply entrenched, so as to even take the time and energy of great men of the church such as Francis Xavier. Amidst all the instructions that he would write to his subordinates in the East on matters concerning Jesuit life and the perpetuation of Christianity, he took time to establish marital alliances. He wrote for instance to Paulo and Antonio Gomes at Goa, asking them to look into the possibility of an alliance for a rich man he had met in Malacca (Christovao Carvalho) with the orphan of a certain lady (children who did not have a father, although they had a mother were considered orphans) whom he called *may* [*sic mãe*] or mother.[31] Such an alliance, he went on to say would benefit both parties as Christovao Carvalho was interested in settling down and the widow would have rest that at least one of her daughters would be married. In this context, Xavier asked the Fathers to remember '. . . the great hospitality and charity which we always received from our mother, and that you and the *veador da fazenda* should take up this matter and carry it through so that this honorable widow is freed from her concern and her daughter is well treated and protected'.

In a society where religion played a crucial role, virtuous women were also those who worked towards the perpetuation of the Catholic faith. Thus the women of Cochin were noticed and written about by the local priest because of their sense of duty and devotion to the religion.

With regard to the doctrine of Our Holy Faith in this city there are many women who know the commandments and articles of the faith, works of charity as well as other things that are necessary for our salvation.

They confess and request for mass to be said and they are *confrades* of Our Lady of Rosario and they donate oil for the lamps and now a lady, Briatiz de Quental by name orders to make a house in this City of Cochin dedicated to Our lady of Conception, which will cost her, according to the master, 300 *cruzados*. . . .[32]

There was also great pressure placed by the church establishment on Portuguese colonial society to be exemplary Christians because of the operation of an important binary, that of Christian and 'pagan'. It was hoped and very important for many that the Portuguese lived honourable lives in order that the 'pagans' would be drawn to Christianity. In the context of women, especially those without spouses, whether widowed, divorced or unmarried it meant that they were not to be immoral. There was a close link between perpetuating a Christian example and the cloistering of women as can be seen in the statement of Friar António de Santa Maria who informs us that the reason behind Archbishop D. Aleixo Menezes establishing a convent for women was so that 'the Gentile kings who place their happiness in the multitude of women for the use of their sensuality' would be confounded in seeing virtuous and chaste women and thereby '. . . the strength of Our Holy Faith; also in order that with the virtue of Holiness of those alms [that the women brought as dowry to the convent] the same Lord would be obliged to conserve that state and the Christianity of the Orient, so surrounded everywhere by enemies. . . .'[33]

The creation of a good colonial society, which could be attractive to non-Christians, it was felt depended on how women were controlled, and respectable women made. If they were married, they were to keep company with their husbands in state-building and if single, they were to be in convents praying for the well-being of the state and its peoples. In sum, the assumption seemed to be that culture resided in women and they therefore had to be controlled or protected, as the case may be.

The understanding of women and their place in colonial society can best be understood in the nature of the convents that came to be established in Portuguese India and the politics of their maintenance, management and organization. The presence of convents were important to colonial society because of the understanding provided by the church that they played an important role in the establishment and fulfilment of a spiritual kingdom. This was consistent with the understanding of institutionalized female spirituality in Catholic Europe. In recent works on women in medieval Europe, the ability of women especially nuns to shape religious experiences has been pointed out. Thus,

As spiritual centres, convents performed a wide range of sacred and liturgical activities like the performance of the Divine Office. Nuns were also valued and revered for their proximity to the divine and the prayers that they could offer for those outside of convent walls. Additionally, convents were institutions that honoured and protected the culturally esteemed practice of female chastity. To a society that expressed order and control through the regulation of female sexuality, convents were symbolically, as well as spiritually, significant.[34]

While this was said of medieval Spain, the same can be imputed for the Portuguese in India as well, being a part of the same Iberian catholic society. This is also corroborated by a Portuguese work of a late seventeenth century writer. The author applauded Archbishop Menezes (the person who began the campaign to establish convents and actually saw to their foundation) for taking great pains to establish the Convent, despite opposition. He considered its establishment appropriate because the efficacy of the prayers of the nuns was already becoming well known. He wrote,

The prayers of the Religious of the Convent of Santa Maria (one of the convents Menezes established in Goa) was very effective in being able to receive from God the answers, for those who solicited it; because in any work and adversity, in which they would take recourse to them (the nuns), their prayers would deliver them (the supplicants), and in adversity (one could be) certain of their consolation and comfort. With these experiences many people asked their help in their prayers, and the effect showed what they would be worth.[35]

The author mentions an instance of their spiritual role in the context of the threat that the Portuguese faced from the Dutch in the form of blockades in Goa. According to the author, the Prioress of the Convent and the other religious of the House sensed the danger and 'with their many prayers for the defence of the town obliged God to answer'. It was said,

. . . they would, as a community as well as individually say their prayers; in the beginning they said a Mass for the said purpose and several others for the same end. They called for a procession with litanies sung each week, and one of Our Lady each day, in which the entire Convent assisted. Almost all the nuns worked [for it] and their prayers were so effective that the enemies would not reach Goa; and it was certain that the Lord was kind to that city because of the virtues and merits of those, his servants (*servas*).[36]

Convents were also important institutions because in them were manifest all the characteristics of the medieval mentality, which sought to restrain and restrict, control and channelize women. Two kinds of institutions emerged that catered to such concerns—the Recolhimento (shelter) and the Convent. With progeny and the perpetuation of a colonial society being the chief concern, the provision of suitable wives was often seen as the responsibility of the state. One of the ways in which this was provided for was by making orphan girls under the care of the State to be given in marriage to officials in the Indies. These *orfás* were also provided with suitable dowries in the form of posts in the Indies, which their suitors would inherit. In Lisbon, *orfás* under state protection had been housed at the *Recolhimento do Castelo de São Jorge* (Shelter of the Castle of St. George).[37] These *orfás*, by the mid-sixteenth century began to be sent to the Indies. The first contingent left Lisbon in 1545, the system functioning intermittently until the eighteenth century. According to Boxer there may have been an average of 5 to 15 each year who were sent, despite the claims of Correia for a much larger number.[38] The crown suspended sending the girls in 1595, but promptly reverted to the old system although never to such a scale as to have a large white population in Portuguese Asia.[39]

In the early stages of the functioning of this system, *orfás* who arrived in India had been housed by prominent families of the city until they were married off. However in 1598, a shelter for the orphan girls was formed, on the lines of the one in Lisbon called the *Recolhimento da Nossa Senhora da Serra*, and sanctioned by the *Misericórdia* of Goa.

Over and above these orphan girls who were sent from Portugal there were also other women from various backgrounds in the city of Goa, whom religious authorities felt, needed to be kept in institutions. In 1610, another shelter called the *Recolhimento de Santa Maria Magdalena* was established for 'disgraced women', who showed inclinations for reform and who were willing to settle down to a married life. This was yet again a transit accommodation for them before they moved into a settled life.[40]

Most important, however, was the foundation of the Convent of Santa Mónica of Goa, a cloistered convent for women, established sometime between 1606–10. There had been attempts earlier, mainly by the Franciscans to establish such a convent for women. However, it was Archbishop Menezes, who succeeded in establishing one which adopted the Augustinian rules. In the course of the seventeenth century the convent drew much flak from the Camâra of Goa and the local authorities, while the initial opposition of the crown to its foundation gradually changed to support. The accusation against the convent was not only its financial wealth, a common complaint against all religious institutions in the seventeenth century, but mainly that it minimized the number of women available for marriage. The Convent of *Santa Mónica*, with a total of 104 in 1650, seemed to many a 'waste' of women who could instead have been married to eligible suitors. According to Coates, 65-70 per cent of the novices who entered the House were either *castiça* ('pure-white' women, but born in India), *mestiça,* or native Goan women. About 2-3.6 per cent were *reinõis* ('white' but from the kingdom) and a handful of them were from outside of Goan control.[41]

These *recolhimentos* and convents played out in many ways the hierarchies of colour and rank that were so prominently a part of Portuguese society. Thus candidates for the *recolhimentos* and the convents were to be 'white' and *bem nascidas,* 'well born', but in the true fashion of Portuguese colonial society of the tropics, exceptions to the rule abounded.[42] The Recolhimento de Santa Maria Madalena from the very beginning had a majority of Eurasian women. In the late seventeenth century, a royal decree had to be issued chiding the authorities as the *Serra* admitted daughters of Hindus, Muslims and African women. No heed seemed to have been paid to it. There was even a prohibition to the entry of illegitimate children to the *Serra*, this rule was also flouted even as illegitimate children of officials had often to be admitted. In course of time, admission of one or more daughters to the Convent of Santa Monica also became a symbol of social position, as daughters were admitted to the convents only after very large dowries.[43]

CONCLUSION

The essay has tried to trace the evidence of women in Portuguese India, largely through sources pertaining to the sixteenth century. A lack of mention of women in our sources does not mean their absence. Where available, sources are careful to categorize women and thereby their depiction on the basis of race and rank, which were very prominent markers of social hierarchy in Portuguese India. Also interesting was the manner in which *mestiças* were understood and represented. Portuguese colonial society in the tropics also replicated institutions of the metropolis that bound women through the establishment of *recolhimentos* and convents. Moreover, women who were favourably considered were usually those who worked towards the maintenance of a colonial society and its religious aspirations. Conditions away from the metropolis did not necessarily prevent the perpetuation of ideals that were considered consistent with Iberian religiosity and cultural conditioning.

NOTES

1. Cf. K.M. Panikkar, *Asian and Western Dominance*, London, 1954 repr.
2. Cf. M.N. Pearson, *Merchants and Rulers in Gujarat: The Response to the Portuguese in the 16th Century*, New Delhi, 1976.

3. Elaine Sanceau, *Portuguese Women during the First Two centuries of Expansion Overseas*, Separata do, vol. V das Actas do Congresso Internacional da História dos Descobrimentos, Lisboa, 1961, p. 2.

4. Donald F. Lach, *India in the Eyes of Europe: The Sixteenth Century*, Chicago and London, Phoenix Books, The University of Chicago Press, (1965) 1968, p. 337.

5. Ibid., p. 338.

6. J.B. Harrison, Five Portuguese Historians, in C.H. Philips (ed), *Historians of India, Pakistan and Ceylon*, 1961.

7. Georg Schurhammer, 'Desenhos Orientais do Tempo de S. Francisco Xavier in *Garcia de Orta*', Numero especial (1956), pp. 247-56. Also reprinted in *Orientalia*, Rome IHSI and Lisboa, CEHU, 1963.

8. P.A. Tiele, Introduction in *The Voyage of J.H. Van Linschoten to the East Indies*, vol. 1 (1935), Delhi, 1997.

9. Manuel da Cunha Maldonado and Joaquim Antonio Roncon, *Usos e Costumes da India, 1846-1847*, in Maria de Jesus dos Martires Lopes, *Tradition and Modernity in Eighteenth century Goa*, Delhi, 2006. Also see for example the work of A. Lopes Mendes, *A India Portugueza*, (1886, Delhi, AES reprint, 1997) a work of the nineteenth century with ample illustrations (382 of them and only seven maps), wherein images began to take on a form that has been more associated with British India. Sketches of individuals from different caste groups, both male and female, were becoming an important exercise for classification.

10. Cf. Boxer, *Portuguese Women during the First Two centuries of Expansion Overseas*, Separata do vol. V, das Actas do Congresso Internacional da História dos Descobrimentos, Lisboa, 1961.

11. E. Sanceau, op. cit., p. 6.

12. C.R. Boxer, *A mulher na expansao Ultramarina Iberica*, Livros Horizonte, Lisbon, 1977, p. 82.

13. E. Sanceau, op. cit., p. 7.

14. Ibid., p. 6.

15. A.M. Mundadan, *History of Christianity in India,* vol. 1, 1989, Bangalore, 1989, pp. 364-5.

16. Ibid., p. 437.

17. 'By 1512, there were two hundred such marriages in Goa and a hundred in both Cannanore and Cochin. In 1529 there were eight hundred of these in Goa....' Footnote in M. Joseph Costelloe, *Letters and Instructions of Francis Xavier,* Anand, 1993, pp. 148-9.

18. Mundadan, op. cit., p. 438.

19. A. da Silva Rego, *Documentaçao para a história das Missoões do Padroado Português do Oriente,* Lisbon 1947-58, vol. 1, Doc 110, pp. 232-9.

20. P. Malekandathil, (ed), *Jornada of D. Aleixis de Menezes: A Portuguese account of the sixteenth century Malabar,* Kochi, (2003) 2004, pp. 249-59.

21. Boxer, op. cit., p. 100.

22. M.N. Pearson, 'The Crowd In Portuguese India' in M.N. Pearson (ed), *Coastal Western India,* New Delhi, 1981.

23. Mundadan, op. cit., p. 438.

24. Luillier, *Voyage aux grandes Indes,* 1705, cited in Linschoten, op. cit., (1935)1997, by the editor, p. 209.

25. *The Voyage of François Pyrard of Laval,* Tr. Albert Gray, London, Hakluyt Society M.DCCC.LXXXVIII, vol. 2. I, pp. 113-15.

26 Linschoten does not explicitly mention the above characteristics as *mestiça*-specific. However, the general context in which it was written makes me infer that he was referring to *mestiças.*

27. Linschoten, op. cit., p. 209.

28. Arie Pos, 'A Stranger's testimony, some of Jan Huygen van Linschoten's views on and from Goa compared with Portuguese sources', in Ernst van Veen and Leonard Blussé (eds.), *Rivalry and Conflict: European Traders and Asian Trading Networks in the 16th and 17th Centuries,* Leiden, CNWS publications, 2005.

29. Lach, op. cit., p. 444.

30. M. Joseph Costelloe, S.J., op. cit., pp. 411-12.

31. M. Joseph Costelloe, op. cit., p. 286. In this case the woman concerned was Violante Ferreira, widow of Diogo Frois.

32. Silva Rego, op. cit., vol. 1, Doc 144, p. 341.

33. Fr. Agostinho de Santa Maria, *Historia da Fundação do Real Convento de Santa Monica da Cidade de Goa,* Lisboa, Na Officina de Antonio Pedrozo Galram, 1699, p. 70.

34. Allyson M. Poska and Elizabeth A. Lehfeldt, 'Redefining Expectations: Women and the Church in Early Modern Spain', in Susan E. Dinan and Debra Meyers (eds.), *Women and Religion in Old and New Worlds,* New York and London, 2001, pp. 27-8.

35. Santa Maria, op. cit., p. 409.

36. Ibid., pp. 412-13.

37. Timothy Joel Coates, 'Exiles and Orphans: Forced and State-sponsored Colonizers in the Portuguese Empire, 1550-1720', Unpublished thesis submitted to the Faculty of the Graduate School of the University of Minnesota in partial fulfilment of the Requirements for the Degree of Doctor of Philosophy, July 1993, p. 260.

38. According to Correia more than a million (*milhão*) of both sexes came to India between 1580-1640, cited in Boxer, op. cit., p. 79.

39. Boxer, op. cit., pp. 83-4.

40. Ibid., p. 87.

41. Coates, op. cit., pp. 432, 300.

42. Boxer, op. cit., p. 87.

43. Ibid., pp. 89-90.

REFERENCES

C.R. Boxer, *Portuguese Women during the First Two centuries of Expansion Overseas*, Separata do, vol.V das Actas do Congresso Internacional da História dos Descobrimentos, Lisboa, 1961.

————, *A mulher na expansao Ultramarina Iberica*, Lisbon, Livros Horizonte, 1977.

M. Joseph Costelloe, *Letters and Instructions of Francis Xavier,* Anand, 1993.

Timothy Joel Coates, Exiles and Orphans: Forced and State-sponsored Colonizers in the Portuguese Empire, 1550-1720, Unpublished thesis submitted to the Faculty of the Graduate School of the University of Minnesota, in partial fulfilment of the requirements for the degree of Doctor of Philosophy, July 1993.

J.B. Harrison, Five Portuguese Historians, in C.H. Philips (ed.), *Historians of India, Pakistan and Ceylon,* 1961.

Donald F. Lach, *India in the Eyes of Europe: The Sixteenth Century,* Chicago and London, [1965], 1968.

The Voyage of François Pyrard of Laval, (Tr. Albert Gray), London, M.DCCC.LXXXVIII, vol. 2.

The Voyage of J.H. Van Linschoten to the East Indies, vol. 1, [1935], Delhi, 1997.

P. Malekandathil (ed), *Jornada of D. Aleixis de Menezes: A Portuguese account of the sixteenth century Malabar*, Kochi (2003), 2004.

Fr. Agostinho de Santa Maria, *Historia da Fundação do Real Convento de Santa Monica da Cidade de Goa*, Lisboa, Na Officina de Antonio Pedrozo Galram, 1699.

A. Lopes Mendes, *A India Portugueza, Breve descripção das possessões Portuguezas na Asia*, Lisboa, 1886, Delhi, 1997.

A.M. Mundadan, *History of Christianity in India*, vol. 1, 1989, Bangalore, 1989.

K.M. Panikkar, *Asia and Western Dominance*, London, 1954 repr.

M.N. Pearson, *Merchants and Rulers in Gujarat: The Response to the Portuguese in the 16th Century*, New Delhi, 1976.

————, 'The Crowd in Portuguese Indian' in M.N. Pearson (ed.), *Coastal Western India*, New Delhi, 1981.

A. da Silva Rego, *Documentação para a História das Missoões do Padroado do Português Oriente*, (12 vols) Lisbon, 1974-8.

Manuel da Cunha Maldonado and Joaquim Antonio Roncon, *Usos e Costumes da India, 1846-1847*, in Maria de Jesus dos Martires Lopes, *Tradition and Modernity in Eighteenth Century Goa*, Delhi, 2006.

Arie Pos, 'A Stranger's testimony, Some of Jan Huygen van Linschoten's views on and from Goa compared with Portuguese sources', in Ernst van Veen and Leonard Blussé (eds.), *Rivalry and Conflict. European traders and Asian Trading Networks in the 16th and 17th centuries*, Leiden, 2005.

Allyson M. Poska and Elizabeth A. Lehfeldt, 'Redefining Expectations: Women and the Church in Early Modern Spain', in Susan E. Dinan and Debra Meyers (eds.), *Women and Religion in Old and New Worlds*, New York and London, 2001.

Elaine Sanceau, *Portuguese Women during the First Two centuries of Expansion Overseas*, Separata do, vol. V das Actas do Congresso Internacional da História dos Descobrimentos, Lisboa, 1961.

Georg Schurhammer, 'Desenhos Orientais do Tempo de S. Francisco Xavier' in *Garcia de Orta*, Numero especial (1956), pp. 247-56.

P.A. Tiele, 'Introduction' in *The voyage of J.H. Van Linschoten to the East Indies*, vol. 1. [1935], Delhi, 1997.

Disease, Morbidity and Mortality in the Indian Ocean World, 1500–1800

Abhay Kumar Singh

It is a medical and anthropometric truism that hunter-gatherers (4.5 million years ago) were healthier than modern humans. Their lifestyle spared them disease and illness which are prevalent today.[1] They hardly segregated and adopted a sedentary lifestyle in scattered groups of 50 to 100 individuals. Possibly, the sparse densities of the populations diminished the incidence of contagious diseases transmitted by viruses and bacteria such as smallpox or measles. The restless and fast mobile lifestyle of hunter-gatherers eventually exposed them minimally to pollute water and other sources of disease-carrying and human-transmitting bacteria, viruses and insects.

Two principal sources of disease transmission among hunting, fishing and gathering people have been recognized. First, wild animals: infections (zoonoses) were acquired by eating or contacting wild animals by any mode whatsoever. Trichinosis, sleeping sickness, tularaemia (rabbit fever), tetanus, schistosomiasis (bilharziasis) and leptospirosis (Weil's disease) are the zoonotic diseases. Various forms of typhus, malaria, and yellow fever can also be included into this category. Encounters with these zoonotic diseases would have been largely incidental and seldom adversely affected many members of a group 'especially in view of the mobility of hunter-gatherers and their tendency to abandon areas when food become short'.[2] Second, the disease organisms present during pre-hominid period and continued to evolve with humans. Numerous disease transmitting micro-organisms, worms, lice, virus and bacteria such as salmonella and *Treponema* (the agents of yaws and syphilis) have been classified into this category.[3]

The symbiotic relationships between men and scavenging, decoy hunting and taming animals originated the processes of social parasitism during pre-Mesolithic age. The symbioses are, generally, unequal and involuntary, in nature, involving certain degrees of coercion operated through transplanted social medium of the more intelligent species. Mutual life supporting motive made this social medium operative and sustainable. The social parasitism has two operative bio-mechanisms: first, the guest exploiting host mechanism and second, host exploiting guest mechanism. This offered sufficient precondition for the evolution of pests and parasites.[4] Tamed animals helped humans to create civilizations with their meat, hides, milk, eggs and bones and simultaneously, also transmitted diseases.

The systematic domestication of nomadic animals vital for seasonal migration occurred in the late-Mesolithic period, earlier than the true Neolithic age. The motive behind their domestication was supplying required meat in addition to hunting to almost permanent human habitation. Some other animals, like cattle, had been domesticated for maintaining settled human life. Finally, other categories of animals had been domesticated to serve as transport animals.

The transition from hunting-gathering to domestication and systematic domestication of animals had numerous biogenic, bio-ecological and biosocial environmental transformations. Guest exploiting host is a spontaneous and automatic and uneven and unequal biosocial environmental mechanism of biogenic transformation operating on social medium provided by host exploiting guest mechanism through biophagous processes in different bio-climate and bio-ecology at different times. One of these biogenic transformations embraces and buttresses the many realms of parasites. Pathogens of domesticated animals and self-appointed domesticates (social parasites) like mice and rats, mosquitoes and blood-sucking insects found their ways in human bodies and began adapting them as well. Hence, human economic needs during upper Paleolithic and Neolithic periods exposed humans to infectious diseases. Mechanisms of this can be tentatively demonstrated through Fig. 1.

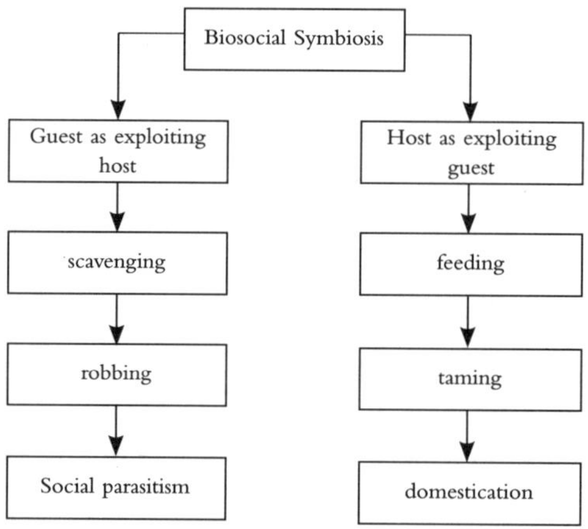

FIG. 1: BIOSOCIAL SYMBIOSIS ORIGINATING DISEASE PATHOGENS

Source: F.E. Zuener, 'Domestication of Animals', op. cit.

The transition from hunting-gathering to farming and domestication of plants had numerous bio-climatic, bio-environmental and bio-ecological downsides. These were anthropogenic and bio-geographical processes. Humankind engineered the epoch of agrarian revolution (Neolithic economic revolution) by domesticating plants and animals—'a process of selecting and propagating beneficial traits in wild crops', occurred independently in the subtropical zone at different times.[5] The advent of farming inculcated intensive and fresh opportunities for developing the guest host relation, since cultivated lands are ideal feeding and breeding grounds for a large numbers of herbivorous and gregarious animals.[6]

Each subtropical zone evolved a dependence on different staple crops: wheat and barley in the Middle East and South Asia; millet and rice in China and South East Asia; maize in the New World. They also domesticated animals to facilitate the Neolithic agrarian revolution. The domestication of animals promoted a 'process of selective breeding'. Thus, sheep and goats, native to West and Central Asia were domesticated for their meat, milk, hides, and wool. Cattle were domesticated all over Eurasia to draw ploughs. Food surpluses supported sustenance of settlements in permanent villages surrounded by material goods and equipment, enabling the evolution of craft skills, experiments with technology, such as pottery kiln, gold, and copper metallurgy. The technique of hand-modelling and firing clay pots evolved independently in many regions. Moulds, wheels and kilns were later innovations and recognized as the 'province of specialized craftsmen'. This became instrumental in initiating social differentiation.[7] Through agricultural revolution humankind embarked upon manipulating earth to arrange its eco-system, not to mention the genes of the plants and animals within those systems. With this, an ever-perpetuating enterprise of undoing a self-regenerating nature—an unthinking process appeared, boosting recurrences of bionomic, bio-climatic and bio-meteorological downsides, partially responsible for biological imbalances and epidemiological outbreaks.

Irrigation farming[8] in early river valley cultures of the Indian Ocean world, especially the flooding of lands for rice cultivation had microbial and epidemiological effect. In the warm, shallow waters of the paddy fields lurked bacteria potent enough to penetrate the skin and 'enter the bloodstream of wading human rice farmers'. *Schistosoma*, a blood fluke that uses acquatic snails as an intermediate host through successive stages of breeding, feeding and development was the result. This micro-organism causes breeding and spreading most debilitating and often deadly disease namely schistosomiasis (or bilharziasis). Evidence of the presence of this disease has been discovered in the kidneys of 3000 year old Egyptian mummies in Egypt.

Slash-and-burn agriculture (jhoom cultivation) created niches that stimulated breeding and fast multiplication of relatively small populations of parasites. This typology of cultivation had been blamed to be the reason for the proliferation of the mosquito *Anopheles gambiae,* communicating the most lethal type of the malaria, falciparum malaria.

Hence, the agrarian revolution revolutionized new and intimate human contact with numerous insects and worms, not to mention bacteria, viruses, protozoa and the rickettsias (micro-organisms intermediate between bacteria and viruses) carried by ticks, fleas and lice.

Sedentary coexistence before 5500 BC exposed humans to infectious disease. The settlement of Catal Huyuk, for instance, was plagued by malaria.[9] William McNeill postulated that humans share some 65 diseases with dogs, 50 with cattle, 46 with sheep and goats, 42 with pigs, 35 with horses and 26 with poultry.[10]

The establishment of permanent settlements attracted other self-appointed domesticates like mice and rats, mosquitoes and blood sucking insects to cohabit with human habitations and surroundings and share and eat the same food. Several thousands of co-evolution, rapidly changing environments, climates and local resources compelled these small, furry animals to evolve a wide range of adaptations and adaptability. They developed the high potency of survival without human cohabitation. These commensalities, all too often, have helped spread diseases.

Permanent settlements also provided breeding sites in forms of forest clearings and stagnant water near dwellings, as well as feeding human bodies to mosquito and blood-sucking insects. Faeces-feeding house flies nourished in these new settlements, polluted human food with their 'dirty feet' and reached in human abdomen to insure a variety of self-perpetuating diarrhoeal disease and bacillary dysentery. Fleas and lice colonized outside of the human body while amocbas, hookworms and countless other parasitic worms moved into its interior. Human proximity promoted proliferation of the pathogenic-producing squalor.

A further vigorous global ecological and climatic manipulation reproduced microbial mutability with lightening speed catalyzing pathogenic parasitizing of humans. The roundworm (*Ascaris*) which was probably acquired from pigs and hookworm, both spread by faecal pollution of land are cases in point. These worms live in the gut and compete with their human hosts for protein, causing anaemia.

In this apparently uneven struggle for survival, humans developed a defence system to combat parasitism of micro-organisms. Four mechanisms of this defence system have easily been detectable. First, the natural defence system developed by those who had survived a disease. The disease survivals were left with some immunity against that disease. Humans hence began evolving a sophisticated immune system to enable them to live with their invaders. Pathogens cooperated in the biological evolution and mutation of this immune system which operated through a compromise mechanism between the invader and the invaded: the host survived by passing on the micro-organisms to other hosts. Second, mothers transmitted those immunities to infants acquired during pregnancy. These immunities delivered across placenta to newborn, which partially immunized them against the inevitable invasion of germs. Third, some humans are genetically protected from disease. Finally, community immunity acquired by people against those parasitic worms surrounding their habitat.

The ravages of two types of diseases are however most lethal and far more difficult to live with. First, a new incarnation of an old disease in a new and more potent form has been supported sufficiently by fast human population progression. The speculation about detecting the time and space of the first manifestation of these new plagues is fascinating but hardly more than guess work. The emergence of cities with populations as large as 50,000 in Asia and Africa offered ideal breeding and feeding sites and hosts to these disease pathogens. Cattle herds of the first urbanism boosted the breeding and hosting of many pathogen-types. The endemicity of smallpox in South Asia is attested by the erection of a smallpox deity in ancient Indian temples with a view to escape its visitation.[11] Contemporaneous to this, smallpox inoculation had been practiced in ancient India.[12]

Second, the invasion of novel pathogens created massive epidemics with an immediate consequence of a spectacular fall in susceptible population. The survivors would have then commenced the painstaking and slow process of population recovery through biological reproduction, only to be eliminated by another new disease, and yet another. The disease holocaust themselves disappeared, only to reappear after a large gap with lethal ferocity against few non–immune individuals. While one human community enjoyed an epidemiological respite, another neighbouring large community experienced the vitriolic voracity of the new diseases. The novel diseases became powerful instruments of controlling a population explosion while "immunologically tempering those that survived them".

In this new pathogenic crucible, micro-organisms incubated, combined, altered, perished, and prospered. Their prosperity often required intermediary hosts for serving as staging areas for later assaults on humans. Similarly, other micro-organisms took relatively longer time to acquire vectors to shuttle them about from intermediary host to people, as well as from person to person. Some diseases so perfectly adapted to humans became so contagious that they no longer needed their old–host or hosts to complete their life cycle. These were crowd diseases like smallpox and measles that were transmitted by small parasitic micro-organisms that either killed or produced lasting immunity. Humans were their carriers. Finally, there were filth diseases like typhoid least suffered by the hunter-gatherers due to their regular migratory life style.[13] Nutritionally impoverished people became ubiquitous hosts to disease, opening the door to a 'synergistic union of malnutrition and pathogens'.

Transoceanic and transcontinental exchangeability of pathogens or micro-organisms were speeded up by the modern world system especially after 1500. The hegemonies and the cores of the modern world system promoted not only the transoceanic European expansion for proto-industrial exchanges but also for the exchange of its biological or bio-cultural consequences. Maritime trade in proto-industrial goods of the Indian Ocean world for bullion from America mediated through the Indian and the Atlantic Oceans by the Europeans during the seventeenth and the eighteenth centuries played a dynamic role in biological or bio-cultural exchangeability.

Physicians and surgeons (Egyptian, Greek, Indian, Chinese, Roman, Sidhha, Arabic, and other medical systems) from earliest times to AD 1800 constantly waged concerted wars against emerging infectious diseases globally to eradicate or even eliminate them from the face of the planet Earth. Traditional medical therapeutics and procedures of the pre-scientific phase to manage these diseases failed miserably, rather aggravated the health problem by making the patient anaemic through bloodletting, depleting them of fluids and valuable electrolytes via faeces, and poisoning them with compounds of such heavy metals as mercury and lead.[14] Very recently, modern medicine with specialism and super-specialism, extreme medical engineering and high-tech medical technology after AD 1800 made ambitious claims of controlling most of the infectious diseases. But tragically, with optimism came a false sense of security, which has helped many diseases to spread with alarming rapidity.

This is a perennial process still persisting and causing extreme damage to world health as observed by epidemiological authentic reports, prepared by meticulous researchers and sponsored by world class health organizations. The recent global epidemiological pessimism revolves fundamentally round two types of diseases.

First, resurgence of major diseases such as malaria, tuberculosis; and reappearance of plague, diphtheria,[15] dengue, meningococcal meningitis, yellow fever and cholera as public health threats in the Indian Ocean world after many years of decline. Antibiotic and drug resistance diseases are other threats to global human health.

Tuberculosis has become a holocaust in Asia, accounting for two-thirds of all the tuberculosis in the world, metaphor as MDR-TB hot zone and usually a death sentence in the developing world.[16] Pertussis (whooping cough) responsible for infecting broncho-pneumonia in children is another child killer disease with 15 per cent mortality rate in the developing countries and 4 per cent mortality rate in the industrial countries. Its infection is on the rise among adults. Adults have now become reservoirs and transmitters for pertussis. The children are especially fatally vulnerable to dengue, hemorrhage fever killing globally around 30,000 people each year. The mortality rate in this disease is around 5 per cent.

Measles remains one of the major vaccine-preventable child killer diseases. 98 per cent of the total deaths occuring in the developing countries have been caused by it. Meningococcal meningitis, a disease covering the brain and spinal cord, accounts for rapid deaths and permanent brain damage—especially in young children. The case fatality rates in Africa range from 8 to 20 per cent whereas in US it ranges from 5 to 15 per cent. Typhoid fever remains a serious global health problem infecting 16 million people worldwide annually with 600,000 deaths, mostly under the age group of 3-19 year. The morbidity rate in this case is much higher—keeping patients away from work for several weeks.

Second, previously unknown highly infectious microbes are emerging at an unprecedented rapidity and tumultuous voracity, propagated in stupendous numbers, infecting new hosts by the medium of animal transmitters or through sexual contact or through aerosol transmission. Thirty such viruses have been identified in the last 3 decades around the world. Germ-friendly ecosystem has been offered by high human density populations in the mega-cities, by stockyards, potentially susceptible animals crowded together, etc. They are supreme parasites—infecting plants, animals and even each other and adapted to sustain in molten rock, at the bottom of the sea, or in strong concentrated acid solutions. They mutate instantly in environmental and climatic changes as well as changes in host's immune system or the latest vaccine. Microbial tenacious selection and reproductive efficiency is both impressive and humbling.

The recent global alert and global response towards infectious diseases concentrate on devising high-tech bio-chemical antimicrobial, biotechnological, genetic engineering,[17] super medical technological and extreme engineering tactics and strategies for disease monitoring, disease surveillance and disease control. The fatal failures of these tactics and strategies only increase severe stress on the biosphere, human health and present epidemiological pessimism.

Health planners and government policy framers realized the need for intensive historical explorations of the realms of bio-geography, bio-anthropology, bio ecology, bio environment, bio-climate and bio-ethics to offer bio-feedback to formulate bio-cleaning and bio-controlling biosphere-friendly strategies for conserving, rather even improving, existing global and regional bio-rhythmicity, bio-compatibility and bio-diversity. Such historical researches are fashionable in America after the researches of A.W. Crosby at least since the 1970s. The researches of this genre are still in infancy in India.

The biological or bio-cultural consequences of the transoceanic European expansion have drawn considerable scholarly attention in recent years. The most influential works in this genre have, undoubtedly, been A.W. Crosby's *The Columbian Exchange* and *Ecological Imperialism* which comprehensively analysed the exchange of plants and animals across the Atlantic after AD 900.[18] There are other historians, too, who have emphasized the importance of the disease factor of the European expansion. Contemporaneous to Crosby's *Columbian Exchange*, are Emmanuel Le Roy Ladurie and W.H. McNeill who perceived a similar concept of a 'unification of the globe by disease'. This unification, according to McNeill, is the product of a 'confluence of the civilized disease pools' or in the opinion of Ladurie is a 'microbial unification' of the Eurasian land mass, a lengthy and slow process, spread over centuries. The process became more unifying and homogenizing, and repeated more frequently with devastating effect after the sixteenth century. Ladurie concluded that neither Africa nor Asia experienced any comparable 'bacteriological genocides'.[19] This discriminatory bio-historiography about the Indian Ocean has been further followed and buttressed by historians of epidemiology like Philip Curtin[20] and Kenneth F. Kiple.[21]

Only in the 1990s Ralph Shlomowitz and his associates, and David Arnold seriously challenged this tradition of the bio-history of the Indian Ocean.[22]

An attempt will be made in the following pages to analyse the roles of the modern world system in intra-Indian Ocean world transmission of diseases, its instrumentality and morbidity and mortality rates associated with specific diseases.[23]

INFECTION AND TRANSMISSION OF DISEASE

Evolutionary biology suggests that morbidity and mortality differed ethnically because Asians, Africans and Europeans had desperate biological reaction to different disease (micro-parasite) environments that evolved across the oceans.[24] Biological agents of disease are the products of evolution.[25] It must have undergone mutations in the past and became environment specific. Diseases, therefore, evolve in specific environments. The essence of bacterial ecology and evolution is the utter expandability of the individual, the clone or the species. The utter expandability of the bacteria practically made it hard, rather impossible, to define a bacterial species. In a specific environment, a definite pattern of structure and function in a bacterium may be adequately defined. Any environmental modification, human made or natural, results in the origination of some slightly different but usually more potent and well adaptive to new environment bacterium than the original. If such a bacterium or micro-organism is available either as a casual contaminant from somewhere else, or as a mutant of the originally dominant form, the rate of multiplication is very rapid.[26] The more the environment is modified, the fewer the surviving animal and plant species and the greater the population density of those opportunistic bacterium and micro-organisms which can easily adapt.[27] The new bacterium or micro-organisms will dominate the situation while old ones will vanish. In a changing world, its successor, too, is bound to be replaced shortly by something more powerful and more adaptive ones, depending on a wide potential for genetic variation with sufficient genetic stability to exploit advantages. In bacteria and micro-organisms, the speed of the process of genetic variation is vastly greater. 'Doubling time is measured in minutes, population numbers in billion per mililitre'.[28] Manmade environmental degradation has the effect of controlling infection where the vector is a 'wilderness' specie, or promoting it where the vector is an opportunist.

Human populations also evolve and are ecology and environment specific. Most human societies, in course of time, normally, develop some sort or rough equilibrium with their pathogens or bacteria or micro-organisms—the specific causes of their disease—adapting to their disease environment in a variety of bio-cultural ways.[29] The natives of one ecology and environment developed resistance to the diseases that predominate that ecology and environment and susceptible to distinctive disease environment often at distant locations. Hippocrates was the first 'great naturalistic physician' who in his book, *Airs, waters and places* laid the earliest foundation of the relationship between environment and health.[30] The *Airs, Waters and Places* had two principal purposes to serve: first, it enabled the doctors to predict the disease of specific ecologies, and second, it served as a guide to the Greek colonists.

Micro-organisms or microbes are very small organisms including bacteria, rickettsiae, fungi, algae, protozoa and viruses. Some micro-organisms are plant-like and some animal-like, but some have the property of both. Viruses are even more complicated because sometimes they are classified as living while sometimes as inanimate. Bacteria and blue-green algae are classified in a specialist group of organisms called prokaryotes and are easily distinguishable from other organisms called eukaryotes.

Bacteria which produce respiratory infections are expelled in large numbers during sneezing or coughing and transported some distance. Air-transported plant pathogenic (disease bearing) fungi such as *Puccinia graminis* release millions of spores to improvise their survival potentiality.[31] Some micro-organisms need a vector in form of an individual of one species functioning as transmitting propagules of another from place to place. It may function as a mechanical transmitter, or a simple carrying agent. It may also work as mediums of some physiological or pathogenic association with the cell, which it transports. When the identity of vector is detectable, the modes of dispersal, transmission new favourable habitat, etc. are established. Viruses and rickettsiae need appropriate micro-environment to survive. *Fungi imperfecti* produce diseases such as 'ring worm' and 'athlete's foot'.

Bacteria are smallest organisms common to almost every available habitat and cause large numbers of diseases. Some even cause acute disease in the host, which may kill it, while others produce immunity by activating the defence of the host body and bring destruction to the bacteria itself. Sometimes bacteria are contained in hosts' body so that it performs the duty of carrier, capable of infecting another member of its species without itself being very sick. The host killer bacteria produce crisis for its own subsistence and survival. Optimum opportunities of symbiosis are needed for optimum growth. Human lower intestine is the repository of massive amounts of bacteria performing specific tasks without either killing hosts or fellow bacteria. Many bacteria maintain their viability for longer periods up to 2000 years in case of actinomycetes' endospores. Some pathogenic bacteria may survive in dust, dirt and occupation deposits for sometime and can infect a new host. Varied dispersal methods had been and even have been adopted by bacteria: aerially dispersal in case of *Bacillus anthracis* spores and droplet infection. The later is the potent means of spreading the viruses, for instance, mumps, measles, influenza and tuberculosis.

Over 300 types of viruses producing diseases in a wild variety of organisms had been identified. Since viral diseases normally affect the soft tissue and leave no skeletal and ortho-pathological evidence, it is pertinent to deduce information from historical references.

Viruses are a heterogeneous group of submicroscopic, sub-cellular, filterable 'infra-microbes' or 'essentially nucleic protein', surrounded by a lipoprotein membrane in nature capable of producing disease not by multiplication but by fabrication from the active host cell. The nucleic acid of an individual virus is of one type either DNA, or RNA, the later is more common among viruses pathogenic for humans. This was the result of an autocatalytic action of the cell metabolism either by a non-specific stimulus, or by a minute quantity of the end product—the virus itself. Energy generating and biosynthetic mechanisms are lacking in viruses. The virus only provides the genetic material for replication. An essential characteristic of virus is constancy of size and form: spherical, rod-shaped, brick-shaped, bullet-shaped, etc.[32]

Viruses are multiplicative only in living cells of the organs of infected organisms. Viruses are antigen-specific agents capable of autocatalytic causation of a trans-naturation in the host cell's protoplasm protein. Hence, viruses are multiplicable transnaturase of protoplasm protein.[33] The structural change in protoplasm protein was a corollary to the chain reactions of virus proliferation in the protoplasm, or a change in the structure of protoplasm protein surrounding newly generated viruses. Protoplasm, a fibrous structure consists of proteins, and lipids usually constitute one-third of the total cell mass. The lipid contents of mitochondria and microsomes have been in the proportion of 24 and 40 per cent respectively.

Viruses are identical with elementary bodies of protoplasm, can act and assimilate other assimilates with weaker action than the viruses themselves, and can synthesize the protein but cannot proliferate in media without cell. The virus protoplasm contact site changed to become identical with the virus and then propagated to other organs, up to polar groups in the whole protoplasm. Viral host cell surface attachment through physical contact is the complementary function of specific viral receptors. Attached viruses are irremovable by washing and remain susceptible to inactivation by antiserum. Cells without receptors for a specific virus are not susceptible to infection by that virus. The process of attachment operates independent of temperature. When the protoplasm, in its structure, becomes identical with a virus, the virus multiplies in it.

The synthetic phase is crucial for the production of many copies of genome and viral messenger RNA (mRNA). Messenger RNA is then translated into viral proteins including capsid proteins by host ribosomes. The reaching of critical concentrations of two viral components, genome and capsid protein result into virion assemblage. Viral mRNA and host mRNA are two different types of proteins performing an anti-viral mRNA function through interferons that 'Induce virus inhibitory protein(s) in the host cell'.

The assimilatory action cannot be performed without the orderly polymerization of protein molecules of the same structure, i.e. incomplete or loose polymerization refutes a virus to act as a typical virus. This disturbs the structure of protoplasm without inducing exact replica. Replica formation is only a result of multiplication of a virus but not a necessity. Virus multiplication is not always accompanied by the manifestation of any pathological disorder. Pathological disorders are visible only at the time of cell disintegration, whereas, in some cases, pathological changes are more remarkable with limited virus multiplication. During virus multiplication, the injurious effect upon the host cell may be effective but clinically 'inapparent' or 'subclinical' infection.[34] Virus properties are occasionally alterable and transmissible to newly produced viruses through analogous phenomena of mutation of organisms. Serial propagation of a virus in a host different from the customary gradually increases its virulence towards the host. Virulence enhancement is the byproduct of enhancement in the combining force between a virus and the host cell, i.e. passing of a virus in succession through a series of the same kind of host.[35]

The whole organism practice parasitism. Parasitism may be operated at the molecular, cellular, tissue, population, macro-ecological to micro-ecological and bio-chemical levels of organization. Parasitism is defined as an intimate and obligatory relationship between two hetero-specific organisms, during which the parasite, normally the smaller of the two partners, is metabolically dependent on the host. The relationship may be permanent or very temporary—an extremely common biological phenomenon.[36] Parasitism generally involves two biogenic processes: First, the host exposure of antigenic substances of parasitic origin, the body molecules of the parasite (somatic antigen) or molecules secreted or excreted by the parasite (metabolic antigens) and second, antibody synthesis by the host against the parasite or host immunological response.

External and internal parasites yield bio-information of palaeo-environmental interests. Parasites affect both animals and man while others are host-specific. Endomorphic parasites depending on animal vector have human relatives and produce specific disease in animals. Parasitic organisms are highly adaptive to complex bio-environment, bio-ecology, bio-climate and bio-meteorology by completing a complex life cycle. Examples of more complex cycles may be seen in the development of malaria and yellow fever, the later being transmitted to humans from *Haemogagus* mosquitoes, which have collected it from certain species of monkey. The liver fluke (*Fasciola hepatica*) undergoes a complex life cycle involving a number of hosts and vectors. They can be utilized as indicators of past environments by palaeo-pathologists and hence, have great epidemiological, pathological and medical value. Unlike diseases producing ortho-pathological traces on skeletal material like leprosy, arthritis and tuberculosis, parasites are identified accurately and adequately if soft or decayed tissues are present, or recovered from coprolite.

Nematoda parasites are communicated to humans by the consumption of contaminated pork by *Trichinella Spiralis* (a small cylindrical worm). *T. Spiralis* infects the host with cysts. Other nematode parasites of man causing chronic diseases are the hookworm (*Acyclostoma doudenale*) and thread worm (*Ascaris lumbricoides*). Various types of flukes have a complex life cycle, involving the water snail as an intermediate host, although man may be directly affected by drinking contaminated water. Blood flukes are productive of very serious health problems such as the blood flukes of the genus *Schistosoma*, and the Chinese liver flukes *Clonorchis* (*Opisthorchis*) *sinensis*. The most common of the parasites recognized in microbiological archaeology are different types of tapeworms that produced health problems but rarely became fatal. It, certainly, reduced resistance power to other diseases and produced severe weight loss.[37]

Various types of parasitism are identifiable.

- Fucultative parasite, i.e. capability of an organism to adapt but partially.
- Obligatory parasite, i.e. having complete dependence on microbes on host during a segment, or all of its life cycle—a physiological relationship.
- Incidental parasite, i.e. accidental, or unnatural dependence on host for survival.
- Erratic parasite, i.e. a parasite that wanders into an unusual organ.
- Periodic or sporadic parasite, i.e. intermittent visitor to its host to secure some metabolic requirements.
- Pathogenic parasite, i.e. a causative agent of a disease state in the host. The disease may be chronic or acute.

Parasites can be classified into two classes on basis of their habitat locations.

- Endoparasites, i.e. those that live within the organs of their hosts in locations like the alimentary tract, liver, lungs, gall bladder and urinary bladder.
- Ectoparasite, i.e. those that attached to the outer surfaces of their host, or are specifically embedded in the host's body surface.

Homologous to this is the categorization of hosts.

- A definitive or final host, i.e. where the parasite attains sexual maturity.
- An intermediate host, i.e. when it serves as a temporary environment but is necessary for the completion of the parasites' life cycle.
- A transfer or parential host, i.e. when a host is utilized as a temporary refuge and vehicle for reaching an obligatory host.
- Reservoir hosts, i.e. when infected animals serve as a source for further animal infection.
- Hyperparasite, i.e. when an organism parasitizes another parasite.[38]

The establishment of a host–parasite interaction results into infection—sometimes as overt infectious disease and sometimes as covert or subclinical infectious disease. When the injury to the host is manifested clinically, an overt infectious disease ensues. The host–parasite cohere, collide and ultimately, parasitic multiplication and its maintenance without reaction is connoted as colonization. When host–parasite interaction does not ramify perceptible deviation from good health, but does not inflict sufficient injury to the host to provoke a response specific for the parasite, the result is designated as covert or subclinical infection.

This is evident from Fig. 2.

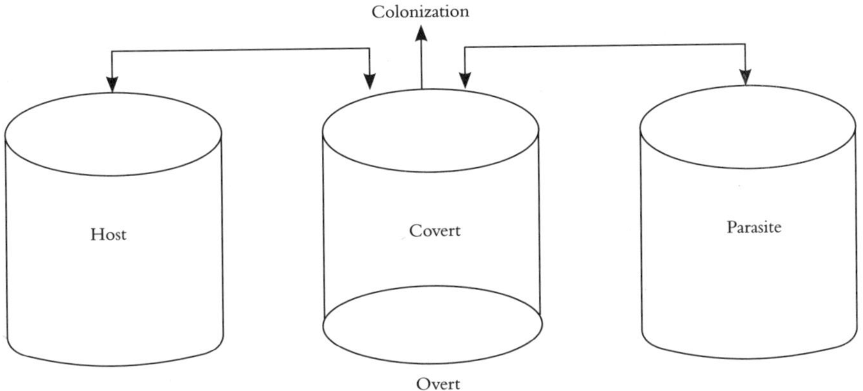

FIG. 2. INTERACTION BETWEEN HOST-PARASIT

Source: Paul D. Hoeprich, 'Host-Parasite Relationships and the Pathogens of Infectious Disease', in Paul D. Hoeprich (ed.), *Infectious Diseases: A Modern Treatise of Infectious Processes*, p. 51.

A typical parasite when hosted by a typical species does not necessarily and universally provoke identical disease in all infected individuals. The variability and typicality of infectious diseases is substantially dependent on the inoculating dose of the parasite. One of the innate adaptations of the species is to develop variability regarding the virulence of the parasite as well as the resistance of the host. The spectrum of intensity of interaction between host and parasite can broadly be delimited for specific species. Clinical inapparent infections outnumber apparent illnesses in cases of most infectious microbes or agents—an epidemiological and medical fact of urgent criticality to the dissemination and perseverance of parasitic micro-organisms.

Host-parasite synchronicity or conjunction is a probability or chance phenomenon leading to an infectious disease—an accidental infection afflicting a previously healthy person between neonatal and geriatric groups. Accidental infections can be designated as the host-parasite conflicting relationship. Such infections are essentially transpired by accidental host-parasite encounter rather than endogenous microbiota or of commensal interactions. More numerous cases in the young and in the old are host-abnormality-specific or obligatory infections. Anatomic, metabolic, neoplastic, collagen vascular, therapeutic, devitalized tissue abnormalities are productive of infections. Such infections are often dreadful, even leading to death. Endogenous microbiota or a quasi commensal relationship to the host is found to be involved in this infection. There are localized or generalized infections attested by signs or symptoms of disease. Inapparent involvement elsewhere other than localization is another distinction of localized infection.

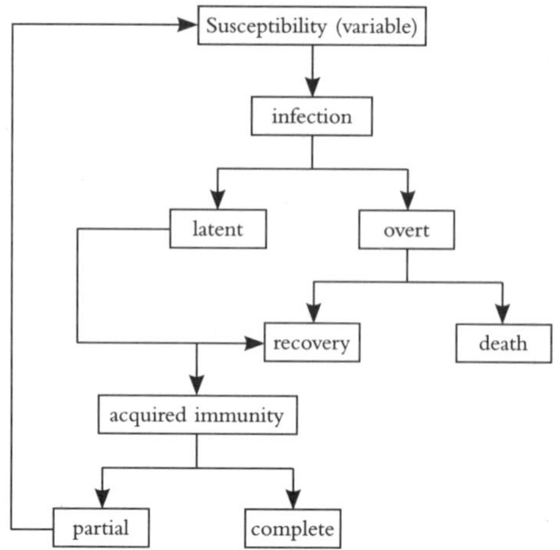

FIG. 3. RELATIONSHIP BETWEEN SUSCEPTIBILITY, INFECTION,
DISEASE AND ACQUIRED IMMUNITY

Source: E.H.R. Harries, *Clinical Practice in Infectious Diseases* (Edinburgh, 1951),
p. 14.

Pathogenicity is variable in virulence and virulence is alterable. Many generations' growth in inhospitable environment frequently results in attenuation of virulence. Reversely, virulence enhancement may be transpired in a parasite by its survival in the dangerous environment of the living host. Survival and replication in a hostile environment is virulence potency amplifier—a type of pathogenicity occurring through the selection of spontaneous mutants. Invasiveness does not guarantee virulence or pathogenicity.[39]

Well established parasites, therefore, have generally acquisitioned or reached a predicament of balanced pathogenicity in the host—a predicament of minimal damage compatibility or parasitic multiplicability. In balanced pathogenicity, some degree of tissue damage becomes compulsory for the effective shedding of micro-organisms to the exterior.[40] Different microbes have different balanced pathogenicity. Many infections have yet not attained this ideal predicament for different reasons. First, each micro-organism metamorphoses occasional virulent variants, causes extensive disease and death before disappearing after 'all susceptible individuals have been infected settling down to a more balanced pathogenicity'. Second, some of the microbes precipitate perilous human diseases which had originated in one bio-ecological zone of the world, weeding out of genetically susceptible individuals and a move and settlement in the direction of a more balanced pathogenicity. Subsequent transference of the microbes to a new bio-geographical and bio-ecological zone of a different human population is productive of more severe infection because of great genetic susceptibility or risk. Examples of this sort are available in the history of the Indian Ocean world profusely. Third, for some micro-organisms the human host is obviously irrelevant for survival. Micro-organisms of this type subsist on other regular species, often with an anthropoid vector, for their maintenance in the atmosphere. Finally, in the course of the history of exploration, human kind had encountered occasional microbes from exotic animals of almost unique bio-ecology that quite accidentally cause serious or lethal human diseases. Contemporary Indian Ocean world is persistently encountering such diseases.[41]

Disease communicability depends on the conflict between disease organism and the host's defence system. Invading organism is annihilated by immunity or defence system except in the carrier state. Intensity of communication is determined by pathogenicity, communicability, invasiveness and the dosage of the organism. Contagion during pregnancy usually proves fatal leading foetal death or cause developmental defects.[42] Auxiliary to this are the host factors of defence, age, and state of health.[43]

Organisms and micro-organisms are expelled from a host in different forms: discharge, droplets or excreta, and communicated to a new host through different transmitting mediums. Direct transmission occurs when the infector and infectee come into actual contact. Indirect transmission involves mediating means of contamination of inanimate objects by the infector and conveyance to the infectee residing at a considerable distance from the infector. Fomites form another medium of disease transfer from infector to infectee. Clothes, handkerchiefs, bedclothes, etc. come into this category. Food, drink, and hands are concerned in mediate transfer of infection. A special type is aerial transmission. It is almost always intra-mural; the air is contaminated by respiratory droplets, their suspended nuclei and dust. Infectivity is intense and fatal at the beginning of the disease, at a new bio-environmental zone, in a new community and declines fairly rapidly.[44]

Infectious disease transmission takes the following paths of transference. First, source of infection, i.e. almost always a human being, either suffering from the disease or carrying the micro-organisms. Rarely is an infected animal responsible for transmitting disease. Detection of the original or primary source is impossible, whilst intermediate or secondary source may be readily discoverable. The identification of the human carrier as a reservoir of disease pathogen helps in coherently explaining anomalies of the disease. Second, routes of excretion, i.e. micro-organisms leave the body of an infector in the discharges or excreta. Micro-organism content of discharges or excreta depends largely on the site of the disease. During coughing, sneezing, laughing and even conversation, the infector expels fine pathogenic droplets from the nasopharynx, nose and mouth. Most of them are invisible and transmit through upper respiratory tract. In common infections, each droplet carries large numbers of infectious micro-organisms. Third, direct and indirect contact communicates disease to infectee not only by touching but also by meeting and coming into close proximity with the infector. Inhalation, ingestion or inoculation, etc., constitute aerial infection—a potent mode of direct transmission. But, American Commission

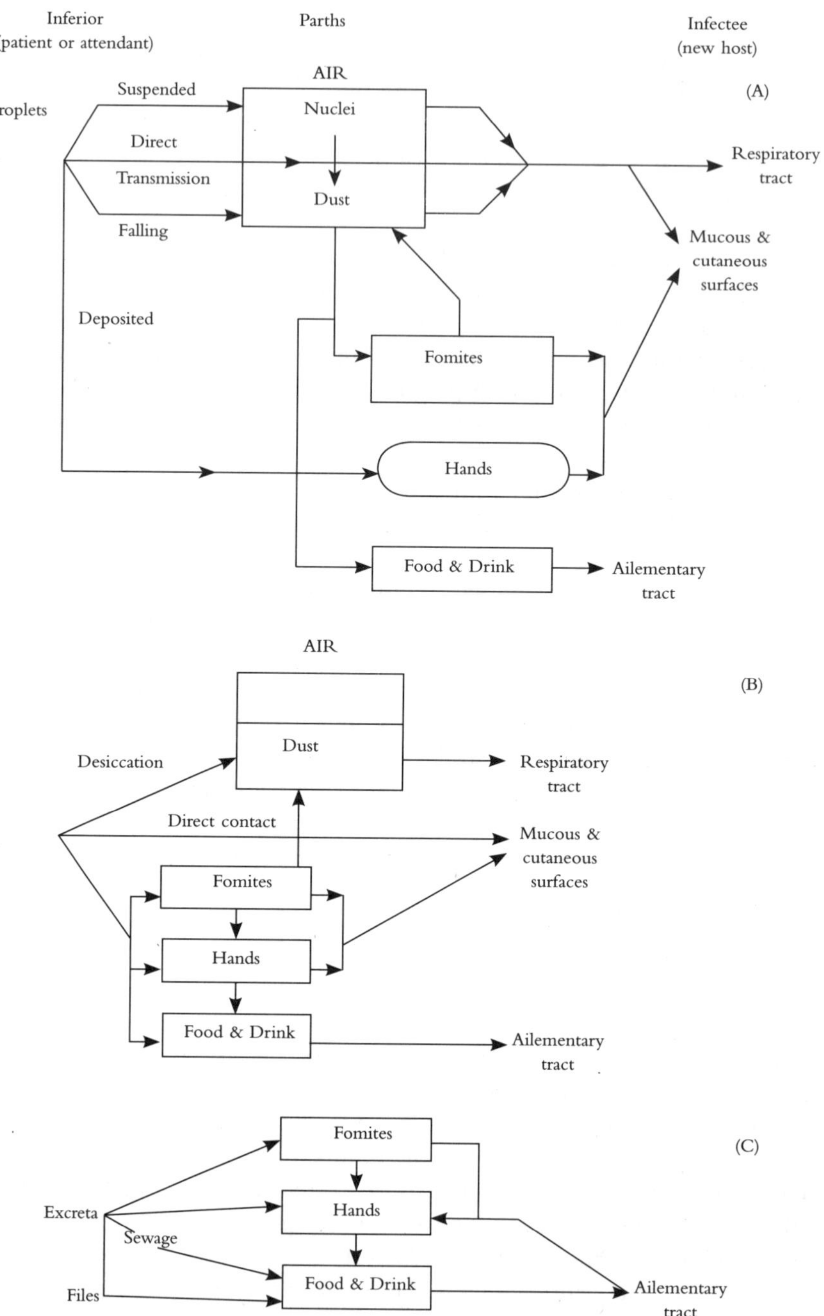

FIG. 4. MODES OF TRANSMISSION OF INFECTIOUS DISEASES

Source: Prepared on basis of E.H.R. Harries, *Clinical Practice in Infectious Diseases*, pp.48-9:

on Airborne Infection (1947)[45] excludes droplets from the category of direct infections. Direct contact is restricted to physical contact, i.e. kissing and sexual intercourse applied in venereal diseases. Indirect contact consists of mediate transfer by contaminated hands, food, instruments, toys, etc.[46] The transference of disease by droplet nuclei through air currents within the range of 3 to 4 feet distance from the infector is connoted as droplet infection while beyond 3 to 4 feet from the infector as air-borne infection by Flugg (1897, 1899). The greater the dilution and the smaller the mass of the droplet nuclei the milder the infection if a new host inspires the air. Dispersal of droplet nuclei depends on location. In the open wind, droplet nuclei are rapidly and effectively dispersed with reduced rate of infectivity, while in an enclosed space, especially a badly ventilated or overcrowded room or ward or ship, the atmosphere remains contaminated with droplet nuclei and highly infectious at least for sometime. Droplet infection can be controlled by personal hygiene whereas air-borne infection by environmental control is the most daunting proposition.

Dust also contaminates air. Micro-organisms survive to a variable extent in the dust. In rooms, wards and ships thus deposited dust subsequently raised by sweeping, bed-making and other human activities in the rooms, wards and ships and the re-infected air may be roamed by currents in the room. In the process of drying, especially sunlight drying, many of the micro-organisms either die or loose their virulence as a consequence of bactericidal action of the sunlight, though nuclei are shielded from this action partially they preserve the toxicity of bacteria. Diffused daylight was found germicidal in 1944 by Garrod. Medical researchers established the fact that respiratory pathogens are more resistant in air than intestinal ones.

Jean Razoux in 1767 listed 6000 diseases,[47] many of which were infectious and spread through specific hosts. Occasionally some succeeded and transferred from man to man. Thus, the stage for human pathogen was established. Measles spread from dog, rhinoviruses causing common cold from horses, the human type of the tubercle bacillus causing respiratory tuberculosis probably resulted from mutation by the bovine type and transmitted by wild cattle, leprosy from water buffalo, diphtheria from cow and syphilis from monkey. There are many other diseases whose original hosts are not certainly known, although they must have arisen from related conditions in other animals. Mumps and small pox can be categorized into this category. Few diseases were capable of spreading as global pandemic by rapid transmission from one host to another. The momentum of the spread of such diseases was quite faster as a result of their airborne infection. Plague and influenza can be put in this class.

The first group of infectious disease has been comprised of those that infect primarily the digestive tract, usually concentrating on the lower part of the bowel. These diseases spread by the dissemination of faecal material containing the responsible pathogens. Water, food, milk, eating utensils, etc., were contaminating mediums infecting men in direct and indirect fashions. Fingers and flies were probably the most potent agents of indirect transfer.

Typhoid and paratyphoid fever, cholera, dysentery due to wide variety of bacteria, summer diarrhoea of infants, bacterial food poisoning, etc. were some of the diseases spread through this route. More than a hundred distinct viruses such as echo, Coxsackie and advenovirus groups can multiply in the intestinal wall generally without producing any symptoms. A variety of symptoms may result from other intestinal infections such as fever, rash, headache, neck rigidity or sore throat, painful cramps of the respiratory muscles, etc. Polio virus (a member of these 'entrovirus') has unique capacity to paralyze. Infectious hepatitis is another entrovirus.[48]

Disease leaves lasting effect on the human body by stimulating the development of disease specific immune system. The immunological effect of an infectious disease can be seen in the production of disease specific antibodies. Through this mechanism, the disease itself produces immunology that diminishes the chances of injury or death or relapsing of that disease. Some diseases provoke a virtually complete and lifelong immunity while others produce partial or transitory or briefer immunity, protecting the individual from further assaults from the specific strain of the infectious agent encountered but not against all strains of the same disease.[49] And still others offer little or no immune response although the immune system may be disordered or even a source of illness, the immune response generally is beneficial. Historically, immune system protected generation after generation from the onslaught of infectious diseases including smallpox and diphtheria. Immunity can also be developed by genetic defenses over generations such as sickle-cell trait which provides some protection against malaria.[50] Cultural responses in form of some typical functional standards of personal and community cleanliness are equally important. Medical technologies have been developed and introduced to control the incidence and severity due to different diseases. People learn to avoid certain sources of danger.

Group immunity determines community epidemic disease patterns. In a previously unexposed community, an acute infectious disease can infect the entire population. If the disease has the property of life immunity, for a time, the community will have limited susceptiblity. As children are born, as unexposed persons migrate to the community and as persons with immunity die the number of susceptibles rise. A recurrence of the disease will infect largely the susceptible, principally those children born after the previous epidemics. Most communities are constructed of individuals with varying degree of exposure to specific disease organisms. Some individuals are immune while others have partial immunity. The rest may have no immunity. The proportion of a population with little or no immunity determines societal susceptibility to a particular disease. The proportion of societal susceptibility is constantly changing as the susceptible individuals are infected, acquire immunity, recover and in the case of some diseases, lose immunity again.

Genetic engineering offers an appropriate explanation for the devastating impact new diseases had upon the new populations. The immune system in human depends on genetic homogeneity and heterogeneity. Genetically homogeneous population develops weaker immune system than genetically diverse population. DNA evidence supports this. The locas (chromosomal position) of part of the immune system and the major histocompatibility complex antigens (termed as class I and class II MHC glycoproteins) has been identified. The chromosomal position clearly demonstrates that class I and class II MHC glycoproteins are transmitted genetically. An individual immune system depends upon the glycoproteins to identify and assist in the destruction of micro-organisms.

Class I and class II MHC glycoproteins function as the body's defenders against cellular invasion. Evolutionary selection favours viral epitopes (and other pathogenic infections) that survive the immune system's counterattack. The greater the variation in MHC glycoproteins the greater the resistance to invasion.[51]

An individual can have up to 6 different kinds of MHC glycoproteins. Populations, in general, have maximum variation in MHC glycoproteins than individuals. Different populations differ spectacularly in the number of types of MHC glycoprotein presence. For example, 40 different types of class I glycoproteins have been identified in sub-Saharan African population, 37 in Europeans, 34 in east Asians but only 17 in a population of North American Indians and 10 in a population of South American Indians. By implication, the lack of genetic heterogeneity among North and South American Indians catalyses the rapid spread of pathogens among these people with great fatality.

This evolutionary biological postulation is one of population genetics rather than individual genetics. Two individuals from different populations may face the same risk of morbidity and mortality but the population with greater genetic heterogeneity will be at less risk. It can, therefore, be simply derived from this genetic biological argumentation that a South American Indian population (with 10 class I MHC glycoproteins) have 32 per cent chances of not encountering a new type of class I MHC glycoprotein as 'A virus passing between two people'. Among sub-Saharan Africans (with 40 different class I MHC glycoproteins), the possibility of a virus not encountering a different class I MHC is 0.5 per cent. This implies that micro-organisms are much more virulent among American Indians with minimum class I MHC diversity than among Africans, Europeans and Asians with a comparatively large number of different class I MHC glycoproteins.

Malnutrition, unhygienic and unsanitary conditions and personal uncleanliness have an adverse effect on individual as well as on the population immune system. Malnutrition increases the frequency and fatality of the infectious diseases and World Health Organization (WHO) prescribed an adequate diet as the most effective 'vaccine' against most diarrhoeal, respiratory and other common infections. The infection rate of measles are high in all social classes but seriousness and death depends largely on the health of the child and is much increased in the poor.[52] The ultimate effect of an infection depends to a large extent on the nutritional state of the animal at the time of exposure to the agent. An acute deficiency of almost any of the essential nutrients markedly influenced the response of the host to the effect of the infections agent. When an animal is well nourished, the same infection may be mild or even unapparent, but when malnourished, the infection may be virulent and sometimes fatal. Malnutrition influences infection through four ways: by effects on the host facilitating initial invasion of the pathogens (the agent of infection); through an effect on the agent once it is established on the tissues; through an effect on secondary infection; and finally, through an effect of reducing convalescence from infection.[53]

Environment, too, has either positive or negative impact on the immune system of a person. Defective social and personal hygiene, difficult environmental and insanitary living conditions speed up the transmission of an infectious disease from one host to another.[54] Infective organisms have been continuously exchanged between man and other animals. The consequences of this exchange can be determined by climatic, environmental, meteorological, hygienic, social, bio-cultural and other conditions.

MARITIME TRANSMISSION OF
MICRO-ORGANISMS AND DISEASES

Maritime populations were much more at risk of death than land based populations.[54] This was primarily due to three reasons: Maritime sickness, shipwreck and naval and piratical engagements among different maritime competing nations. To this must be added three other most potent maritime mediums promoting the multiplication of micro-organisms at a faster rate: Overcrowding of the ship, unsanitary and unhygienic ship environment, and finally, frequent ecological changes where exposure to extremes of heat and cold, storms and the movement of the ship had direct effects on health. Consequently, most of the trans-Atlantic bound Euro-mercantile and naval ships of the seventeenth and the eighteenth centuries appeared as 'floating slums' in which different kinds of diseases spread rapidly.[55]

Oceanic and transoceanic transportation of merchandize and men produced ideal circumstances to facilitate the intercontinental transmission of diseases. The embarkation point in maritime organizations became the first site of carrying infectious diseases by passengers which once introduced would spread and multiply rapidly among the passengers. Maritime morbidity and mortality were further aggravated by the enclosed nature of shipboard society.[56] Chronic overcrowding with appallingly lack of elementary hygiene were the worst instinctive characteristics of the seventeenth and the eighteenth centuries merchant ships. When such merchant ships changed their ecological and climatic zones and entered into a new environment, normally, to tropical zones, they became 'pest-houses' near the equator. For the rest of the journey, they were just the 'floating slums' and 'their occupants died like flies'. They had to endure extreme hot, bitter cold, tropical rain and seasonal storms that wet and chilled them to the bone especially those who had no place to sleep except the upper deck.[57] Hence sudden transitions of climate and ecology are productive of certain diseases in races and individuals.[58] Lack of constant ventilation, lack of personal cleanliness and hygiene in all its branches, unsanitary living conditions and the difficult environment on board ships were ideally suited for the generation and multiplication of micro-organisms.

Undernourishment, due to inadequate ration scale, consisting chiefly of biscuits, salted meats, dried fishes, wines, water, etc., on board ship, dramatically, increased the frequency and fatality of infectious diseases. Chronic lapses were reported in distributing rations. Fresh fruits, fresh vegetables and fresh milk were naturally lacking after the first few days. But even when there was an abundant supply of fresh greens, the feeding conditions on board remained largely susceptible to infectious diseases of some kind. Fresh meat possibly could only be had at port. It was possible to carry a few cows and kill them whenever the need arose. Normally, no provision was made for keeping cows on board ship, so, they were kept on the open deck and soon suffered from broken legs and exterminated. Chickens taken on ships were kept along the sides of usually half-flooded decks. They, therefore, were either drowned or contracted some specifically pullarian disease and died. Their remains were consumed by the sailors. Cheese was available for consumption but in the tropics, its purity was maintained for a limited time. If consumed, it became a source for quite a number of rarely fatal 'Unpleasant gastric ailments'. Morbidity rate, therefore, increased considerably and incapacitated the patients to perform their duties. The famous French technique of 'potage' had not been adopted by the Dutch, English and the Scandinavian ships. The European seamen except the French received dried vegetables, either peas or beans in the form of soup, besides their daily ration. Its nutritive qualities were vanished due to faulty cooking methods. That was why, they called it the 'vegetable water'.[59]

To supply sufficient uncontaminated, drinkable drinking water to the ship board society was another reason promoting the spread of infectious diseases. After a week of the onset of ships from the port, the drinking water speedily putrefied by mutation and multiplication of varieties of bacterial life in it. No matter what substances were added to counteract the bacteria, and to keep the water fresh, sooner or later a greenish or brownish slime began to ooze from the spigot of the water barrel. The situation became worse by unsanitary habits of the crewmembers using the water. Typhoid fever seems to have been endemic on most of the Indiamen due to contaminated water. At Asian and African ports the Euro-maritime people consumed large amounts of polluted water that contained billions of microbes that invariably led to outbreaks of dysentery and typhoid fever. Despite use of a microscope, the Euro-ship surgeons failed to discover the tiny microbes responsible for these dreadful scourges. They prescribed drinking wine or gin or rum as an alternative to drinking water. The beers of those days were useless.[60]

Initially alcohol was recognized as resisting the ravages of tropical fauna and flora and hence moderate alcoholism was prescribed by maritime surgeons. The French navy especially supplemented wine for water but the experiment was a ghastly failure. The vinous experiment, therefore, had to be discarded almost as soon as introduced. Later maritime surgeons unanimously declared that even moderate doses of alcohol tended to lower the resistance power of the maritime labourers and when taken in excess it certainly reduced the resistance power. This was found particularly true in cases of pneumonia and tuberculosis, to which many alcoholics fall victims. Alcohol of poor quality and the inflammatory kind was distributed on board the Indiamen as daily allowance that killed seamen and natives directly by poisoning, promoting chronic nephritis, stertorous breathing and indirectly as 'Being a cause of quarreling and sickness.'[61]

The maritime surgeons forbade alcoholism in warm climates. Excessive wine consumption, coupled with excessive heat of the tropical climate, little food of briny nature, and the intense cold and dew of the nights increased a suicidal tendency in the patients, often compelling them to go 'raving mad, and oft-times running in large crowds towards the sea and rivers' to destroy themselves. Post-mortem examination by Dutch surgeons confirmed the location of a worm in the brains of such patients.[62] Dr. William Twining recorded a degree of venous congestion in the brain and spinal canal arising as a result of drinking wine up to a degree of intoxication in the tropics while A. Balfour and H.H. Scott noticed vile decoction.[63] The heat of the tropical middays alone was sufficient precondition to cause fatal affections of the brain.[64]

Maritime exchanges carved ideal environment for waxing and waning nutritional diseases. The lack of knowledge of the maritime people about the constitution of good nutrition worsened the health situation on shipboard. The possibility and potentiality of dietetic variableness on trans- and inter-oceanic bound ships were relatively limited. The maritime labourers, therefore, became shorter and suffered considerably from anaemia. Vitamin and protein energy deficiency (malnutrition) contributed to the soaring rate of maritime mortality.

The classic shipboard disease, however, was scurvy, triggered by vitamin C deficiency. Approximately 30 weeks of vitamin C deprivation was productive of the classic scurvy with bleeding gums. Lack of fresh vegetables, fresh fruits and milk were responsible for acute vitamin C deficiency from which half of all the patients on board died.[65] It remained the scourge of seamen for more than 300–400 years.[66]

Ascorbic acid is essential for many oxidation reactions in the body. Ascorbic acid also plays an important role in formation of hydroxyproline, an integral constituent of collagen which, in turn, is essential for growth of subcutaneous tissue, cartilage and bone. It enhances the removal of iron from cellular ferritin, thereby, increasing the concentration of iron in the body fluids. Ascorbic acid contributes significantly in increasing intercellular cement substance between the cells, the formation of bone matrix and the formation of tooth dentin.[67] Acute efficiency of scorbutic acid for 20 to 30 weeks cause scurvy. One of the most serious effects of scurvy is the failure to heal. This is because of the failure of cells to deposit collagen fibrils and intercellular cement substance. Another consequence of vitamin C deficiency is the cessation of bone growth. The cells of the growing epiphyses continue to proliferate but no new matrix is laid down between the cells and the bone fracture generally easily occurs at the point of growth because of the failure to ossify. Ossified bone failure reduces osteoblast secretion, a new matrix for the deposition of new bone. Consequently, the fractured bone does not heal. This was precisely the reason that European ship surgeons generally operated and cut the leg of crew members whose legs were smashed by a bullet or a runway gun in the hold. Hence, surgical skills improved extraordinarily among the surgeons serving on European merchant ships.[68] The blood vessel walls become extremely fragile in scurvy, presumably because of failure of the endothelial cells to be cemented together properly and failure to form the collagen fibrils normally present in the vessels walls. The capillaries especially are likely to rupture and as a result, many small petechial hemorrhages occur throughout the body. In extreme scurvy, the muscle cells sometimes fragments; lesions of the gums with loosening of the teeth occur; infections of the mouth develop; vomiting of blood, bloody stools and cerebral hemorrhage occur, and, finally, high fever develops before death. After a few weeks the patient may suffer from pneumonia, or kidney diseases or some other complications and dies.

Beriberi is a vitamin B deficiency disease more specifically thiamine (vitamin B1). The first reference to the disease occur in Portuguese documents from Maluku in the mild sixteenth century and that region continued its reputation for this scourge even in the seventeenth century.[69] Rice husks contain plenty of thiamine, but in course of centuries people adopted methods to strip away the husk to make rice more palatable, appealing and amenable. Traditional hand-milling technique of rice produced the neurological and cardiovascular symptoms of 'dry' and 'wet' beriberi for many people throughout the world. This led to infantile beriberi, which was practically always fatal for nursing babies and toddlers with thiamine-deficient mothers. The disease was also prevalent among those whose diet was confined to white bread. The disease was more particularly prevalent among institutionalized populations—such as slaves on plantations, prisoners, children in orphanages, those at sea for long periods of time, etc.[70] The vitamin and protein deficiency diseases gradually caused a loss of strength, severe out burst of mental depression, pain in the muscles, pneumonia, kidney disease and some other complaints.

Maritime transportation, therefore, remained the most potent conduit of inter-oceanic as well as trans-oceanic transmission of diseases. Different regions of the Indian Ocean can well be defined on basis of its specific diseases. The same applies to trans-Atlantic regions. When different regions of the Indian Ocean evolved interacting networks, and frequently, developed closer maritime connectivity throughout history, region, climate, weather, environment, ecology, and gene specific diseases were transmitted to other regions of the Indian Ocean, initially with a high fatality rate. During 1500-1800, interacting networks were developed by the maritime powers of Europe to connect directly or indirectly India with the trans-Atlantic regions of the world. This connectivity became the channel of trans-Atlantic transmission of new fatal diseases.

Intra-Indian Ocean Transmission of Diseases

Transmissibility is a biological process of capacitating the micro-organisms to transfer themselves to a new host. It varies greatly from pathogen to pathogen. The modes to transfer of pathogenic agent are quite different. Epidemiological historians demarcated four general means of pathogenic transmission: contact transmission (direct and indirect contact); common vehicle transmission (such as food and water); airborne transmission by droplet nuclei or dust; and vector transmission usually by insect vectors that transmit a pathogenic agent from one host to another of same or different species.[71] The intensity of the transmissibility of a disease depends upon two factors: the period of communicability and the period of incubation. The time interval required by a disease pathogen to be transferred to another host is known as the period of communicability. The period required to shed disease organisms into the environment and to transmit it to new host is called the incubation period. During the period of incubation, the new host is often asymptomatic and incapable of transmitting the disease to others. Hence, the incubation period is followed by a period of communicability as a general rule during which the new host experiences morbid effect of the infection and sheds organisms into the environment. Disease symptoms are generally and most clearly observable during the period of disease communicability. Some hosts may have subclinical infections in a subconscious state and shed disease organisms that infect others. There are many diseases which have a long incubation period. In cases of such diseases the period of communicability begin a little before the diseases become clinically evident. These diseases spread great distances in relatively shorter time intervals. Smallpox, measles, etc., are such diseases.

The transmissibility of a disease increases by its ability to adopt a new environment and to live outside its hosts for longer periods. Such micro-organisms develop great survival ability in droplet nuclei or in faecal material. Such micro-organisms' capacity to infect a new host varies considerably with different diseases and with environmental circumstances.

The virulence of a disease is a measure of its ability to injure its hosts. There are two aspects of virulence: first, pathogenicity, which refers to the severity of a disease as measured by morbidity and mortality; and second, invasiveness, which refers to the capacity of the parasites to invade tissues. Smallpox, for instance, was a very virulent disease as it caused high morbidity and mortality in its host population. The virulence of a disease, however, has considerable variability from one host to another. The virulence of many important diseases has changed in the course of time. Virulence is also affected by environmental conditions. The Indian Ocean world has been marked by diverse ecological zones with its definite disease syndrome offered ideal conditions for intra-zonal transmission of diseases. When diverse ecological zones of the Indian Ocean interacted with each other through human activities, the bio-spheres of the interacting ecologies were transformed pathogenically. This stimulated intra-Indian Oceanic transmission of diseases with fatal consequences initially on human population. The intra-zonal interaction was established through direct and indirect mediums. India, during the sixteenth and seventeenth centuries, had direct expanding maritime traffic with some of its Indian Oceanic neighbours while with the rest it had indirect maritime traffic through inter-zonal medial ports. The cosmopolitanism of the medial ports also enabled local as well as intra-zonal pathogenic exchanges. Major, middle and minor international emporiums of the Indian Ocean can be recognized where large numbers of people from different local communities and nationalities assembled producing ideal conditions for the transmission of infections and their dissemination when people returned to their native places.[72]

The Indian Ocean, well before 1500, had been a significant arena for the transmission of diseases. Unfortunately, only meagre qualitative data is available. Our treatment of the subject, therefore, has to be inevitably highly sketchy, and selective. For a variety of reasons, including the early domestication of animals and plants, their dense populations, their ecological and social diversity, constant interregional maritime contacts, India, Africa, the Middle East, China, middle Indian Ocean, etc., had long been the direct or indirect source of transmitting epidemic diseases that periodically penetrated into other parts of the Eurasian landmass.[73]

Some idea about the diseases suffered by crew members of Indian ships bound to western Indian Ocean regions can be gained from a sea-travelogue in Persian, the *Anis-ul-hujjaj* of Safi bin Wali Qazwini.[74] *Anis-ul-hujjaj* warns the ship's passengers especially those with bilious (*safārwi*) humour to avoid staying on the outer side of the deck, since it is productive of giddiness. Staying near the main mast was prescribed to minimize the impact of all movement and to avoid accentuating sea-sickness.

Provision for fresh water was another problem discussed in *Anis-ul-hujjaj* in some details. Stored water in the *fintas,* in course of time, often became distasteful and contaminated. Stored water was made palatable by the addition of sour juice of some fruits or syrup of pomegranate. On Indian ships, fresh water was procured through buckets or leather bags. Euro-surgeons forbade drinking water which had been 'inside a mussuck or any leather vessel'. It is hard to clean it properly once the vessel becomes tainted with foul water.[75] Such contaminations were capable of spreading gastro-intestinal infections. Not surprisingly, many sailors developed inguinal hernia as a result of handling heavy weights of water although *Anis* had not noted such patient. This ailment was more frequently prevalent among European sailors who had to handle heavy weights.

The preservation of provision was primarily a Euro-maritime phenomenon, a long duration voyage problem from Europe to India, often, lasting not less than eight to nine months. Indian ships, normally, undertook relatively shorter voyages some times lasting 40 to 50 days. The problem of preserving salted and smoked fish, imported from Thatta, in Sind and Maldive islands respectively was undeniably acute on Indian ships also. Diverse preparations of meat also faced the same problem on Indian ships. Our contemporary sources were silent in this respect. Another major risk relating to the preservation of provisions on Indian ships went unnoticed by contemporary chroniclers. The harm often done to cereals by water especially during bad weather, though productive of gastro-intestinal, liver, kidney and other diseases, found no mention in our sources. The menace caused to provisions, timber and human health by infesting rats on Indian ships has also been largely neglected.

In terms of cleanliness, Indian ships may have been the worst. To avoid foul odour and terrible stench, which were further intensified by vomiting and retching all round, affluent passengers on Indian ships carried perfumes of different kinds, especially those derived from *gulab* (rose) and *bed-I mushk* (*salix syqostomon*). There are inferences to cleanliness or washers as part of the crew in the *Anis–ul-hujjaj* but no details have been given.[76]

Maritime diseases such as scurvy, diarrhoea and nausea have not been mentioned by *Anis-ul-hujjaj*. Abul Fazl's list of the crew has no place for a maritime surgeon. *Anis-ul-hujjaj* emphasized the need of a *tabib* (physician) and *fassad* (blood-letter) aboard an Indian ship. We have no solid clue to suggest that Indian ships were necessarily provided with the services of a maritime surgeon as was the case with the Euro-mercantile marine. Perhaps this explains why our author advises travellers to carry some medicines with them, a long-standing maritime tradition in India. Sour fruits and juice were prescribed for the bilious humour while sweets such as honey and sugar for phlegmatics. Guard against 'empty belly' and 'overeating' were prescribed to overcome conditions producing giddiness and retching due to the rolling and pitching of the ship.

Indian officials and wealthy passengers suffered least from maritime diseases as a result of their keeping personal medical luxuries. Even, the ordinary Indian sailors hardly succumbed to scurvy, at least, not in its most hideous form, because they generally undertook short duration voyages. Even a long voyage, for instance, from Bengal to Persian Gulf or vice versa, was broken often at regular intervals by calling at ports on the east and west coasts of India, on western Indian Ocean ports and so forth for natural botanizing, and for loading the ships with locally produced fresh fruits and vegetables. The same applied with return voyage. Nicholas Downton, arriving from the Persian Gulf in 1611, found limes, plantains, water melons, radishes and cucumbers available at Swally Road (Surat). Similar articles were also available at Dabhol.[77]

Calenture, a tropical malady described by Keeling as fever and delirium, in which the victim imagined the sea to be green fields, into which he was impelled to jump. *Anis-ul-hujjaj* referred to a form of sickness which brought about a violent change of temper (*taghaiyur-i dimāgh*). This can be taken as equivalent of the occurrence of calenture.[78]

The other major health hazards of the Indian Ocean conveyance were natural and manmade. Tempests, severe storms at the setting of the north-east monsoon and at times, in the south-west monsoon[79] calms, etc., were natural Indian Ocean health hazards. During tempests and severe storms accidents happened and injuries were received by seamen and passengers. Among man-made health hazards, piracy was foremost. In piracy, both professional buccaneers and amateur pirates were involved. High sea piracy seldom came from Indian pirates; the Indian pirates were confined mostly to coastal areas. Europeans spared outgoing Indian ships but plundered incoming Indian ships from foreign ports laden with vast treasures of silver, gold, copper and tin. When such piratical acts were resisted, wounds were received by both parties. Sometimes bullet injuries were shared as Abul Fazl and *Anis-ul-hujjaj* insisted the presence of *banduqchis* (gunners) on Indian ships. To treat injuries of maritime accidents required surgical aids. Maritime piracy and the growing use of gun powder, lead bullets and other projectiles increased the need of surgical operations on Indian ships. General amputations and amputations of thighs were also required on Indian ships. Unfortunately, we have no mention about Indian maritime surgery in our original sources.

Any speculation based on guesswork because of the dearth of archaeological data about the timing of the appearance of new killer pathogens has to be frequently contradictory. Historians of epidemiology have still speculated. Their reasoning implies that with the emergence of big cities with populations as large as 50,000, and substantial cattle herds around 3000 BC in Mesopotamia and Egypt and perhaps later India (Harappan culture), new pathogens including smallpox spread to humans.[80] William McNeill identified 500 BC as the time when pathogens started influencing the growth of civilizations in Asia and Europe. Micro-parasites of smallpox, chickenpox, mumps, influenza, diphtheria and numerous others quickly and directly transmitted from human to human. Marauders, merchants, porters, missionaries and marching armies, moving from one ecological set up to another also linked their pools of pathogens. Thus, one people's familiar disease became fatal for another people. The invasion of a novel pathogen would have been epidemic with a dramatic reduction in population as the most susceptible individuals were eliminated. With the increase of mass immunology, the illness themselves would have disappeared, and attacked other populations, that had a larger non-immune population, often near the limits of their food supplies.

With passage of time, there developed disease centres in the Indian Ocean from where diseases rotated intra-Indian Ocean regions. The earliest disease centre of the Indian Ocean was the Sumeria that appeared about 5000 years ago in Southern Mesopotamia in the flat lands around the lower reaches of the Tigris and Euphrates rivers. The Sumerian farmers were also the first to cultivate unintentionally the villains of the animal world, their competitor for the food and shelter. Lice, fleas, internal parasites, mice, rats, roaches, houseflies, worms and a plethora of others were the villainous competitors of the humans who carried diseases. Modern medical technologies attest that rats are carriers of plague, typhus, relapsing fever, and other infections, and the other varmints are carriers of hosts of infectious diseases. The Sumerian peasants and herdsmen also stimulated the incubation of worst invisible micro-predators: fungi, bacteria and viruses. Crowd and filth diseases made their roads in fields, flocks and cities of Sumeria. They shared the diseases of the domesticated beasts as a consequence of sharing commonly the water, air, environment and habitat. The synergistic effect of these was the production of new diseases and more potent variants of the old ones. Hence pox viruses produced smallpox and cowpox; distemper, rinderpest and measles were produced by viruses exchanged among dogs, cattle and humans or by a combination of different viruses; humans, pigs, horses and domesticated fowl in contact with wild birds shared and still share influenza and so forth.[81]

With urbanization in different ecological zones of the Indian Ocean, disease centres multiplied. In the evolution of the disease centres in the different ecological zones of the Indian Ocean, Sumerian culture made considerable contribution at least since 3000 BC. The process started intensifying since 2500 BC when Sumerian ports developed maritime trade with Harrappan and Mediterranean peoples.[82] The Mesopotamian contribution to the evolution of diseases, of new disease centres in the Indian Ocean was two-pronged: first, it transferred many originally Mesopotamian diseases to new disease centres in its original form; and second, highly adaptive Mesopotamian pathogens of diseases combined with different viruses, bacteria, and other micro-organisms of the new disease centres to produce more potent new diseases or variants of old ones. The maritime and overland intermingling of Mesopotamian peoples with the peoples of the other disease centres of the Indian Ocean jeopardized and exposed immunologically innocent people to chronically obsolescent disease microlife. The process was not unilateral, rather bilateral, affecting both the cultures unequally and unevenly. Maritime peoples of the disease centres of the Indian Ocean suffered the most. Literary and anthropological sources from ancient Middle East especially from the Philistines and Hebrews, Israelites, Egyptians, etc., would attest the concomitant spread of communicable diseases.[83]

India emerged as the major centre of transmitting some killer diseases to intra-Indian Ocean regions from very early times especially the western Indian Ocean region up to the tips of east Africa and through this to the interior of Africa. Africa, too, served as a dynamic centre of transmitting disease in the reverse direction. The whole of the Arab world evolved as one of the intermediate centres of transmitting disease in all the directions especially in western disease environments and disrupted the existing equilibrium, the breakdown of isolation in Europe, South Asia and the middle Indian Ocean region. Similarly, the middle Indian Ocean region was the other intermediate centre spreading diseases in South-East Asia, East Asia, some parts of South Asia and so forth. Maritime connectivity between India and east Africa, India and east Mediterranean and west Mediterranean regions was regulated by Arabic maritime entrepreneurs up to AD 1500.[84] Intensifying maritime trade of India with western Indian Ocean regions breached previously isolated areas, and the disruption of the equilibrium simply meant the introduction of new pathogens and new strains of bacteria, protozoa, and viruses in the Indian Ocean region.[85]

The process of systematic and regular transmission of diseases directly to intra-Indian Ocean regions and indirectly to Europe started taking a definite shape from 500 BC. The micro-parasites of smallpox, diphtheria, influenza, chickenpox, mumps and numerous other diseases passed directly from human to human and needed no intermediary carrier. Maritime, military, diplomatic, religious and cultural missions became the major channels of transferring these diseases directly from one disease zone to another by linking their pools of pathogens. An invasion by novel pathogens often resulted into a massive epidemic that dramatically reduced the population by eliminating the most susceptible individuals. The famous green historian Thucydides comprehensively describes the epidemic that begun in Africa, spread to Persia and reached Greece in 430 BC. He claimed that it killed initially 25 per cent of the Athenian forces; then lingered in southern Greece for the next four years, killing up to 25 per cent of the civilian population. What the epidemic was it is hard to classify, but on the basis of symptoms described, plague, smallpox, measles, typhus, or even syphilis and ergotism seem to be the most probable candidates.[86] The Roman empire between AD 165 and 180 suffered massively from some epidemic most probably from 'Antonine plague' killing from a quarter to a third of the populations in the infected regions. The second epidemic struck Rome between AD 211 and 266.

The population of South Asia, the Middle East and East Asia also shrank by widely shared pools by diseases that rotated outwards to incorporate other intra-Indian Ocean regions into their vortices. Japan presented a classic example in this regard. Before AD 552, Japan seems to have escaped the epidemic diseases. Smallpox ravaged Japan in 522 after the Korean Buddhist missionaries visited the Japanese court. In 585, either smallpox or measles appeared in Japan to kill new, non-immune Japanese population. During 700-1050, Japan's 'age of plagues' developed. During eighteenth century, Japan was rocked with 34 epidemics, in the ninth with 35, in the tenth with 26 and in the eleventh with 24. Smallpox was well represented.[87]

The process of systematic and regular transmission of diseases to intra-Indian Ocean regions operationalized more and more rigorously after AD 1500, when Eurasians actively engaged in intra-Asian trade. Military and cultural conduits, too, contributed considerably in this pathogenic transfer. The emergence of India as one of the giant centres of world trade during the seventeenth and the eighteenth centuries, transformed its role as a disease centre of the Indian Ocean. Since the seventeenth century, India acted as a terminal not only for the transshipment of merchandise but also for the transfer of diseases. During 1670-1800, Bengal acted as the most dynamic disease centre with Calcutta as its pathogenic capital.

The hegemonic power of the modern world system during their respective tenure of hegemony contributed remarkably to the intra-Indian Oceanic exchange of micro-parasites of disease by intruding commercially and pathogenically the unknown regions of the Indian Ocean. Cores of the modern world system also played equally an important role in intra-Indian Oceanic epidemiological exchanges. Cape Town, Mauritius, Japan, China, East Africa, middle Indian Ocean regions, South China Sea, etc., were linked to Bengal epidemiologically by Euro-maritime entrepreneurs participating in Bengal's Indian Oceanic trade. Maritime cities and major and minor port cities and emporiums of the Indian Ocean functioned as major and minor centres of transferring and receiving diseases.

The origins of the variola virus (smallpox) are obscure and lost in the mists of the past. Smallpox was not described by Hippocrates and probably not known to Galen or his successors either. The earliest western references seem to come from early medieval chroniclers, notably Gregory of Tours most probably during the 540-94. The great physician and alchemist Abu Bakr Muhammad ibn Zakariya' al Razi (835-923) is still considered to be the authority on the variola and measles. His *Kitāb al-jadarī wa'l—hasba* is the landmark on smallpox in the history of medical science.[88]

Unsurprisingly, smallpox traveled to intra-Indian oceanic regions in either an active or inactive form through migratory populations such as mariners, maritime merchants, porters, pilgrims, soldiers, etc. The resilient virus of smallpox managed to have precipitated epidemics even after lengthy oceanic voyages. Persons infected by the disease might sicken and die, or recover during a long voyage, but the virus has the capacity of persisting even in clothes and beddings. Smallpox broke out in ports where ships without infectious persons on board docked and sent laundry ashore to be washed.[89] As a consequence, smallpox reached to Japan quite early in their history from East Asia most probably from China, when Japanese contact with China was infrequent.

Mariners, maritime traders and porters became the main agent if not the exclusive ones to spread smallpox, cholera and plague to East Africa from India. By AD 1600, the Africans encountered smallpox. Smallpox was the most dangerous epidemic often called 'murderous': the choice obviously was for the 'variola major.'[90] Smallpox was introduced in South Africa from India in 1713 caused epidemics there. The Dutch infected ship returning from India introduced smallpox at Cape Town in 1713. Smallpox reappeared there in 1755 and 1767. Maurice, the French slaver, who visited Kilwa in 1775, postulated that smallpox was an epidemic there and people knew of inoculation.[91]

Smallpox appeared in Siberia in 1630 by crossing the Urals from Russia. From the sixteenth century, smallpox was the most feared disease because of its 'swift spread, the high death rate and the permanent disfigurement of survivors'. The death rate in a single epidemic could soar past 50 per cent. When it first struck Kamchatka in 1768-9, it killed two-thirds to there-fourths of the local population.[92]

Three most probable routes of transferring smallpox to Russia, and then to Siberia, via Urals can be demarcated. First, in the sixteenth century, men from the west went to Siberia via the Urals in search of furs demanded by upper class and rising bourgeoisie of Western Europe where smallpox was transmitted in all probability from the East Mediterranean region.[93] Second, from eastern Mediterranean smallpox was transmitted to Astrakhan to Siberia via the Urals. Third, from India it was transshipped to Isphahan to Astrakhan and from there to Siberia. After 1556, India emerged as one of the principal trading partners of Russia. From now onwards the focus shifted from the Ottoman empire to Iran and Turan and India.[94] Smallpox remained epidemic in Siberia and its periodic depredations, every ten or twenty or thirty years was so murderous that the 'young were totally susceptible, and an entire generation could be lost in a few weeks'.

Even during the fifteenth and sixteenth centuries, smallpox was widely prevalent in India, killing more people here than in other countries. Dacca's civil surgeon Dr. Wise noticed that out of 16 maharajas of the Tipperah family, five died of smallpox between the fifteenth and the eighteenth centuries. During the eighteenth century, smallpox was widely prevalent in India and killed people in larger numbers here than elsewhere.[95] In London, in the eighteenth century, the annual death rate was, on an average, 4,000 per million whereas in Calcutta, in 1833, it was 11,796 per million; in 1838, 6976 per million; in 1844, 13,537 per million; in 1849, 7,981 per million; in 1850, 10,721 per million; in 1857, 8,237 per million; and in 1865, 29,250 per million. In London, smallpox was responsible for about 12 per cent of the total deaths in the eighteenth century whilst in Calcutta, between 1 January and 31 December 1849, it was responsible for 13 per cent of all the deaths; and 47 per cent of all the deaths between 1 January and 31 May 1850.[96]

Cholera was another epidemic transmitted ocean-wide from Bengal where it was endemic for at least 2000 years. It became pandemic and transferred from Bengal to all over the Indian Ocean. About 50 per cent of those affected by cholera usually wiped out in these great pandemics.[97] The severity of cholera was much more intense in India in the seventeenth century than in the eighteenth. It was attributable to moist heat, a damp season, repressed perspiration, excessive intake of chills, indigestible articles of food, chiefly to vegetables, though occasionally to flesh and fish.[98] Some doctors found insanitation chiefly in the cities like Calcutta as the root cause of the outbreak of epidemic cholera. At places of pilgrimage cholera broke out primarily due to bad food, perpetual prevalence of insanitation, unscientific drainage system and so forth.

In 1882, Dr. Koch, the German scientists, diagnosed the stool specimen in Egypt at the site of a cholera epidemic and found a vibrio-like organism. Cholera micro-organisms produce a toxin that damaged the glut lining, leading to diarrhoea and rapid, often fatal dehydration. *Vibrio cholerae,* a bacterium, whose survival is dependent upon its ability to hibernate, shrivelled to one hundred and fifty times smaller than its normal size, beneath the mucous coating of algae. Cholera is, therefore, endemic around the shores of Bengal and Bangladesh. Recent epidemiological and ecological researches have shown that algal reservoirs for cholera are very responsive to temperature, rainfall and nutrient loading in waste waters. As the nutrient load to the onshore water in the Indian Ocean was increasing, at a faster rate, at least, since AD 1600, from various sources, algal blooms were becoming more extensive. Global warming has also catalyzed this growth.

Euro-surgeons and physicians who came to India in the sixteenth and seventeenth centuries described cholera in their reports and books. Correa, in 1543, in his *Lendas da India* and Gracia da Orta, in 1563, described cholera as they saw it at Goa.[99] Charles E. Rosenberg is of the opinion that prior to the 1830s, physicians in India, had only seen cholera.[100] Only scattered reports of the occurrence of cholera, therefore, are available till the last quarter of the eighteenth century. It is assumed that cholera in epidemic from did occur, but the cases were not recognized as such because of the unfamiliarity of the Euro-surgeons and physicians with it. That was why, Macnamara wrote: 'Our possessions in India prior to 1781, was surrounded by large provinces regarding whose inhabitants, we had literally no knowledge whatsoever; in these territories the course of the epidemic could not possibly be traced.'[101] The earliest notice of this epidemic has been given in a letter from Dr. Tytler, the Civil Surgeon of Jessore on 23 August 1817.

The extreme dampness of the months from the end of June to the end of September or early part of October gave rise to cholera in Bengal, Assam, Bihar, Orissa, etc.[102] Because the defective elimination of hydrocarbon by the lungs, under diminished heat generation burdens the liver of the elimination of these elements.[103] Usually, Europeans were comparatively more resistant and suffered less from cholera.[104] Reports of the Marine Surveys of India (1876–80) are full of cholera outbreaks with fatal effectivity.[105]

Cholera radiated outwards in almost every direction from lower Bengal. The disease advanced along three main routes. First, overland route lying along northern India from Bengal to the Punjab, thence via caravan routes into Afghanistan and southern Russia, Poland, Germany, France and Britain, where it occurred in the early 1830s for the first time. Second, overland but least frequented route was from Bengal to Nepal, Tibet and China. And third and more commonly followed route was the expanding maritime traffic between India and its Oceanic neighbours. It moved with mercantile and troopships eastwards from Bengal into Burma, Malay, Thailand, Java and rest of middle Indian Ocean and South China Sea islands, and on to East Asia, reaching up to Japan in 1822.[106] In the southern direction, it reached to Sri Lanka through peninsular India. It swept southwest from Calcutta or Madras to Mauritius. On the East African coast, cholera was recorded for the first time in 1821. It reached there from India to Macca, to the Gulf of Aden and Red Sea ports through coastal shipping. In the nineteenth century, cholera was introduced into the East African interior.[107] It also transmitted to Egypt from Bombay via the Red Sea and southern Arabia. From Egypt, it reached to the eastern Mediterranean and Western Europe. In 1819, Burma came under its influence. In 1819, Siam and Malacca were inflected when a mercantile ship reached there from the Bengal coast. Subsequently, Penang, Singapore, Java and the Phillippines were in the grip of cholera. Some Dutch mercantile marines coming from Calcutta had anchored at Malacca spreading cholera there.

Three important factors were under operation in disseminating cholera in different regions of the Indian Ocean. First, an important factor in the westward movement of cholera from India to the Middle East was the Hajj, the Muslim pilgrimage to Mecca. In a sense cholera was starting its tolls where the bubonic plague had barely left off. In fact, in many places the two diseases followed identical trails—the customary lines of pilgrimage and trade.[108] Through pilgrims cholera was transposed to Mecca in 1831 and then to other parts of the Middle East. 13 per cent of the population of Cairo had been wiped out by cholera. After 1831, cholera visited Mecca many times by Muslims from south and South East Asia, who reached there in large numbers mainly by maritime routes. Some Muslims followed the more traditional overland routes. Between 1831 and 1912 cholera broke out among pilgrims at Mecca 27 times, on average, once every three years—though never so severely destructive as in 1831, 1839 and 1865.[109]

The second factor in the transit of cholera was the movement of troops around the Indian Ocean region. The troops frequently served both as victims of the disease and the agents of its wider dissemination. The British military raids during the early nineteenth century coincided with the spread of cholera. Arrival of troops at Malacca in 1819 coincided with the spread of a cholera epidemic. In 1840, the movement of troops from India to China brought cholera during the Opium War. According to Ackerknecht troops have always played a major role in spreading cholera.[110]

Another important factor was the rise of Calcutta as a trading metropolis. Epidemiological significance of the maritime cities has so far been neglected although these cities performed much the same role in epidemiological terms as they did in maritime commercial ones. They were disease entrepots: the principal points of entry for incoming micro-organisms which were then transposed into the port's hinterlands by roads and rivers. They were also terminal points from where pathogens were exported to other parts of the ocean.

By virtue of their geographical position, ecological set-up and social environment, maritime cities might also be singularly suited as reservoirs and incubators of infectious diseases. Nowhere had this trait loomed larger than in the case of Calcutta, where water supplies were easily and repeatedly infected by cholera. Calcutta was also situated close to the centre of a zone of endemic cholera from where epidemic repeatedly erupted. The transposition of cholera from Calcutta by soldiers and sailors, pilgrims and migrant labourers to intra-Indian Ocean regions reflected the wider patterns of pathogenic mobility. Calcutta, 'the city in swamp', also served as the epidemiological capital of Bengal.[111]

In intra-Indian Oceanic transfer of *kala-azar,* India played a significant role, though its history in India itself is obscure. Philip H. Manson-Bahr has asserted that *kala-azar* is a widely distributed disease which occurs in India, especially in Assam, Madras and along the Ganges and Brahmaputra; in China, north of the Yangtse between the coasts of Peking and Hankow and Western Abyssinia, etc. Genus *Leishmania* is responsible for its spread. Two stages can clearly be discernible in its development: intracorporeal and extracorporeal. Possibly these represent respectively asexual and sexual forms. The former were normally traceable in man though some other vertebrates were also liable to the parasites of *kala-azar. Kala-azar* parasites in human were almost invariably intracellular.[112] Sand flies *(Phlebotomus argentipes),* a human blood feeder became instrumental in intra-Indian ocean spread of *kala-azar.* During the early nineteenth century, *kala-azar* became an all–India phenomenon. *Kala-azar* transferred to intra–Indian Oceanic regions by maritime traders, troopships, and so forth.

Dysentery was another disease in different forms that carried off more victims in Bengal and Assam than all the other diseases put together. Chronic and scorbutic dysentery almost invariably spread towards the beginning of the cold weather.[113] Diarrhoeal diseases are primarily water and food–borne diseases caused by organisms living in drinking water or water coming into contact with humans. Micro–organisms including bacteria (*shigellae*), viruses and protozoa, many of which might remain viable in water for months, were largely responsible for diarrhoeal diseases. These genetic characteristics were largely responsible for not only its long survival, but also for its transportation from place to place and from person to person during the monsoon.

Different pathogens of diarrhoeal diseases were productive of various affections. Flux (dysentery), bloody flux (blood mixed with stool), alvine fluxes of Bengal, hepatic affections, liver affections, chronic enlargement of the spleen and the livers, etc., are some of the most burning examples.[114]

Skin diseases such as leprosy, *Pityriasis versicolor*, pemphigus, chilblains, moist eczema, dermatitis, scabies, leukoderma were all present in India from ancient times. Most of the skin diseases were common among the poorest people. Ancient Indian physicians have discussed seven major and eleven minor forms of leprosy.[115]

Filariasis (elephantiasis) is extremely common in India where almost 60 per cent inhabitants were affected. It had been transported in all the directions of the Indian Ocean from India mainly to Africa, south China, Samoa, etc.[116]

There were other epidemic centres in the Indian Ocean with their specific pathogens that were transmitted to intra-Indian Oceanic regions by maritime transportation. One of the most dynamic epidemic centres of the Indian Ocean was the Middle East. It was from these regions that many disease micro-organisms were transmitted to intra-Indian Ocean regions including India. One such disease was the plague.

A disease called *pali* plague or *maha mari* with high death rate and sloughing of the cornea was known in India.[117] The plague originated along the edge of the Himalayas where India and China share a common border. It was here that the species of black rat (*Rattus rattus*) carrying plague seems to have originated. Bubonic plague is an example of a four-factor disease (host, agent, vector and reservoir). *Yersinia pestis* was the rodent of plague which reaches to humans through frantic infected fleas searching for another living rat after their infected host has died. After the flea bites its new host, the bacillus enters the lymphatic system and heads for the nearest lymph node giving rise to bubo in the groin, armpit and the neck, depending on the location of the bite. Historically, bubonic plague brought death to about 60 per cent of those infected. The most deadly form of all was the pneumonic plague, which usually develops from bubonic plague. This requires no insect vector for transmission because it spread quickly and directly from person to person through breath and clothes.

It is believed that pre-fourteenth century plagues transmitted from the Sino-Indian border to Egypt from where into Russia. The earliest reference to a certain kind of plague occurred in Kalhan's *Rajatarangini* (between 751–81), caused by the *luta* disease.[118] Plague's spread normally followed two routes: First, from the Sino-Indian border to the south-east, thence to India and from India through the Caucasus and up the Volga; and the second from the north-west by way of Pskof and Novgorod.[119] Early fourteenth century plague was spread from Yunnan to the rest of China by Mongol armies. The plague of 1331, reportedly caused havoc in some parts of north-east China by killing about 90 per cent of its population. By 1346, following one of the above discussed routes from where bubonic plague reached the Black Sea. Muslim merchants brought it to south-west Asia, Egypt and North Africa. While Indian merchants carried it west to Italy and then to northern and western Europe.[120]

The second cycle of plague was one of the greatest holocausts in the known world. It began to rampage in Asia around 1300 and continued until about 1800. India experienced the second cycle of plague in 1342. Another cycle of plague broke out in Bengal in 1575 and in Kandahar in 1598.[121]Many parts of northern India experienced the virulent impact of the second cycle of plague during the early seventeenth century. Salbank recorded the great ravages of 'a wonderful great plague raging in the city' of Agra in a letter dated 29.10.1616. According to the letter, the plague lasted for three mouths killing 'no less than a thousand people a day'.[122] On 14.1.1617, Master Crowther reported that this virulent plague had lasted for 100 days.[123]In 1616, a plague broke out in Bijapur. In 1684, another plague occurred in the south, a place of seldom occurrence of plague which most probably reached to Bijapur in 1689. In 1702, Nepal was attacked by plague.[124]Its ravages covered the whole country and it lasted for eight years.[125]

The third cycle started in Asia about the middle of the eighteenth century.[126] The third cycle of plague epidemic created havoc in Gujarat first between 1812 and 1821 about which we have detailed reports. It first broke out in Kutch in 1812 killing 50 per cent of the population of the region. It reached to Kathiawad in 1813 and subsequently to Rajputana. Between 1812 and 1837, it covered Gujarat, Rajasthan and central India.

Dr. Hutcheson firmly believes that the disease has spread into India from the Mediterranean especially from Levant from where Gujarati printers *(chhipis)* of Kutch, Bhavnagar and Surat have imported the pathogens of plague. Information gathered from local medical traditions of the disease infer the prevalence of two distinct diseases. In local parlance, they were termed as *ghaut no rogue* (the knotty disease), *kogla no rogue* (the expectorating disease) and *tao no rogue* (the fever disease).

The local medical traditions identified the following symptoms for *ghaut no rogue:* great and general uneasiness of the frame, pains in the head, lumbar region and joints on the day of the attack, knotty, and highly painful swellings of the inguinal or axillary glands, the parotids were affected in 4 or 5 hours; these symptoms went on increasing in violence in course of time. Suppuration of the glandular swellings occurred on the fourth or fifth day. *Kagla* and *tao no rogue* was identifiable as high fever with burning and excruciating pains, hiccough with deep and oppressing breathing ensued, considerable pains in the chest and joints and about the navel. The former most probably was bubonic plague while the later was the pneumonic plague.[127]

Contemporaneous to this, was the occurance of a *mahamari (gola* or *futkiya rog)* in Kumaun and Garhwal in the Himalayas. The first record of the *mahamari* is dated 1823 at Kedarnath in Garhwal. Dr. Hutchesion recognized this plague as endemic of Kumaun and Garhwal. John Strachey in a report, dated 2 December 1849 clearly wrote that 'for about 34 years past, there has existed in Kumaun and Grahwal a disease known by the local name of *mahamari,* apparently identical with the plague of Syria and Egypt, the characteristic symptoms being violent fever of the most contagious nature, always accompanied with swellings in the armpit, the disease ordinarily seeming to reach its crisis on the third day.[128] From the following report it is possible to postulate that plague was present in nineteenth century India in more or less endemic form in the Kumaun and Grahwal district of the Himalayas. It was from here, that plague was disseminated.

Later plague (1890 and 1900) caused the greatest mortality, originated in Yunnan, the southern China and reached to India through multiple maritime linking networks. The speed, with which plague and its vectors could travel the ocean highways and sustain a chain of infection from port to port depended largely on the sophistication of the maritime technology. In the age of steamer traffic, the migration of plague from one part of the ocean to the other was very rapid.[129]There were other routes too, particularly from the Middle East, a likely source of outbreaks in the past. It reached from Middle East to western India and from there to intra-Indian centres depending on the nature of interaction. In some instances plague was carried by ship-borne rats and their fleas, moving with cargoes of grain from one port to another.

The whole of the middle Indian Ocean region served as the medial centre of transmitting plague in the different parts of the Indian Ocean since early times. Pneumonic plague may possibly have the major reason of the terrible Java epidemic of 1625-6. The bubonic and pneumonic plague had ravaged China 11 times between 1300 and 1400.[130]Some of these epidemics must have disseminated to South-East Asia through the middle Indian Ocean as happened with the 1890 plague. This plague (1890) originated in Yunnan, in southern China and reached India via Canton, Hong Kong, and Bombay.[131]

The Middle East performed two functions since ancient times; disease centre and medial centre. Micro-predators such as fungi, bacteria and viruses raged through the packed crowds of the cities of the Middle East. As disease centre, many of these pathogens were transferred to India. *Bilharziasis* was one of them. Evidence of the presence of this disease has recently been traced in the kidneys of 3000 year old Egyptian mummies.[132]Bilharzia caused by fluke worm is today still the curse of Egypt and Africa. There are three species of Bilharzia: *Bilharziam mansoni, Bilharzia hoematobia* and *Bilharzia japonica*. The first distributed throughout the continent of Africa, being most abundant in Egypt, the Congo, French West Africa and Nigeria, while the second was almost entirely confined to Africa. *B. Japonica* occurred most commonly in China, especially in the Yangtse valley, Southern Japan, upper Burma and the Southern Philippines.[133] The parasite has a complicated life cycle. In Huang He and the Nile valley, especially the flooded lands, the disease had killed competing plant species where fluke worm lived on various species of water snails.[134] In the warm, shallow waters of the paddy fields, lurked parasites which possibly became capable of penetrating the skin and entering the bloodstream of wading human rice farmers. Foremost among these is a blood fluke called *schistosoma*. *Schistosoma* produces the debilitating and often deadly disease called *schistosomiasis*. One common symptom of the disease was hematuria, i.e. the discharge of blood with the urine but hematuria can occur also in the course of other diseases. *Schistosoma* causes far reaching destructions into various tissues especially in the urinary tract.[135] Finally, the parasite led to epilepsy, kidney failure and malignant cancers. It was through Middle Eastern medial centres that *Bilharzia mansoni* and *Bilharzia hoematobia* had been transported to India. Middle Indian Ocean functioned as medial centre for transmitting *Bilharzia japonica* to India either from China or from Japan. There was another possibility of transmitting *Bilharzia japonica* over land or through river route to Assam, Bengal, etc. Cases of epilepsy, kidney failure were recorded by the late eighteenth and early nineteenth century by European surgeons practising in India.[136]

Measles is probably the result of rinderpest or canine distemper. They are highly contagious and require no old hosts to complete their life cycle. They spread with remarkable ease from human to human. The microlife of measles reached India with Perso-Arabic maritime merchants engaged in Indo-Western Indian Ocean trade.

Middle East also functioned as medial centre of transmitting diseases from one disease zone of the western Indian Ocean to the other. Africa was a disease zone of the western Indian Ocean, the birth place of various diseases, which were disseminated to intra-Indian Oceanic regions via Middle Eastern medial centres. Africa had a whole collection of local diseases that had presumably evolved with man in Africa and remained tied to African conditions. Many strains of malaria, yellow fever, sleeping sickness, various intestinal helminthiasis like *ankylostomiasis*, framboesia or yaws, and the like are African diseases.[137] When the Asian civilizations had developed close enough and long enough, direct and indirect contact through medial centre with Africa, these African diseases fatally affected Asiatic civilizations.[138]

Many of the African diseases are mainly mosquito-borne diseases. Malaria, yellow fever, etc., are some of the examples. Malaria is a protozoan infection, produced in humans by four species of protozoan parasites *(plasmodium)*. *Vivex malaria*, one of the oldest of the malarial types, originated in Africa. It was the ubiquitous malarial type which changed locales to become a scourge of much of the world.

The most virulent is the *Plasmodium falciparum* which kills one in ten people it infects.

Malaria is dangerous because the mosquito is very adaptive to the warmer and more humid conditions. It is adaptive to enhanced global warming and breed at the optimum temperature between 20 °C-30 °C and at temperatures as low as 16 °C. Mosquitoes are very active at relative humidity over 60 per cent. Mosquitoes die at temperatures above 35 °C and at relative humidity less than 25 per cent. Hence, even a small swelling in temperature will accelerate the incubation period of the parasite from three weeks to one week. A monsoon of three or four months is likely to enhance breeding by the different species of anopheles.

Bengal presented an ideal ecological set-up for the breeding of mosquitoes. Large development works of the eighteenth century cleared forests. Newly cleared land created abundant breeding sites for mosquitoes. Epidemics of malaria followed.[139] The endemic of miasm in Bengal killed 300 sailors out of 1000 at Calcutta and 200 out of 1000 at Hooghly yearly during the second half of the eighteenth century. In 1757 and 1762, severe epidemics of miasm (malaria) in Bengal had been recorded.[140]

Yellow fever, a four factor disease whose animal reservoir was the monkey, emanated from Africa. The disease causes acutely debilitating fever, chills, jaundice, internal haemorrhaging, coma, etc. The incubation period for yellow fever lasts from three days to several weeks, depending upon temperature. Its principal vector, the *Aedes agypti* mosquito thrived at more than 24 °C temperature and high humidity. It ceased to spread when the thermometer sank below 75 °F and it stopped abruptly as an epidemic when freezing point is reached. It was most prone to occur and spread during rainy season. Humans picked up the disease when they entered forest areas harbouring the monkey reservoir. When yellow fever infected people returned to settlements, the *Aedes* mosquito then retransmitted the disease from person to person in the community. Yellow fever, therefore, was a disease both of forests, and of adjacent urban settlements. In the past, yellow fever had been transmitted to intra-Indian Oceanic regions by ships. The favourite haunts of the disease, thus, were the sea coast towns, the banks of rivers and flat delta country.[141]

Yellow fever can be introduced to an area by either an infected individual or an infected mosquito which typically breeds in urban areas. An extrinsic incubation period of 10–12 days is required to transmit the disease to another victim. Poor residents faced increased risks because of overcrowding and insanitary living conditions.[142]

During seventeenth, eighteenth and early nineteenth centuries, yellow fever became a regular visitor to the ports of India. Dr. William Twining, a medical practitioner of Bengal in the early nineteenth century wrote, 'The yellow fevers, so justly dreaded in Western Europe, is hardly to be accounted an endemic of Bengal; although we every year meet with some patients in whom an intense yellow suffusion of the skin occurs, in fevers of considerable severity.'[143] W.S.R. Hodson in 1853 discussed in some length a type of fever which occurred near Peshawar locally named as *Euzofzai* fever. The toll of the fever between the 1st March and 15th June 1853 amounted to the death of 8352 persons out of a population of 53,500.

According to W.S.R. Hodson, the fever was very similar to typhus but had some symptoms of yellow fever and confined mainly to the natives. It appeared to be highly contagious or infectious.[144] Thus, yellow fever was present in India during the late eighteenth and early nineteenth centuries but not in a high epidemic form. Sleeping sickness, a major epidemic of east and central Africa carried by tsetse flies failed to take a flying leap across the western Indian Ocean to India.

Other African micro-organisms were transferred to India and because of India's insanitary conditions, these diseases spread in certain regions of India. Framboesia or yaws is a disease of this sort. Yaws was the disease of the African race but it was known in India and China at least during the eighteenth and the nineteenth centuries. The disease was contagious and transferred from person to person by contact or the 'absorption of the poison through some abraded surface'. The poison might be carried from individuals by flies.[145] Many of the 'boils and blains', ulcers and sores, 'desquamations, tumours and skin diseases' were exchanged between India and Africa via the western Indian Ocean by direct and indirect contacts.

Ankylostomiasis was another disease which reached India. It was particularly prevalent among plantation workers of Assam and Bengal. It is an intestinal helminthiasis. The slave trade was responsible for the transmission of this helminthic infection in India.[146]

In the East Indian Ocean region, Japan was the disease zone of some importance. *Japanese encephalitis*, an inflammatory brain disease transmitted to India somewhere between seventeenth and nineteenth century. *Japanese encephalitis* is a vector disease spread by a mosquito named *culex annulirostris* which thrived whenever mean daily temperature exceed 17.5 °C. Approximately 30 per cent of infected people end up permanently incapacitated according to a recent estimate.

Infectious jaundice, a severe form of fever, caused by *leptospira icterohaemorrhagiae,* associated though not invariably with jaundice, enlargement of liver and sometimes of the spleen prevailed in Japan during the nineteenth century. The disease had long been known in Egypt. This epidemic has been found in the Andaman Islands and in the Malay states. It infected some parts of India either via the western Indian Ocean or via the middle Indian Ocean.

The home of the seven-day fever is the Japan. It spread due to *Leptospira hebdomadis,* especially during the summer months in Japan. It is found in China (*shaw-si*) but possibly occurs also in India and Dutch east Indies.[147]

The Indian Ocean, therefore, consisted of various and distinctive disease zones, which continuously interacted with each other pathogenically either directly or indirectly through medial centres throughout the seventeenth, eighteenth and early nineteenth centuries. Alongside maritime trade, soldiers, pilgrims, migrant labourers were the other important human element in intra-Indian Oceanic exchange of micro-organisms. This migration traffic, too, had a history of high morbidity and mortality at the depots before embarkation, during intra-Indian Oceanic crossings, and on arrival in new disease environments. In some disease zones of the Indian Ocean, Indian labour was the putative beneficiary of the health hazard wrought by epidemics among indigenous populations elsewhere in the Indian Ocean. More generally, the disease burden, whether imported or endemic took heavy toll on Indian labourers.

There were other factors, too, that contributed positively to continuing high rates of morbidity and mortality from epidemiological links between India and the Indian Ocean territories. The inadequacy of housing, insanitary living conditions lived in by poor populations of the Indian Ocean societies, low level of medical technology prevailing in Indian Ocean countries, limited medical provisions available for the people and so forth increased the fatality of the migrant diseases. The main burden of the migrating infectious disease fell on Indian labour.

These biomedical and environmental deficiencies, industrial slums and polluted water supply, inadequacy of medical technology to fight diseases, inadequacy of medical measures undertaken, etc., did much more than 'virgin soil' explanations to promote certain diseases in certain societies than other. In this sense, India in general and Bengal in particular is significant not only as a recipient or disseminator of diseases within the Indian Ocean,'a broking housing for regional epidemiological transactions, but also as a prime example of the kinds of conditions in which diseases, deadly or debilitating, have continued to thrive.'[148]

Susceptibility to disease might be very high, too, in the rural areas either through persistent poverty and famine, or as a result of environmental change through deforestation, etc., or through the linking of hinterlands and so forth. The movement of labour from centre to periphery and reverse for employing in developmental works, plantations, etc. provided social and epidemiological conditions favourable to disease transmission. The forests and wooded hillsides, cleared for indigo, tea, opium, jute, silk, coffee, etc., created conducive disease ecology in which pathogens could proliferate.

NOTES

1. Kenneth F. Kiple, 'The History of Disease', in Roy Porter (ed.), *Cambridge Illustrated History of Medicine*, Cambridge University Press, 1996, p. 16.
2. Ibid.
3. Ibid.
4. F.E. Zuener, 'Domestication of Animals', in Charles Singer, E.J. Holmyard and A.R. Hall (eds.), *A History of Technology*, vol. I, Clarendon Press, 2nd edn, 1955.
5. Jeremy Black (Gen. ed.), *Atlas of World History*, London, 2004, p.20.
6. F.E. Zuener, op. cit., p. 338.
7. F.E. Zuener, 'Cultivation of Plants', pp. 353-75; Sir Lindsay Scott, 'Pottery', pp. 376-412; Seton Lloyd, 'Building in Brick and Stone', etc., all available in Charles Singer et al. (eds.), *A History of Technology*, vol. 1.
8. M.S. Drower, 'Water-Supply, Irrigation, and Agriculture', in Charles Singer et al. (eds.), ibid., pp. 520-57.
9. Jeremy Black (Gen. ed.), op. cit., p. 20.
10. Kenneth F. Kiple, op. cit., p. 19.
11. W. Crooke, *The Popular Religion and Folk-Lore of Northern India*, in 2 vols, vol. 1, Delhi, 3rd repr., 1968, pp.123-74.
12. Kenneth F. Kiple, op. cit., p. 24.
13. Alfred W. Crosby, *Ecological Imperialism: the Biological Expansion of Europe, 900-1900*, Cambridge University Press, repr. 1986, p.30.
14. Vivian Nutton, 'The Rise of Medicine', pp. 52-81; Roy Porter, 'What is Disease', pp. 82-117; Edward Shorter, 'Primary Care', pp.118-53; Roy Porter, 'Medical Science', pp.154-201; Roy Porter, 'Hospitals and Surgery', pp. 202-45; and Miles Wheatherall, 'Drug Treatment and the Rise of Pharmacology', pp. 246-77; all available in Roy Porter (ed.), *The Cambridge Illustrated History of Medicine*.
15. Edward Shorter, loc. cit., 138.
16. See the reports prepared by Dr. Arata Kochi, director of WHO's global tuberculosis programme; Dr. Michael Iseman of National Jewish Medical and Research Centre.

17. See Kenneth F. Kiple and Friehild Conee Ornelas (eds.), *The Cambridge World History of Food*, Cambridge, 2000, in 8 parts, pt. vii, especially some papers on food biotechnology and genetic engineering.

18. Alfred W. Crosby, *The Columbian Exchange: The Biological Consequences of 1492*, Westport, 1972 and his, *Ecological Imperialism: the Biological Expansion of Europe, 900-1900*, repr. 1995.

19. Emmanuel Le Roy Ladurie, *The Mind and Method of the Historian*, Brighton, 1983, pp. 28-83; W.H. McNeill, *Plagues and People*, Oxford, 1977.

20. P.D. Curtin, *Death by Migration: Europe's Encounter with the Tropical World in the Nineteenth Century*, Cambridge, 1989.

21. K.F. Kiple, *The Caribbean Slave: Toward a Biological History of Black People*, Cambridge, 1984.

22. R. Shlomowitz and L. Brennan, 'Mortality and Migrant Labour in Assam, 1865-1921', *Indian Economic and Social History Review* (hereafter *IESHR*), vol. 27, no. 1 (1990), pp. 85-110; and D. Arnold, 'The Indian Ocean as a Disease Zone, 1500-1950', *South Asia* (hereafter *SA*), vol, 16, no. 2 (1991), pp. 1-21.

23. R. Shlomowitz and L. Brennan, 'Maritime Mortality Revisited', *International Journal of Maritime History* (hereafter *IJMH*), vol. 8, no. 1, June 1996.

24. Philip R.P. Coelho and Robert A. Mcquire, 'African and European Bound Labor in the British New World: The Biological Consequences of Economic Choices', *Journal of Economic History* (hereafter *JEH*), vol. 57, no. 1, March 1997, p. 88.

25. Ivan Polunia, 'Disease, Morbidity and Mortality in China, India and the Arab World', in Charles Leslie (ed.), *Asian Medical System: A Comparative Study*, University of Califormia Press, Berkeley, 1977, p. 122.

26. Macfarlane Burnet and David O. White, *Natural History of Infectious Disease*, 4th edn., 1972, p. 35.

27. Ivan Polunia, op. cit., p. 125.

28. Macfarlane Burnet and David O. White, op. cit., p. 36.

29. K. David Patterson and Geral W. Hartwig, 'The Disease Factor: An Introductory Overview', in K. David Patterson and Geral W. Hartwig (eds.), *Disease in African History: An Introductory Survey and Case Studies*, Durham, 1978, p. 3. Henry E. Sigerist, *A History of Medicine: Primitive and Archaic Medicine*, vol. 1, New York, 2nd edn., 1955, p. 38.

30. H.A. Waldron, *The Medical Role in Environmental Health*, Oxford University Press, 1978, p. 2.

31. Myra Shackley, *Environmental Archaeology*, George Allen & Unwin, 1981, p. 42.

32. Paul D. Hoeprich, *Infectious Diseases: A Modern Treatise of Infectious Processes*, Philadelphia, 3rd edn., 1983, p. 4; and Hideo Moriyama, *The Nature and the Origin of Life*, Tokyo, 1955, p. 3.

33. Hideo Moriyama, *The Nature and the Origin of Life*, Tokyo, 1955, p. 3.

34. Ibid., p. 137.

35. Ibid., p. 163.

36. Thomas C. Cheng, *General Parasitology*, New York, 1973, pp.3-9; P.D. Hoeprich, 'Host-Parasite Relationships and Pathogenesis of Infectious Disease', in Paul D. Hoeprich (ed.), *Infectious Diseases: A Modern Treatise of Infectious Processes*, pp. 45-50.

37. Myra Shackley, *Environmental Archaeology*, p. 155.

38. Thomas C. Cheng, *General Parasitology*, pp. 8-9.

39. Paul D. Hoeprich, op. cit., p. 53.

40. Cedric A. Mims, *The Pathogenesis of Infectious Disease*, Academic Press, London, pp. 3-5.

41. Ibid., pp. 4-5.

42. E.H.R. Harries and M. Mitman, *Clinical Practice in Infectious Diseases*, Edinburgh, 1951, p. 7.

43. Franklin H. Top, *Communicable and Infectious Diseases*, 4th edn, 1960, p. 25.

44. E.H.R. Harries and M. Mitman, *Clinical Practice in Infectious Diseases*, Edinburgh, 1951, p. 49; Franklin kH. Top, *Communicable and Infectious Diseases*, 4th edn, 1960, pp. 27-8.

45. *American Commission on Airborne Infection* (1947), *American Journal of Public Health*, vol. 37, p. 13.

46. E.H.R. Harries, op. cit., pp. 49-51.

47. James C. Riley, *Sickness, Recovery and Death: A History and Forecast of Ill Health*, London, 1989, p. 105.

48. Macfarlane Burnet and David O. White, *Natural History of Infectious Disease*, Cambridge, 1972, pp. 105-6.

49. James Riley, op. cit., p. 116.

50. K. David Patterson and Gerald W. Hartwig, 'The Disease Factor: An Introductory Overview', in K. David Patterson and Gerald W. Hartwig, *Disease in African History: An Introductory Survey and Case Studies*, Durham, 1978, pp. 3-4.

51. Philip R.P. Coelho and Robert A. Mcquire, 'African and European Bound Labor in the British New World: The Biological Consequences of Economic Choices', *JEH*, March 1997, p. 91.

52. Thomas Mckeown, *The Origins of Human Disease*, Oxford, 1988, pp. 52-3.

53. Ibid., p. 53.

54. Robin Haines, Ralph Shlomowitz and Lance Brennan, 'Maritime Mortality Revisited', *IJMH*, vol. VIII, no. 1, June 1996, p. 140.

55. Frank E. Hugget, *The Past, Present and Future of Life and Work at Sea: A Documentary Inquiry*, London, 1975, p. 11.

56. Mark Staniford, op. cit., p. 121.

57. C.R. Boxer, *The Tragic History of the Sea 1598-1622*, The Hakluyt Society, 2nd Series, Cambridge, 1959, p. 15.

58. Kenneth Mackinnon, *A Treatise on the Public Health, Climate, Hygiene and Prevailing Disease of Bengal and the North West Provinces*, Cawnpore, 1848, p. 353.

59. Hendrik William Van Loon, *Ships and How they Sailed the Seven Seas, 500 BC–AD 1935*, p. 203.

60. Ibid., pp. 206-10. For purified water see C.R. Boxer, *The Tragic History of the Sea*, p. 16.

61. Andrew Balfour and H.H. Scott, *Health Problems of the Empire: Past, Present and Future*, pp. 335-6. For chronic nephritis and stertorous breathing see Madhavkar's— *Madhavnidanam*, Sudarshan Shastri (trans.), Varanasi, 1960, p. 361.

62. Phillipus Baldaeus, *Beschrijving der Oost Indische Kusten Malabar en Choromandel der Zelver aangrenzende Ryken en het Machtige Eyland Ceylon Nevens een Oustandige en grondigh doorzochte outdekking en wederlegginge Van de Afgoderye den Oost Indische Heydenen* (trans.), 'A True and Exact Description of the Great Island of Ceylon' by Pieter Brohier, *The Ceylon Historical Journal*, vol. VIII, nos. 1-4, July 1958 to April 1959, 1st edn., 1960, p. 378.

63. For congestion in the brain and spinal canal see William Twining, *Clinical Illustrations of the More Important Diseases of Bengal with the Result of an Inquiry into their Pathology*

and Treatment, vol. 1, Calcutta, 1835, p.46; and for vile decoction see A. Balfour and H.H. Scott, op. cit., p. 336.

64. R. Montgomery Martin, *The History of the Indian Empire,* vol. 1, Delhi, 1st repr, 1983, p. 491.

65. H. Willem Van Loon, op. cit., p. 197.

66. Kenneth Kiple, 'The History of Disease', Roy Porter (ed.), op. cit., p. 46.

67. Arthur C. Guyton, *A Textbook of Medical Physiology,* 6th edn., London, 1981, pp. 910–11.

68. Hendrik Willem Van Loon, *Ships: And How They Sailed the Seven Seas, 5000 BC-AD 1935,* New York, 1935, p.197.

69. Anthony Reid, *Southeast Asia in the Age of Commerce: the Lands below the Winds,* vol.1, New Haven, 1988, p. 60.

70. Kenneth F. Kiple, op. cit., p. 46.

71. Ann Bowman Jannetta, *Epidemics and Mortality in Early Modern Japan,* Princeton University Press, 1987.

72. Ivan Polumia, op. cit., p. 125.

73. David Arnold, 'The Indian Ocean as a Disease Zone, 1500-1950', *SA,* vol. XIV, no. 2, 1991, p.4.

74. I have used A.J. Qaisar's translation of *Anis-ul-hujjaj,* 'From Port to Port: Life on Indian Ships in the Sixteenth and Seventeenth Centuries', in A.D. Gupta and M.N. Pearson (ed.), *India and the Indian Ocean,* pp. 330-49.

75. Louis Tarleton Young, *The Carlsbad Treatment for Tropical and Digestive Ailments and How to Carry it out Anywhere,* London, 1899, p. 206.

76. A. J. Qaisar, loc. cit., p. 341.

77. Ibid., p. 342.

78. Ibid., p. 343.

79. James Horsburgh, *India Directory or Directions for Sailing to and from the East Indies, China, New Holland, Cape of Good Hope, Brazil, and the Interjacent Ports: Compiled Chiefly from Original Journals at the East India House and From Observations and Remarks, Made during Twenty One Years Experience Navigating in Those Seas,* 2 vols., vol. 2, London, 1826-7, p.13.

80. Kenneth F. Kiple, op. cit., p. 24.

81. Alfred W. Crosby, *Ecological Imperialism: The Biological Expansion of Europe, 900-1900,* p. 31.

82. Jeremy Black (Gen. ed.), *Atlas of World History,* p. 24.

83. For details see Alfred W. Crosby, op. cit., pp. 31-4.

84. Gerald W. Hartwig, 'Social Consequences of Epidemic Diseases: The 19th Century in East Africa', in K. David Patterson and Gerald W. Hartwig (eds.), *Disease in African History: An Introductory Survey and Case Studies,* Durham, 1978, p. 25.

85. K. David Patterson and Gerald W. Hartwig, 'The Disease Factor: An Introductory Overview', ibid., pp. 7-8.

86. Kenneth F. Kiple, op. cit., p. 25.

87. Ibid., p. 51.

88. S.J. Nidham, *Science and Civilization in China,* vol. 6, *Biology and Biological Technology,* pt. 6, *Medicine,* Cambridge University Press, Cambridge, 2000, pp. 124-5.

89. Ann. Bowman Jannetta, *Epidemics and Mortality in Early Modern Japan,* pp. 50-64.

90. Johani Koponen, *People and Production in Late Pre-Colonial Tanzania: History and Structure,* Jyvaskyla, 1988, pp. 153-62.

91. Ibid., p. 163, and David Arnold, 'The Indian Ocean as a Disease Zone, 1500-1950', *SA*, vol. XIV, no. 2, 1991, pp.6-7.

92. Alfred W. Crosby, op. cit., p. 38.

93. Ibid., pp. 22-31.

94. Stephen Frederic Dale, *Indian Merchants and Eurasian Trade, 1600-1750*, p. 84.

95. O.P. Jaggi, *Medicine in India: Modern Period*, in D.P. Chattopadhyaya (Gen. ed.), *History of Science, Philosophy and Culture in Indian Civilization*, vol. IX, pt.1, Delhi, 2000, p.143.

96. O.P. Jaggi, 'Disease as Seen by Early European Physicians in India', in D.P. Chattopadhyaya and Ravinder Kumar (eds.), *Science, Philosphy and Culture: Multi-disciplinary Explorations*, p. 87.

97. M.N. Pearson, 'The Thin End of the Wedge: Medical Relatives as a Paradigm of Early Modern Europeans', *Modern Asian Studies*, vol. 29, pt. 1, February 1995, p. 156.

98. K.J. Macpherson, *Annals of Cholera from earliest Period to the Year 1817*, London, 1872, p. 121.

99. O.P. Jaggi, op. cit., p. 56.

100. Charles E. Rosenberg, *Explaining Epidemics and Other Studies in the History of Medicine*, Cambridge, 1992, p. 111.

101. Macnamara, *A History of Asiatic Cholera*, London, 1876.

102. Edward Thronton, *Gazetteer of the Territories under the Government of the East India Company and the Native States on the Continent of India*, vol. II, London, 1854, p.112, and D.A. Macleod, *Sketch of the Medical Topography of Biswanath and its Immediate Neighbourhood with an Account of Diseases Generally Prevailing in Assam*, Calcutta, 1837, p. 26.

103. Charles Morehead, *Clinical Researches on Disease in India*, vol. 1, London, 1856, p.4.

104. Edward Thronton, op. cit., vol. IV, p. 481.

105. A.Dundas Taylor, *A General Report of the Operations of the Marine Survey of India from the Commencement in 1874, to the End of the Official year 1875-1876, 1878, 1879, 1880*, etc., Calcutta, 1876-80.

106. David Arnold, 'The Indian Ocean as a Disease Zone', op. cit., p. 8. A.B. Jannetta, op. cit., p. 47.

107. Ibid.,

108. Nancy Elizabeth Gallagher, *Medicine and Power in Tunisa, 1700-1900*, Cambridge, 1983, pp. 23-44.

109. David Arnold, op. cit., p. 9.

110. Erwin H. Ackerknecht, *History and Geography of the Most Important Disease*, New York, 1965, pp. 24, 26.

111. David Arnold, op. cit., p.10.

112. Philip H. Manson-Bahr (ed.), *Manson's Tropical Diseases: A Manual of the Diseases of Warm Climates*, London, 1929, p. 134.

113. D.A.Macleod, *Sketch of the Medical Topography of Biswanath and its Immediate Neighbourhood with an Account of the Disease Generally Prevailing in Assam*, pp. 26, 28; William Twining, *Clinical Illustrations of the More Important Diseases of Bengal with the Result of an Inquiry into their Pathology and Treatment*, vol. 1, Calcutta, 1835, p. 18; Edward Thronton, *Gazetteer of the Territories under the Government of the East India Company and the Native States on the Continent of India*, vol. IV, p. 711, etc.

114. D. A. Macleod, loc. cit., p. 28; M.N. Pearson, 'The Thin End of the Wedge: Medical Relatives as a Paradigm of Early Modern–European Relations', op. cit., p. 45; W. Twining, loc. cit., p. 227; John H.Van Linschoten's 'Observations of the East Indies,' p.194; R. Montgomery Martin, *The History of the Indian Empire,* vol. 1, p. 491.

115. Ranes C. Chakravorty, 'Disease of Antiquity in South Asia', in Kenneth F. Kiple (ed.), *The Cambridge World History of Human Disease,* Cambridge University Press, 1993, p. 410.

116. Philip H. Monson-Bahr, op. cit., p.205; R. Montgomery Martin, op. cit., p. 491; and K.D. Patterson and Gerald W. Hartnig, 'The Disease Factor: an Introductory Overview', K. David Patterson and Gerald W. Hartwig (eds.), *Disease in African History: an Introductory Overview,* p. 9.

117. Macpherson K.J., *Annals of Cholera from Earliest Period to the Year 1817,* p.112 and James Ranken, M.D., *Report on the Malignant Fever Called the Pali Plague which has Prevailed in Some Parts of Rajputana since the Month of July 1836,* Calcutta, 1836, p. 25.

118. Kalhan's *Rajatarangini,* vol. 1, p.196, quoted in *Extracts from Records of Past Epidemics* (n. d.), p.1.

119. William Ernest Jennings, *A Manual of Plague,* London, 1903, pp. 3-4.

120. Jeremy Black (Gen. ed.), *Atlas of World History,* p. 72.

121. For Bengal *Tarikh-i Daudi* and for Kandahar *Tarikh-i M'asumi* have been referred in *Extracts from Records of Past Epidemics in India,* n.d., pp. 2-3.

122. Extracted from I.O.Records (O.C.No. 568) in ibid., p. 34.

123. Ibid., p. 4.

124. Ibid., pp. 5-7.

125. William Ernest Jennings, op. cit., pp. 3-4.

126. Jeremy Black (Gen.ed.), *Atlas of World History,* p. 72.

127. William Ernest Jennings, op. cit., p.13.

128. Kenneth F. Kiple, op. cit., pp. 26-8.

129. R. Nathan, *Plague in India,* 2 vols., vol. 2, p. 62.

130. Joseph Needham, *Science and Civilization in China: Civil Engineering and Nautics,* vol. 4, pt. 3, Cambridge, 1971, p. 530.

131. David Arnold, op. cit., p. 11.

132. Kenneth F. Kiple, 'The History Disease', op. cit., p. 21.

133. Philip Mansor-Bahr, *The Dysenteric Disorders: the Diagnosis and Treatment of Dysentery, Sprue, Colitis and Other Diarrhoeas,* London, 1944, p. 275.

134. Henry E.Sigerist, *A History of Medicine: Primitive and Archaic Medicine,* vol. 1, New York, 1955, p. 63.

135. Ibid.

136. Kenneth Mackinnon, *A Treatise on the Public Health, Hygiene and Prevailing Diseases of Bengal and the North West Provinces,* Cawnpore, 1848, p. 7.

137. Marston Bates, *The Forest and the Sea: A Study of the Economy of Nature and Ecology of Man,* London, 1960, p. 163; James A. Lee, *The Environment, Public Health and Human Ecology: Considerations for Economic Development,* London, 1985, p. 57; H. Vandyke Carter (M.D.), *Including Notes on Pellagra, Clou, De Biskra, Caneotica and Aleppo Evil,* London, 1876, p. 5, etc.

138. Marston Bates, loc. cit., p.163.

139. David Arnold, op. cit., p. 16.

140. Kenneth Mackinnon, op. cit., p. 89.

141. S. Philip H. Manson-Bahr, op. cit., p. 191.

142. Jonathan B. Pritchett, 'Strangers' Disease: Determinants of yellow fever Mortality during the New Orleans Epidemic of 1853', *Explorations in Economic History* (hereafter *EEH*), vol. 32, no. 4, October 1995, p. 520.

143. William Twining, *Clinical Illustrations of the More Important Diseases of Bengal with the Result of an Indquiry into their Pathology and Treatment*, Calcutta, 1835, p. 201.

144. George H. Hudson (ed.), *Twelve years of a Soldier's Life in India: Being Extracts of the Late Major W.S.R.Hodson*, London, 1859, p.139.

145. H.Vandyke Carter (M.D.), op. cit., pp. 5-9.

146. David Arnold, op. cit., p. 16; Jahavi Koponen, op. cit., p. 155; K. David Patterson and Gerald W. Hartwig, 'The Disease Factor: An Introductory Overview', in K. David Patterson and Gerald W. Hartwig (eds.), *Disease in African History: An Introductory Survey and Case Studies*, p.9.

147. Philip H. Monson-Bahr, op. cit., p.15.

148. David Arnold, op. cit., p.15.

Life and Death: Life of Early English Settlers at Bombay, 1661–1760

Vaibhav Sharma

…For in five hundred, one hundred survive not; of that one hundred, one quarter get no estates; of those that do, it has been recorded above one in ten years has seen his country.   ROE THOMAS and FRYER JOHN, *Travels in India 17th Century*

BOMBAY AS A PORT and settlement had a dual character: it was both welcoming and forbidding to Europeans. The site of Bombay had two major features: a natural harbour and a partly rocky and partly marshy terrain. The harbour itself was excellent, deep, without shoals, and sheltered from storms as it faced the mainland across Bombay creek, instead of facing the Arabian Sea directly. In fact, the Portuguese Viceroy of Goa had opposed the cession of Bombay to the British on the ground that it was the best port in Portuguese possession, 'with which that of Lisbon is not to be compared'.[1]

The negative side was the unhealthy marshy site with feverish swamps, which took heavy toll on European lives, and gave rise to the famous proverb: 'two monsoons are the ages of a man'.[2] Apart from the unhealthiness of the climate, intemperance of the early English, war with the political rival on the coast such as the Marathas, Sawantwadis, Sidi's and other European companies, there was also a problem of finances for the maintenance of the island. All this added to the growing mortalities on the island to the extent that many Company servants were not willing to serve on the island.

The other effect can be seen in terms of the declining population of the island which remained more or less the same in the first century of English rule on the island. Fryer in 1675 remarked that the population of the island was 60,000, but by 1715 Cobbe remarked that the population had dwindled to 16,000. This may have been the result of the pestilential climate, coupled with disorders, thus giving rise to the commercial rivalry and the hostility of the Marathas, the Mughal's and the Portuguese. The population remained approximately 60,000 in the 1750s.[3]

The central purpose of this essay is to show the various problems that the English faced on the island, and steps taken to overcome these problems such as constructing a hospital on the island.

MORTALITY ON THE ISLAND: VARIOUS CAUSES

Each year the monsoon winds of July and August bring rains on which the Indian population depends. But for the early European travellers it was very unhealthy. The mortality rate was very high by modern standards. The sea voyage itself claimed a fair share of victims during the six or seven months journey. By the time they reached India contagions and other unknown causes of death awaited their arrivals. For instance, Ovington in 1690 writes: 'We arrived here, at the beginning of the rains, and buried of the twenty-four passengers which we brought with us above twenty before they ended; and of our own ships company above fifteen.... The most sovereign, nor the safest prescription in the physical art, could restore the weakness of our languishing decay'd natures and that thoroughly confirmed to us the unhealthfulness of the place....'[4]

Nor was this mortality confined to particular epidemic years. It was a regular annual occurrence. The survivors used to hold thanksgiving banquets towards the end of October to celebrate their deliverance. 'What use sending trusty factors, and hardy soldiers there?' writes Anderson 'They breathed the poisonous air, but a few short months after their lives and services are lost to the employers forever. Three years was the average duration of European life.'[5] Such instances found continuous references in the early official records and travel writings during the closing years of the seventeenth and first quarter of the eighteenth century.

HEALTH CONDITIONS AND DISEASES

There exists little evidence to throw light on the health conditions on the Bombay island during the period preceding its cession to the English crown by the Portuguese. Heitor da Silveira named it the 'Island of Good Life', which he would scarcely have done, if the climate had proved very deleterious.[6] While Fryer in 1673 spoke of the country people and the Portuguese in the old days lived a long life, which he believed to be largely due to their temperate habits.[7] But Bombay under the English acquired an evil reputation mainly because of the mortality which rose at an alarming pace.

'I reckon the people of Bombay live in the charnel houses,' wrote Fryer in 1673, 'the climate being extremely unhealthy, as first thought to be caused by bubsho, rotten fish; but, though that be prohibited, yet it continues as mortal. I rather impute it to the situation which causes an infecundity in the earth, and a putridness in the air, what being produced seldom coming to maturity, where by what is eaten is undigested; whence follow fluxes,[8] dropsy, scurvy, barbiers, gout, stone, malignant and putrid fevers, which are endemical diseases.'[9]

More fatal was the disease known as the 'morshideen'[10] to the Portuguese, which was choleric in nature and also called 'Chinese death'. It was divided according to their system into four types. The first type was simple choleric[11] and its only symptom was severe griping. The second was attended with diarrhoea as well as pain; the third with pain and vomiting. Purging and intense pain were symptoms of the last type referred by Thevenot to be 'cholera morbus' and generally brought its victim to a swift end within twenty-four hours.[12]

The remedy upon which the Europeans relied was applied to the ball of the sufferer's foot. If he winced, it was expected that he would recover but if he showed no signs of pain he was given over by his attendant.[13] Manuchi, a Venetian physician had discovered another cure by which he gained a vast reputation at the Mughal court, where he resided for forty years. His infallible remedy was this: 'Take an iron rod about an inch and half a meter in diameter and then heal it with red hot fire, extend the patient on his back, and apply the ring to his naval, in such a manner that the naval may be as the centre to the ring. As soon as the patient feels the heat takes away the ring as quickly as possible, when so sudden a revolution will be brought in his intestines that his pain immediately ceases.'[14]

The native remedy mentioned was to apply a twin red hot rod under the heel until the patient screamed with pain and then to slap the same part with the sole of the same shoe—to drive the pain centre of the body to an extremity. Thus, the treatment of the diseases was more terrible than the diseases themselves. In the case of flux, the remedy was to take 'two branches of the torrified rhubarb and a branch of cummin seed; all must be beat in to lime water or (if in that wanting) in the rose water. The common people in the Indies have no other remedy for this distemper, but the rice boiled in water till to be dry, they eat it with milk turned sour and use no other food as long as distemper lasts; but the same they use for the bloody flux....'[15]

The climate instead of improving became more pestilential year by year. In 1668 many of the soldiers were sick of 'the flux' (dysentery).[16] The Company records regularly refer to the sickness of the place, which was aggravated by the lack of supply of medicines, the need for which is constantly referred to in the factory records. For instance, a letter of 1668 noted, 'This island we find much more sickly then other places, the rather (as we conceive) for that the water hereupon is very bad, where for that which we commonly drink is fetched from Salsette, there being no spring upon the island. The private soldiers pay extends not to that charge and, therefore, they are incident to several diseases (for the relief of which we have supplied them with part of our store of wine, etc.), in so much that the doctor's complaint is that the medicines are very scant and not proportionate to the maladies, representing also to us the necessity of two able surgeon to reside on this place.'[17]

The quality of water was so bad that the letter of 1668 again mentioned and the officials asked for some rose water from Persia for the entertainment of strangers. They also asked for some Shiraz wine 'to encourage our water, by the badness whereof and the malignity of the air we have had diverse sick, towards whose recovery we have on all occasions furnished them with wine according to doctor's directions'.[18]

Seeing the mortality and sickness in Bombay there was an urgent need of a surgeon. Accordingly Bombay received its surgeon, Dr. John Bird.[19] Unfortunately, the island continued to have disease, problems of the lack of fresh water and medicines. This was so severe that even Dr. Bird did not want his wife to stay on the island. This became more evident from the letter sent to the Company 'The surgeon you sent for Bombay makes a humble request that you would permit his wife to take her passage in the next ship that come forth ... the air or water or both, do not very well agree with the soldiers. The general disease of this island is the flux and looseness, which caused the great expense of the physic, but the surgeon, has sent a list of suitable medicines, which please to let them not fail of by the next shipping....'[20]

Later in the same year, Bombay again required another surgeon as John Bird, the assistant surgeon, was going home. His services had proved useful for the soldiers.[21] Randolph, the surgeon of the ship *Rebecca*, however was willing to remain in India and was induced to accept the post. Randolph was given the permission to stay back since the friend of John Bird, Thomas Farley's conduct was found unsuitable for the job.[22]

In 1670, there was a considerable sickness and mortality among the European community, attributed largely to the late monsoon and lack of sufficient medicines. In July, the Bombay Governor Gray wrote: 'Tis now a sickly time for fluxes and feavours, yet praised be God, wee have buryed but one man these three months past.' But later in that month the deaths took place of Revd James Hutchinson, the Minister, and John Martyn, the Secretary of the Council, whose place (Gray said) it would be difficult to supply with 'a sober person and one that is qualified for it'. In August he wrote: 'We have diverse people sick and tis a hard sayinge when the surgeons tell them they have no medicines to give them, occasioned by the improvidence of those (to whom) our Masters intrust it at home, that neglected sendinge any out by the last three ships, which falls out unhappily to the loss of the Company's servant and alsoe to the greate encrese of these (through having to pay the ships double the amount that the medicines so obtained would have cost at home).'[23]

Unfortunately this sickness and mortality continued in the year 1670; Charles Smeaton, an able accountant died within two months of his appointment.[24] Another loss was the death of Revd James Sterling on 15 December.[25] Babor, member in the Council also died from gout or some other illness.[26] In September there was fortunately an improvement in this respect, Gray having purchased a medicine chest from Dr. Bird for Rs. 44, and received another which the Company had sent out on one of the ships, so that he hoped the supply would be sufficient till the next year's shipping arrived.

Mortality did not get the respite in the coming years and it fell heavily on the new soldiers especially in the first few months after their arrival. Thus, out of forty recruits put on the ships in March 1677, half had died by the following January.[27] The sickness and diseases among the Europeans, especially among the English, continued without any partiality concerning their ranks. For instance, Aungier died of the dysenteric illness that had attacked him in 1673 and eventually caused his death in 1677.[28] Philip Giffard, the deputy governor of Bombay fell ill in early 1676 and in September of that year left for Surat, where he died two months later.[29] John Petit who succeeded Giffard also fell ill in November 1676 and also left for Surat.[30] In the months between October 1675 and February 1776, a hundred soldiers perished.[31]

In 1676 John Child, who later became governor of Bombay, had been appointed accountant of Bombay and second in the council by the President and Council of Surat. He pleaded his apprehension of the diseases and refused to accept the office.[32] Phillip Giffard and John Petit writing in 1676 remarked 'we have buried up to fifty men, most new men; they die of fluxes which for the most part will take all it seized by reason of the bad diet and lodging and ill government of our people in their sickness, and also they living so remote they can not be looked after as they ought....'[33]

In the same year there were seldom less than forty or fifty of the English soldiers sick and totally incapable of doing any duty.[34] Seeing the sickness and mortality in the garrison the officers were finding it inconvenient to disband the old soldiers because each one of them was worth ten raw men as they did not easily adjust to the climatic conditions of the island.[35]

The Company thus seeing the growing mortality of the English on the island put forth the plan of building a hospital, so that the sick might be provided with constant attendance and a regular diet. The proposed building was to be capable of receiving seventeen persons and the cost was not to exceed four thousand rupees. It was estimated that about a thousand rupees would cover the annual expenses of the establishment which was to be under the superintendence of a resident surgeon. But notwithstanding the high mortality rate, no definite steps were taken till 1675.

HOSPITAL

Aungier had during his visit to Bombay in 1670 marked out a convenient place for a hospital, and ordered one to be built at the Company's expense. Gray accordingly employed stonecutters, bricklayers and carpenters in getting materials ready for one, but no actual building appears to have begun during the year.[36] In 1671 the progress was made with the building of the hospital, and in April its completion was reported to be in sight.[37]

This was indeed an urgent requirement. The letters from Bombay contained frequent references to the sickness and mortality prevalent among the European residents. The Council was particularly affected. In January both Cotes and

Captain Burgess were sick. Cotes in fact had a long and a tedious illness that ended in his death in November.[38] The sickness prevailed heavily on the island especially in October. This becomes more evident from the factory letters:

October: Is a very sickly time with us, there being at least 45 soldiers were sick and both Dr. Powell[39] and Dr. Boice[40] given over—so that we have no person that knows how to apply anything to anyone's distemper, which does much to discourage our men.

13 October: We have a very sickly time of it here, not one of us enjoy our health hardly a week together. Dr. Powell is dead, and Thomas Boice given over long ago, so in much want of a doctor.

23 October: We have still many of our men desperately ill and some die almost daily.

8 November: We find no abatement of those malignant distempers among us, which carry off daily one or other away—not two of us in the fort well, but some dangerously ill, which is a great hindrance to business and to the accounts.[41]

It was not till the close of the year that the want of the surgeon was remedied. Dr. Bird, whom it was proposed to re-engage, refused to accept the appointment, unless his emoluments were raised.[42] Once again in 1673 Aungier wrote a letter to the Company relating to some demands affecting the English population. One such demand was relating to the hospital, to be erected partly at the cost of the Company and partly out of voluntary contributions. He explained that the house formerly built for the purpose was not conveniently situated and was more suited for the warehouse. He proposed, therefore, to build another one nearer the fort. The need for this had been stressed by Dr. Bird, the surgeon at Bombay, in his report of 1 January to the Company. Aungier complemented him, saying he had a very heavy task, because of the constant sickness in Bombay, especially in the months of May, June, September and October.[43] He, therefore, welcomed the addition of another surgeon in the person of Dr. John Fryer whom the Company sent out on the *Unity*, one of the seven ships that arrived in December.[44]

In the second half of the year an illness suggestive of influenza had 'swept away a great many people by violent fevers and pains in the head'. Aungier was among the victims, but he had recovered by the end of July. In October and November he was again on the sick list with 'colleck' (colic); and a consultation of 5 November records that the Governor, the Deputy Governor, and some others had been 'of late much indisposed as to health, so that for some time the Council could not all meet. On 24 November, however, Aungier was 'somewhat recovered from his violent distemper'.[45] Aungier had long wanted to have a hospital built[46] but was awaiting the Company's reply to the proposals in his letter of 15 January 1674, where he had specified the four principal 'wants' of Bombay.[47]

Seeing the mortality on the island, the Surat Council wrote to Bombay on 18 December 1675 regarding the need for a proper hospital. The letter read that the Surat Council was determined to erect the necessary building with all speed without waiting for any order from England for the island. The Surat Council further directed the Bombay government to collect materials without further delay on the Company's charge and also proposed to supply the material for building.[48]

The letter further stated that the Surat Council had considered the two models for the hospital for the island but it seemed to the Council that they were too narrow to have a yard room. Even the rooms for the surgeon and his family and for attendants were not as spacious as they ought to be. The Council thus asked for another model to be drawn which was to be larger and airy, and consequently healthier for the sick people. The Council also asked the Bombay government to first look for the place where it was to be built as they were not much inclined to the place formerly decided by Colonel Bake fearing that the air was not so good there and it was too near the sea. The Council advised the Bombay government to choose a place in line of fortification and with space for gardens. They also advised that the foundation of the building to be of good stone and lime and the ground to be raised to three or four feet high. The walls, they recommended, should be built of sandstone and lime and raised to a convenient height.[49]

In 1676, the Bombay Council suffered severely from illness, resulting in the death of two of its members—Giffard, the Deputy Governor, and Captain Ustick. Petit also was ill for several months. Towards the end of January he went for a week's change of air to the mainland, but it did not do him much good, as at the beginning of February he was still ill and 'very weak and much spent'. It was noted later on that month that he and Giffard 'still linger under a tedious flux, and are both now very weak and reduced to that extremity that, if they recover, they will not be capable of performing their duties for some considerable time'.[50]

At the beginning of March Giffard became much worse and in such pain from an internal ulcer that he could not 'stir off his cot. Even when the pain abated, he had to stay in bed and was so weak that he could transact no business, not even by dictating a letter.' This continued to 8 September, when at his request he was allowed to go to Surat, in the hope that the change might save his life, but it ended there on 22 November.[51] Meanwhile, Captain Ustick died on 30 July after a month's illness. The brunt of the work, therefore, fell on Petit, the 'second' of the Council, though he was himself 'under violent distemper for 8 to 9 months, which brought him to the point of death'.[52]

The scarcity of cash and high cost of materials necessarily hampered building operations, and in November the Surat Council limited these to work on the fort and hospital.[53] In January 1676, the Surat Government reminded Bombay about the place of hospital and its design. The Surat Council enquired about the model they suggested and wanted the hospital to have proper wards, surgeons' lodgings and the surgery rooms to be one story high. The Council also asked for the supply of *chunam* through boats in order to keep expenses under control.[54] Surat on 17 January 1676 again sent the plan of the hospital which would accommodate about 70 persons with overall charges not more than Rs.4,000 to Bombay. The Surat Council also wrote that its yearly maintenance would amount to about Rs.1,000 and would be governed by the chief surgeon and his mates who were to reside there.[55]

Giffard, however, wanted to get the bastion work pushed on first, and the proposal languished until he and Petit suggested a feasible alternative.[56] They had jointly built a large house and some shops at their own expense, for which they each received Rs. 1,500 as an advance of salary under the rules for the encouragement of building that had been sanctioned by the Company. They proposed that the Court of Judicature be shifted to this new house[57] and the premises be converted into a hospital.[58]

This proposal found favour with Aungier and his colleagues, and the house and the ground attached to it, which were valued by a committee at Rs.7,400, were acquired accordingly in October. The work of building a wall round the other house then proceeded, but that of fitting it up as a hospital was held over till the end of December, when Dr. Thomas Wilson arrived from Surat. He was a fellow of the London College of Physicians, who had been sent out by the Company in the hope that he would improve the health of its servants. The Surat Council appointed Dr. Wilson Physician-in-Chief at Bombay.[59] It appeared that the proposed building was never erected and that instead a new court of judicature was built in the bazaar, while the old court situated on the esplanade to southeast of present-day Cooperage was transformed into a hospital in 1677.[60]

The hospital that had been provided in 1676[61] proved to be of great use to the 'poor sick soldiers', as the surgeons could give them better attention and prevent them taking 'hurtful things'. The general health of the inhabitants also appears to have been good, and the 'Physician-General' Wilson was accordingly spared in response to Aungier's request for his services in April.[62]

The records say little about the hospital, but Child highly praises Dr. Bird the surgeon in charge, for his skill, care and experience, as well as 'his exemplary quiet living'. A list of medicine was also sent to Surat in March, and a chest of medicine was also sent out by the Company. It also sent Dr. John Staveley and Dr. Bird seem to have had ample assistance as four other medical signatories (including the apothecary, John Kitson) testified in the case against Ensign Hughes.[63]

With the functioning of the hospital the mortality rate started dropping. This becomes evident from the letters of the Bombay Council to Surat: '...our soldiers thanks be to god continue very healthful for where as last year from October to February there died hundred men, this year we have lost only fifteen most of which imposthumation in the liver, much of which benefit we must attribute to the new hospital…. Our Bombay bills of the mortality have swelled to high, where as in the hospital nothing can come in or out without passing the doctors eye, that he have confidence this hospital will save our honours with 100 pounds yearly which the transport of soldiers exacts….'[64]

Unfortunately, this did not continue for long and the rate of mortality again began to increase in the 1680s. Moreover, circumstances were not helped by the fact that the island was often destitute of physicians and consignment of medicines, which was spasmodically sent out by C.O.D., and they often proved to be bad.[65]

In 1682, the hospital suffered from occasional want of doctors and drugs. At the beginning of the year Dr. Bird and Dr. John Stavely were sent to Surat. In April Dr. Bird badly wanted drugs as the quantities and kind of drugs required (rhubarb and poppy heads) did not arrive from Surat before July. In July several men were ill and even Dr. Bird was confined to his chambers. Kitson, assistant surgeon also died during the same year. This made Ward ask for a surgeon for the island.[66]

This need for a surgeon continued for a few more years as evident from the letters of the Company dated 29 December 1686:

enclosed is the list of the English deceased since or last to this instant with those that are living the mortality has this year been exceeding violent, and shall much want recruits by all opportunities; and chief reason that can be given the absolute want of the good Europe medicines that should have been yearly sent out fresh … year 1682-83 which puts us to use the country physician, which serves only to augment the charge of the garrison and does more harm than good . For a supply what is necessary we now send a list or invoice here enclosed under the hand of the chief surgeon, and an account of what is wanting and necessary to the relief yearly of the stores and marine.[67]

The situation in the following year further worsened and the Bombay government asked Surat to send medicines urgently for the garrison. According to the doctors list out of 35 soldiers sent from England that nearly 19 arrived on shore, 17 of them were sent to the hospital at once as they were suffering from scurvy, of which 2 died.[68] One of the reasons for the wretched conditions of these soldiers, who came on the ship 'Worcester', was that they were badly treated and starved on the voyage. Bombay wrote to the Company to send strong men, otherwise most of them would die by the time they came to Bombay. The Bombay government also instructed the captain's of the ships to supply bread and provisions properly to the soldiers on the voyage.[69] In 1690 it was reported that there were only 35 Englishmen in Bombay who were not ill. While in 1691 there were only 5 civil servants alive.[70]

In 1702, Robert Bartlett, doctor from England, offered his much needed service to the Company. It was much relief to the island as there was no doctor since the death of Dr. Skinner. Mr. Bartlett was thus allowed to serve on the island with a salary of $4 a month to be paid in *xeraphins* according to the custom of the island, together with the same allowance for his diet as

Dr. Skinner was allowed, and the usual assistance belonging to the hospital. It was also decided to bear his charges till he arrived on the island.[71]

In 1720, the Bombay government was again asked to reduce the charges of the hospital. The Bombay government thus appointed a committee comprising of B. Midford and O. Phillipps in September to consider the means to lessen the hospital charges. The committee proposed that the quantity of arrack used in drawing cinnamon water and spirit of wine amounts to a considerable sum, so for the future a stock of it for the whole year be laid in by the paymaster. The drugs should be purchased in quantities rather than in retail to reduce the charges. The food for the sick could be kept in stock by the barrack master rather than brought by the surgeon when required.[72] On the recommendation of the committee the Board ordered the land paymaster to provide a quantity of drugs and arrack for making spirits sufficient for a year's expense and that for the future the barrack master cater for the sick people in the hospital as formerly.[73]

The condition of the hospital however remained bad as it needed more servants and assistants and small necessities for the sick. The Board in 1738 thus decided to take all possible care of the sick people agreeing to appoint additional attendants. The list of the additional attendants is as follows[74]:

				Rs.
One Purvoe Book-keeper	...	...	...	5
One Hospital Assistant	...	...	...	6
Seven Ward Servants	...	...	...	28
One Outdoor Assistant	...	...	...	8
One Bedding Servant	...	...	...	5
TOTAL ...			Rs.	52

The board also passed several regulations to improve the condition of the hospital:

The patients were to be entered under the care of each surgeon respectively as they went into the hospital, each surgeon taking one week. Both surgeons were to constantly visit the hospital and consult with each other in cases of danger. The surgeons should preserve good harmony and agreement with each other as the most likely method to contribute to the relief of the sick. The officers of the military and marine should duly visit the hospital as an encouragement to the sick and observe whether proper care was being taken of them and report what they think may be amiss.[75]

A month later, on the 2 August 1738, the Government communicated their decision to Messrs John Neilson and Michael Weston, Surgeons, in the following letter: 'Some irregularities in the hospital having been complained of and sundry conveniences being represented to be wanting for the better care of the sick, the President and Council have taken the same into consideration and thereupon make the following regulations and orders which I am directed to signify to you for your observance.'

These regulations were:

It being represented that the number of servants and attendants at present allowed are not sufficient to take due care of the sick.

The President and Council have given orders to the paymaster for completing the ceilings of two of the rooms in the hospital already begun and to make two dozen screens to keep the sick from the sight of each other.

The paymaster is also to purchase 100 'banian' shirts, as many drawers and caps, fifty beds, fifty quilts, 100 pillows with sheets and pillow cases for them for the use of the sick. That this number may be always kept up, and if possible without further charge to the Company, half a month's pay is ordered to be stopped from every soldier or sailor that shall go into the hospital with the venereal distemper to be applied for that purpose. And when any such soldiers or sailors are discharged you are to sign it to the land or marine paymaster respectively that their pay may be stopped accordingly.

To prevent the ill-effects of people going out of the hospital and changing their diet too soon, a discretional power is hereby given to you in case of necessity to make some small addition to the present allowance of wine and fresh provisions when a patient is on the road to recovery.

It is ordered that as they go into the hospital the sick be entered under the care of each surgeon, each surgeon taking one week, but that you both constantly visit the hospital and consult with each other in all cases of danger.

That each surgeon have a key of the medicines.

As nothing can contribute more to the relief of your patients than preserving a good harmony between you, the same is positively enjoined.

The officers of the military and marine are ordered daily to visit the hospital as an encouragement to the sick, and are to observe whether proper care be taken of the sick and report what they find amiss.

The Moody's account of hospital charges having pretty much increased of late, you are to take particular care that nothing be charged therein but what is actually used for the hospital.[76]

The condition of the hospital remained pathetic with no proper bedding facilities, medicines, or kitchen utensils required for preparing the diet of patients for over two decades. In 1740, on the report of the inspectors of the hospital, the President informed the board about the pathetic condition of the hospital. Thereafter in consultation with the board he agreed to make sundry repairs, alterations and to provide it with other necessities such as bedding, patient's clothes, kitchen utensils, etc., and to make proper arrangements which hitherto had not been present and remained without account. It was also agreed to make arrangements for frequent visits of the doctors and to meet their demands of medicines.[77]

On the basis of the recommendation made by the President to the board, efforts were taken. The dispensary and the medicine rooms were now well provided with facilities and arrangements were made for the frequent visits of the doctors to the hospital at least once in the day to prescribe to each person what medicine and diet was necessary. The hospital was also provided with native doctors who were also entrusted with the duty to visit patients regularly. The diet for patients was also improved and they were given mutton, fowl and bread. Those who could not digest this food were given milk and other light food.[78]

The other problems of the hospital related to various rooms in the hospital such as the distillery, several wards being exceedingly wet and damp, chiefly imputed to a large amount of water in the kitchen, and the compound wall which still did not have proper drainage for carrying off waste liquids. These and other such problems made the place unwholesome for the patient. For instance the quilts, bed sheets, pillows and the supplies which were promised were not made available. The presidency was running short of funds and required a large amount to cater to the needs of the hospital, as a result of which the lives and the health of many Europeans suffered.[79] The President taking note of these problems laid before the Board certain instructions for putting the hospital in order.

On 14 March 1746 the Board read the letter from Thomas Marsh[80] and others that on visiting the hospital they found the visiting hours of the surgeons as six or seven in the morning. This appeared to them an improper time as they visited only once a day and then ordered what was to be given, they could not know whether the prescriptions would be appropriate or not. Neither could the patients judge their own state or any alterations the night may have produced on them. The letter therefore suggested that eight or nine o'clock would be a more proper time. By that time the surgeons could be informed of what had been ordered the preceding day had had the desired effect, and the patients would also be better able to give an account of themselves in all other respects necessary to be known before new prescriptions were given.[81]

It was also proposed that it might not be well to divide the patients between the two surgeons. As most of the doctors differed in their practice, if the same patient was prescribed indifferently by either, it could be harmful for the patient. In dangerous cases the surgeons should consult with each other for the common benefit of the sick under their respective care. Marsh wrote,

Taking notice to the surgeons of the large quantity of port wine used in the hospital and enquiring whether there might not be cordials given to patients when on recovery that would be equally serviceable and be had on much easier terms, and finding that a cordial may be made of Batavia arrack with spices far preferable in many cases and much cheaper than port it induced us to recommend to the surgeons the practice of it in all such cases, in order to cause some saving in the article of wine. This they promised us to do.[82]

The letter further read that on examining the hospital abstract and computing the necessary services they judged three men may be spared and accordingly recommended striking them off their books. 'Here they esteem it their duty to take notice that only two of these are real, the other nominal, whose pay of Rs.5 a month the head surgeon informs us was an indulgence of Mr. Morely's to him. When these are struck off the allotment will stand thus.'[83]

In the surgery 1 mate and 2 assistants				...	3
A purvoe to take account of medicines					
Linen	...	...	...	...	1
Ward Servitors two to each ward		...		...	12
Vietdalman	...	...	...	...	1
Washeman	...	...	...	...	1
Barber	...	...	...	...	1
Halalkhors	...	...	...	...	5
Black Doctors, one at Mahim and one at					
Bombay	....	...	...	...	2
TOTAL					26

They also proposed 'separate cells for lunatics near the back gate of the hospital that may be fitted up at a small expense the better to secure these unhappy and sometimes disorderly patients'. They also thought it very useful if the hospital servants could all be moved within the gates, so that their services could be utilised in time of emergency. They thus asked the Board if the land paymaster may be ordered to allot them ground for their dwellings, and, if a month or two's pay was advanced to them on this account. These suggestions of the Hospital Inspectors were approved and forwarded to the surgeons as standing orders.[84]

However it seemed that though many things were in order, abuses against the sick continued as was evident from the despatch of 1751 which asked for the monthly inspection of the hospital by the military and the marine paymasters along with the major to avoid any abuses that the sick might face.[85]

Other Problems: War, Storms, Epidemics, Fish Manure

The Mughal-Sidi and Bombay conflict were further adding to the problems. Ovington writes:

... the sword has also done its work and so much were the constitutions of Europeans undermined by the deleterious air, that slight wounds were healed with great difficulty. All wounds and contusions in the flesh are likewise healed here; and if they are; tis with difficulty and extraordinary care; they happen generally to be very dangerous and cure admits of more delays and hazards in the healing, than what is usual in other parts. But the corruption of the air has a more visible and immediate effect upon young English infants; whose tender spirits are less able to resist its impression; so that one of the twenty of them live to any maturity or even beyond their infant days, were it otherwise....[86]

Alexander Hamilton writing in the same period reported: 'The number diminished with a rapidly and truly alarming rate, of seven or eight hundred European who have inhabited before the war not more than sixty were left....'[87]

Bombay during this period again grapped with problems, especially of plague, fever, etc.[88] The Bombay correspondence contains the following information about the great plague which wasted western India between 1686 and 1696. Towards the end of 1686 the Bombay Council writes: 'We have abundance of men sick and many of them die. We are finishing the account of our majesty's ship *Phoenix* but by reason of some of her men lying sick in the hospital, and we know not how God will deal with them, cannot close the account to send up, which as soon as we can, shall be done.'[89]

In 1692, Bombay then had scarcely 800 European inhabitants but only 100 survived the fever of that year. The plague bulletin further unfolded a story even more gruesome that there was only one doctor to attend to the sick, and the sufferings were further increased with the death of Dr. Wilson. He too was the victim of the same disease. This deprived the city of its only medical officer. The Council observed at Surat 'for as such as by the death of Dr. Skinner the island of Bombay was deprived of physician and the island very sickly and a many poor people and soldiers laying in danger of perishing for the want of, the help and the advice of a doctor we resolved to entertain Mr. Barlett in the said station allowing him 4 pound a month to be paid in the xeraphins, according to the customs of the island, together with the assistance belonging to the hospital.'[90]

The pathetic condition of the island is illustrated by the fact that the doctor, who was expected to look after the health of the town, was merely paid fifty pounds a year. It was little wonder that the death was knocking at every door on the island. The climate of Bombay was more fatal to the European than the natives. Ovington once jokingly observed: 'As the ancient gave the epithet of fortunate to some islands in the west, because of their delightfulness and health, so the modern may in the opposition to them, dominate Bombay and unfortunate in the east, because of the antipathy it bears to those two qualities.'[91] It was a sorry joke but true.

The earliest storm of which records remains commenced on 15 May 1618, and was thus described by Cordara.[92]

The sky clouded, thunder burst, and a mighty wind arose. Towards the nightfall a whirlwind raised the waves so high that the people, half dead from fear, thought that their city would be swallowed up. Many provision boats, which were lying at anchor off the shore, were dashed to pieces. In the city and the villages, houses were thrown down and rendered unfit to live in. The monasteries and convents of the Franciscans and Augustinians were utterly ruined. The three largest churches in the city, and both the house and the church of the Jesuits were unroofed and gaped in the clefts almost past off the palm groves. Thousands of palms were torn out by the roots, and some the wind lifted through the air like feathers and carried great distances. The whole was like the ruin at the end of all villages.[93]

That it affected Bombay equally is clear from the following account of Faria e Souza in his Asia Portugueza,

In May 1618, a general and diabolical storm occurred in the neighbourhood of Bombay on the fifteenth of that month, and continued with such violence that the people hid themselves in cellars, it continued dread lest their dwellings should be leveled with the earth; and at 2 p.m. an earthquake destroyed many houses. The sea was brought into the city by the wind, the waves roared fearfully, the tops of the Churches were blown off and immense stones were driven to vast distances. Two thousand persons were killed. The fish died in the ponds and most of the Churches, as the tempest advanced, were utterly destroyed. Many vessels were lost in the port. At Bombay sixty vessels with their cargoes and some of their crews foundered.[94]

On 30 September 1696, Bombay was visited by 'a hard gust of wind from the east and south west with thunder and lightning and some rain'.[95] In 1702–3, a fearful mortality prevailed in Bombay. The natives suffered much, and only 26 Europeans were left. Following the calamity was a violent storm, which destroyed the produce of the island and wrecked the greater part of the shipping.[96]

A terrific storm destroyed three grabs (country vessels) completely armed and equipped on 30 November 1740, and it was followed again on 11 September 1742, by a cyclone, the force of which was such 'as has not been expected in the memory of any one now on the spot'.[97] A Government consultation of 11 September 1742, reads: 'This day had an exceedingly hard storm of wind and rain. The ships in the road drove from their anchors, and a large Moor ship parting her cables ran ashore between Cross Island and Dongrie. The *Somersett* and *Salisbury* ran foul, the *Somersett* breaking her main yard and part of the quarter galley and receiving, it is believed, other damages; the *Salisbury's* head was carried away and part of the cut water. The gale was as excessive as has not been exceeded in the memory of many now on the spot.'[98]

Three days later the President lay before the Council reports from the several outposts of the damage done through the late storm. The rain fell in torrents and all the ships in the harbour were forced from their moorings. The royal ships *Somersett* and *Salisbury* were damaged, along with a large vessel belonging to a Muslim. The Mazagon fort was untiled, the thatched posts at Cooley and Sidi bandars were blown down. The Drong battery and houses and sheds at Suri (Seweri) were untiled and the guard rooms to the eastwards very leaky, though paved with stone cemented with mortar. The Candala and Marine batteries were much damaged and the soldiers guarding them were left with no shelters.[99]

The President went on to observe that Mazgaon was of no strength and the Cooley and the Sidi banders were all the same as the Drong battery. These places were of no further use than to prevent boats landing from Salsette, and hindered people from going off. No boats were permitted to land in any of the bays or creeks between Sion and Sewri or Sewri and Bombay, without first having the permit from the customer of Bombay or Mahim. Sewri commanded a large bay and had been thought a necessary security. As the President conceived all the previously mentioned small posts were requisite for answering the end proposed. He noted that making light *cajans* hutches or roofs with ordinary tiles only drew on the constant expense and in the end were much more chargeable then doing them well at first. Provided that they were thought necessary, he propose to direct the engineer to estimate the expense at every place for rendering them firm and secure, subject only hereafter to the turning of tiles, which was but a small matter. To this the Board agreed, intimating that they will form their determination when the estimate was laid before them.[100]

In 1762, a very violent gale considerable damage to the small craft in and about the harbour, blew down coconut trees and 'in other respects damaged most of the "oarts" ("hortas" i.e. orchards or plantations) and houses on the island'.[101] The storm became the basis of a petition from the 'Fazendars' or tax payers of Mahim and Bombay and the oart farmers, asking for the remission of assessment. The 'Veredores' or officials, calculated the total damage at Rs. 14,330, but the Committee of Accounts reduced his estimate to Rs. 8,500, which amount was sanctioned by the government.[102]

Diseases were as usual prevalent at Bombay. Commodore Harland, finding that he could not agree with Sir Nicholas Waite who reached Bombay from Surat in 1704, implored the Court to have his son settled in Bengal rather than Bombay, a place of mortality, without trade and shipping.[103] There were then on the island eight covenanted civilians, including members of the Council and two more persons who could write and two youth taken from ships. The garrison consisted of six commissioned officers and forty European soldiers, civilians and soldiers were suffering from illness.[104] Their condition was so bad that they called Bombay, 'the unhealthful, depopulated and ruined island'.[105]

In 1707, the civilians were reduced to seven and some were invalids. There were but six commissioned officers, two of them frequently ill, and not even forty English soldiers. In May seven civilians had dwindled to six and were deeply depressed by their condition and wrote: 'it will be morally impossible to continue much longer from going underground, if we have not a large assistance out before October'.[106]

In 1710, Burnell wrote that health conditions had not greatly improved as could be seen from his description of the hospital and the cemetery. The former he described as 'Its enough to make a man die with the thoughts of going in to it … a strange ill fashioned contrived things … seldom empty; few above the degrees of soldiers and sailors entered in it … so many have gone in it ill and have come out so well that they never ailed anything after.'[107]

There was no exaggeration in Burnell describing the burial ground as 'a cormorant paunch never satisfied with daily supplies it receives, but is still gaping for more English bodies. It was my strange experience to stand among the closely packed bones that are buried here. The ground is full of bones as the valley that Ezekiel saw in his vision….'[108]

Various reasons apart from diseases for the unhealthiness of the climate were put forth by the Company's representatives in Bombay. Writing in 1671 to the Court of Directors, the Company servants reported that the appalling conditions were due to the habit of manuring the coconut palms with putrid fish. Hamilton too had the same opinion in this regard: 'They being laid to the trees putrefy and is a cause of most unsavoury smell and in the morning there is generally seen a thick fog among those both but little good water on it, and the air is some what unhealthful to the brains and the lungs of the Europeans and breeds consumptions fevers, fluxes.'[109] Similarly Ovington in 1689 remarked that fish manure is the mortal enemy to the lives of Europeans.

In 1673, Aungier had remarked in his report that after the first intermission of the rains in May and after its total ceasing in October, the air and the water were unwholesome by reason of the crude pestiferous vapours exhaled by the violent heat of the sun into the air and vermin created in the wells and tanks, which render those months most sickly to the inhabitants especially the European.[110] Ovington too held the same opinion about the monsoons, 'The prodigious growth of vermin and of venomous creatures, at the time of monsoon do abundantly likewise demonstrates the malignant corruption of the air and the natural cause of its direful effects upon the European. For spiders here increase their bulk to the largeness of the man's thumb, toad are not of a much less size than a small duck.'[111] Burnell, who visited this place in 1710, also validated the fact.[112] Ives too gave the same opinion.[113]

BOMBAY AND THE PROBLEMS OF BREACHES

One of the other chief reasons of Bombay's decline was the gradual silting up of the creeks which divided Bombay into groups of islets. At high tide the sea swept through the breaches, overflowed the major portion of the island and laid the pestilential deposits highly productive of malaria.[114] In 1675 a proposition for draining the swamps was submitted to the Court of Directors by the Bombay Government and subsequently several surveys were conducted apart from this authority and the order to commence the work was given.[115] The Court of Directors between 1684 and 1710 constantly emphasized the need of stopping up the breaches at Varli and reclaiming the drowned lands.[116] Burnell however informs us that the breaches were completed in 1710 especially in the north between Sion and Dharvi between Mahim and Dharvi and between Mahim and Worli.[117]

In 1711, began the work for the great breach between Worli and Mahalaxmi. Brunell however made no mention of the attempts to close the great breach, though he noted how the sea came through it almost to the foot of the hills on the eastern side of the island.[118] He also gave the detailed account of the 'noble large dam' between Sion and Dharvi and a brief account of the dam between Dharvi and Mahim. Between Mahim and Worli there was, he says a small ferry boat. 'When I left the island they were about damming up this breach, designing to go through with all that remains open to oceans invasions.'[119]

THE EARLY EUROPEANS AND THEIR LIFESTYLES

Added to such natural causes was the dissolute life led by the majority of the Europeans during this epoch. 'I can not', wrote Ovington, 'with out horror mention to what a pitch of action and the enormities were grown in this place. Their principles of action and the consequent evil practices of the English forwarded their miseries and contributed to file the air with those pestilential vapours that seized their vitals and speeded their nasty passage to the other world. Luxury, immodesty and prostitute dissolution of manners found still new matters to work upon....'[120]

Intemperance was ripe to the degree that grog shops flourished exceedingly in the city. In a report to the Company at the beginning of 1671, Giffard wrote: 'Several persons of ordinary sort come out of both sexes, whose lives and carriages not being enquired in to, prove when they arrive hear so strangely debauched and factious that they are not only dangerous and troublesome, but are also a disgrace both to their country and religion.'[121] This led the officers to ask for English women for the soldiers on the island in order to put a check on such activities.[122]

The Company in 1669 had sent out single women, not only of a class suitable to become soldiers' wives but also 'some gentlewoman', who had been recommended for such treatment and respect 'as their virtues shall deserve'.[123] But this also did not solve the problem as many women failed to get a husband. This becomes evident from Giffard's remark which shows that some of them did not succeed in getting husbands, viz., 'The ladies that are unmarried begin to despair and desire leave to go home if they were not more successful in the following two months.'[124]

The pleasure seeking habits of the English soldiers seems to have continued, as in 1673 Aungier again suggested the Company should send out English women 'of meaner sort, but of honest reputation, for the soldiers do frequently converse with the Country women, whom we force them to marry for preventing sin and God's judgement thereon'.[125] In response to the Aungier's suggestion, in 1675 the Company had sent out a considerable number of 'sober young women of the meaner sort that may be fit for soldier's wives'. Some of them declared that, 'They had been promised for lodging and diet at the Company's cost for a year and a day....'[126]

Giffard proposed they should be given this, as had been done in 1668,[127] saying they had nothing with them and must either starve or do worse, unless this was done. Also he thought they would not be chargeable long, 'as they goe pretty fast, some married, some sure (and) some in a fair way (to matrimony)'. Aungier and his Council, however, held that the allowance they asked for could not be granted without the Company's order for it, and only sanctioned a charitable pittance of 8 *xeraphins* a month for those who could not subsist without it.[128] Giffard on the other said he could get out one to board them at the rate and feared they would be turned out of their lodgings, 'which will be a new trouble'.[129]

This refers to another source of worry: a few of the women by scandalous behaviour 'not only daily dishonour the nation and their own sex, but declare they will use their utmost endeavour to make their impudence more notorious'. The President and the Council ordered such women to be warned that, unless they reformed, they would be confined and put on bread and water, pending deportation to England.[130] Another difficulty was that women who 'come out yearly, be what they will, at their arrival all pretend to be gentlewomen, high born, (of) great parentage and relations, and scorn to marry under a factor or commissioned officer, though ready to starve'. He wanted 'poor country girls or hospital girls' to be sent out instead, after a strict enquiry into their lives and conversation.[131]

In 1677, the Company sent out twelve young women, intended as wives for the soldiers, saying they had taken care to choose only 'civil' ones and they had not been able to get any 'country girls' of the kind asked for.[132] In view, however, of the trouble these women gave and the scarcity of the English soldiers, both the Surat and Bombay Councils would have preferred fewer women and more men.[133]

It seems the problem of intemperance among English soldiers remained, as in August 1694 John Wright applied for and gained permission to keep a tavern in Bombay and no doubt drove a roaring trade. The authorities tried to check intemperance, but it had grown beyond control. The legislature passed as such during that period of lawlessness had ordained that 'if any man comes in to a victual house to Bombay to drink, punch he may demand one quart of good Goa arak, half a pound of sugar, and if the bowle be not marked with the clerk of the market's scale then the bowle may be freely broken with out paying anything either for bowle or punch'.[134]

In order to check the over indulgence by the Englishmen and the women, the Company officers were demanding again that the English women be sent out to the island. Along with this they also tried to introduce strict measures to control these habits. In 1674 for the moral reformation of the Englishmen Aungier introduced regulation against excessive gaming and keeping punch houses. He also took steps to check brothels and prevented soldiers from keeping wenches or loose women in their house.[135] Giffard tried to check drunkenness by the proclamation prohibiting the supply of drink to soldiers, except for ready money.[136]

In the face of dramatic mortality, it was hardly surprising that the small community of European adopted an 'eat, drink and be merry' attitude which only created yet another source of death:

its true that many Englishmen die here very suddenly, but in my opinion the fault is chiefly their own, they eat much succulent food particularly beef and pork, which the ancient legislation have forbidden for good reasons to the Indians; they drink very strong Portugal wines at the hottest time of the day; in addition they wear as in Europe tight fitting cloth which are useless in these countries since they are much more sensitive to the heat than the Indians with their long and flowing garments.[137]

In 1676, mortality was high amongst the new soldiers mainly after the initial months of their arrival on the island, occasioned chiefly by their immoderate drinking of punch and toddy. Aungier had five files of them sent up to Surat on ships in September, hoping this would 'preserve and inure them to the climate'. The recruits included several artificers and some Germans, such as Aungier had asked for,[138] and the Company called for a report about the behaviour of the foreigners in the corps. Petit said there were but few Germans, who were all civil, quiet persons, as were also (so far as he could learn) the other foreign men sent out. They were, moreover in general more frugal and hardy, and less given than the English, 'who can't live without flesh and strong drink'.[139]

In 1685, the deputy governor attributed the problems of the English to the actions of the native Christians, Indo-Portuguese with whom the soldiers were quartered, who discovered *dhatura* and other poison as the surest medium for revenge for the outrages upon their wives and families.[140] Attempts were also made to check the evil by building barracks and by encouraging soldiers to marry women of the country.[141] But orders of this nature made little impression on this class described by the Council at Surat as composed of 'debauched broken tradesman and renegade seaman'.[142]

The other cause of this unhealthiness of Bombay was largely attributed to the old standing practice of *buckshaw*, i.e. manuring coconut trees with putrid fish.[143] The Company in their dispatch of 22 February expressed the similar view that this was prejudicial to health, and called upon the Surat Council to consider whether it could not be forbidden. Accordingly notice was given to the inhabitants of Bombay to discontinue the practice. But both Aungier and Giffard anticipated there would be difficulties; many years were to pass before the prohibition became effective.[144]

The second decade of the eighteenth century however began to augur well for the island. One of the chief reasons was the policies which the Court of Directors began to take in the middle of the eighteenth century. For instance they advised the Bombay Council to issue orders prohibiting the *buckshowing* of the toddy trees in the Mahim and the Worli woods to allow the free flow of the western breeze to stop the breaches. To burn the continual fires for manure was universally prohibited, dry manure being permitted up to 1766 when it was discovered that the indulgences was turned to bad uses and this practice was also discontinued.[145] (The ill-effect of this practice had been reported by many early travellers, but the Council did not want to diminish the scale of cutting down the trees while the Company derived revenue for Rs. 25 per thousand. Coconuts presented this dilemma of losing the coconut revenue. The Council corresponded with Bombay for many years and finally abolished this practice.)

Despite all these difficulties, the English continued to fight with the unhealthiness of the island that was compounded by their lifestyle. In 1720 a dam was constructed across the great breach at Mahalaxmi and a considerable area of marshy ground was drained. This gradual reclamation of the land from the inroads to the sea coupled with better medical attendance and a more temperate style introduced a considerable change for the better, and by the middle of the eighteenth century Bombay began to lose its reputation of insalubrity and was accounted for as a tolerably healthy station. Cobbe, who visited this place in 1715 also reported the same, i.e. that the climate was healthier compared to the earlier days because of stopping of the *buckshow* and repairing of several sea breaches.[146]

Another reason that could be cited as the improvement in the European's diet and manners of living and also to the place being provided with more skilful physicians than earlier who were also well paid.[147] Grose in 1750 remarked that, 'the climate is no longer fatal to the English inhabitants as it used to be and is incomparably more healthy than that of many of our settlements in India ... this is no longer to be dreaded on account, provided any common measure of temperance be observed without which the tenure of health's in any climate be hazardous'.[148] Edwards Ives visiting Bombay in 1754 also held the same opinion about the place: 'the island of Bombay of late have been healthier than it was formally and ... could be justly styled as the grand store house of all the Arabian and Persian commerce'.[149]

In spite of Grose and Ives' remarks Bombay had still not come out of the clutches of mortality. For instance in 1757, there was serious a epidemic among the labourers employed on the fortifications which necessitated the appointment of the country doctors, whose medicine met with great success.[150] In fact according to the consultation of 18 November the town had became so dirty that the Bombay government decided to appoint a member of the Board to the office of the scavenger and to defray the cost of the sufficient number of labourers and scavengers carts by a tax upon the town people.[151] The mortality rate was high till this date but still the island was much safer than what it was in the seventeenth and early eighteenth century which led to the saying 'two monsoon were the ages of man'.

NOTES

1. Da Cunha, *Origin of Bombay*, Bombay, 1900, p. 5.

2. Ovington, *A Voyage to Surat In The Year 1689*, ed. H.G. Rawlinson, London, 1929, p. 87.

3. S.M. Edwards, *Gazetteer of Bombay City and Island* (hereafter, *GBCI*), Bombay, 1909, vol. I, pp. 150-4.

4. Ovington, *A Voyage to Surat*, p. 86.

5. Anderson, Philip, *English in Western India*, London, 1856, p. 62.

6. *GBCI*, vol. I, pp. 29-30.

7. Roe Thomas and Fryer, *Travels in India 17th Century*, New Delhi, 1993, p. 241.

8. Emphasis is mine. Thevenot who visited western India in 1666 also, talks about it, i.e. 'flux alone is a common and a dangerous distemper in the Indies. For many die of it.' Sen Surendranath, *The Indian Travels of Thevenot and Carrie*, p. 151, fn. 15, p. 336, Linshotten also found this distemper quite common at Goa.

9. Roe Thomas and Fryer, *Travels*, p. 240.

10. Morshideen was doubtless a corruption of Marathi modsi, from modane to break, in allusion to the internal agony which characterized the attacks. It was further corrupted by the English into mortane chine or Chinese death; *GBCI*, fn.3, p. 161.

11. The emphasis is mine. Cholera was first commonly used by Europeans at the beginning of the nineteenth century, it was previously commonly called 'morteshien'. The following are the symptoms which were given by the Portuguese writer in Goa: 'and this malady attacked the stomach, caused as some experts affirmed by the drill; though later it was maintained that no cause what ever could be discovered. The malady was so powerful and evil that it immediately produced the symptoms of the strong poison for example vomiting, constant desire for water ,with drying of the stomach, and cramps that contracted the hams and the soles of the feet, with such pains that the patient seemed dead with the eyes broken and the nails of the fingers and toes black and crumpled and for this malady our physician never found any cure and the patient was carried off in one day, or at the most a day and night, in so much that not ten in a hundred recovered and those who did recover were such as were healed in haste with medicines of the little importance known to the natives'; Cf. Wilkinson's *Two Monsoons*, pp. 171-2.

12. Sen Surendranath, *The Indian Travels*, p. 151.

13. Ibid., p. 62.

14. Ibid., fn. 2, p. 62.

15. Ibid., p. 151.

16. Flux was the epidemical disease of this place, for which the best and the most approved remedy is the marmalade of Bussora. E.F.I., 1668-9, p. 68.

17. Ibid., pp. 72-3.

18. Ibid., p. 75.

19. Ibid., p. 68.

20. Ibid., p. 86.

21. Ibid., p. 95.

22. Ibid., p. 89.

23. E.F.I., 1670-7, p. 19.

24. Ibid., p. 5.

25. Ibid.

26. Ibid.

27. Ibid., p. 177.

28. Ibid., xv, p. 5.

29. Ibid.

30. Ibid.

31. *GBCI*, III, p. 180.

32. Campbell, *Material towards a statistical account of the town and island of Bombay,* Bombay, 1893, III, p. 544.

33. *GBCI*, III, p. 162.

34. Ibid., p. 154.

35. Ibid., p. 155.

36. Ibid., p. 16.

37. Ibid., p. 26.

38. Ibid., p. 27.

39. He was appointed at Bombay in July 1668. E.E. Sanisbury, *Court Minutes of East India Company, 1668-70* (hereafter *Court Minutes*), Oxford, 1938, pp. 37-8.
40. He was formally employed in Persia. EFI, 1668-9, p. 41.
41. Ibid., p. 28.
42. Ibid., pp. 40-1.
43. Ibid., p. 81.
44. Fryer, op.cit., p. 30.
45. EFI, 1670-77, pp. 81-2.
46. Ibid., pp. 16, 42 and 81.
47. Ibid., pp. 129, 141. A clear instance of Aungier's desire to improve the conditions for Europeans at Bombay was his request to the Company to send more medicines to that place than to Surat. He pointed out that the former had nearly 300 Englishman on shore and ships as against some 30 at Surat. In February he engaged two surgeon mates, in view of the large amount of sickness in the garrison. The principal surgeons at this place were Dr. John Fryer and Dr. John Bird, ibid., p. 131.
48. G.W. Forrest, *Selection from the letters, dispatches and other State papers, Home Series*, Bombay, 1887, vol. I (hereafter *FHS*), p. 74.
49. Ibid.
50. EFI, 1670-7, p. 144.
51. Ibid., p. 272.
52. Ibid., pp. 144, 151.
53. Ibid., p. 161.
54. *FHS*, vol. I, p. 78.
55. Campbell, vol. III, p. 543.
56. Letter of Surat Government to Bombay, 19 July 1676, 'We like well your proposition of making the present Court of Judicature and hospital and the Deputy Governor's and Mr. Petit's house the Court of Judicature. At the same time we would be further satisfied how the Deputy Governor and Mr. Petit mean by sparing their house to the Company whether they design renting it annually to them or selling it outright; if the first, then what rent they demand, for we would not put the Company to too great a charge. If they wish to sell, then how will they oblige themselves to the Company for the money they have received on account of their arrears of salary, for at present we look upon the house as sufficient security which is also agreeable to the Company's order. But let the house be valued by any indifferent persons and the valuation thereof sent us and then we shall be able to give a more definitive answer.' *FHS*, vol. I, pp. 98-9.
57. This was 'a fair large brick house tiled, built in the line of the town in that part where the English colony is to be' which was acquired in 1675. Ibid., fn.
58. Ibid., p. 162.
59. Ibid., *FHS.*, vol. I, p. 101.
60. Ibid., p. 178. Bombay 24 January 1677; *GBCI*, vol. III, p. 106; *FHS*, vol. I, pp. 98-9.
61. Ibid., p. 162.
62. Ibid., p. 278. Petit and a member of his Council were exceptions. Mansell Smith during the second half of the year suffered from 'a bone', and thought he was not long for this world if he stayed on in Bombay. Petit's case was much the same. His health had not improved. Ibid., p. 186.
63. EFI, 1678-84, p. 78.

64. Ibid.

65. The first physician that was sent was Dr. Wilson, appointed in 1676. In 1680 a surgeon on 45 shillings a month and an assistant on 30 shillings were sent from England; Edwards, *Rise of Bombay*, Bombay, 1902, p. 67.

66. EFI, 1678–84, p. 122.

67. *FHS*, vol. I, p. 148.

68. Campbell, vol. III, p. 77.

69. *FHS*, vol. I, p. 150.

70. *GBCI*, vol. III, p. 162.

71. *FHS*, vol. I. p. 245.

72. Public Department Diaries (hereafter PDD), 1-A of 1720, pp. 233-4.

73. Ibid., p. 235. For their garrison and military, the Government had to purchase Goa and Batavia arrack. Entries in the Diaries show the following prices: (1) Batavia arrack, Rs. 90 the leaguer of 150 Gallons in 1731; Rs. 120 in 1735; Rs. 95 in 1736; Rs. 120 in 1737; Rs. 150 in 1744; and Rs. 225 in 1773. (2) Goa arrack, Rs. 14 and Rs. 16 the hogshead of 50 gallons in 1735; Rs. 15½ in 1737; Rs. 20 and Rs. 26 in 1742; Rs. 27 in 1744; Rs. 26 in 1745 and 1748; Rs. 25 in 1749; and Rs. 23 in 1751. Campbell, vol. III, p. 546.

74. Campbell, vol. III, pp. 547-8.

75. PDD, 10-A of 1737-8, pp. 244-5.

76. Campbell, vol. III, p. 548.

77. PDD, 13-A of 1740, p. 370.

78. Ibid., pp. 384-7.

79. Ibid.

80. Hospital inspectors.

81. PDD, 19-I of 1746, pp. 70-1.

82. Ibid., p. 72.

83. Ibid., p. 73.

84. Ibid., pp. 74-5.

85. PDD, 24-B/1751, p. 578.

86. Ovington, *A Voyage to Surat*, p. 89.

87. Hamilton, op.cit., p. 237.

88. The plague appeared in Aurangzeb's camp in 1684 and in 1689, at Surat in 1684 and 1690, at Bassien in1690, and in Bombay at intervals from 1689-1702; *GBCI*, vol. III, fn.2, p. 164.

89. *FHS*, vol. I, p. 155.

90. Ibid., p. 245.

91. Ovington, *A Voyage to Surat*, p. 86.

92. *GBCI*, vol. I, p. 97.

93. Ibid. This account refers chiefly to Bassein, and the ruin the storm wrought there was still visible several years later. Cited from, *GBCI*, vol. I, fn. 2, pp. 97-8.

94. Faria e Souza in his Asia Portugueza, vol. III, Cf. *GBCI*, vol. I, fn. 2, p. 97.

95. Campbell, p. 111.

96. Ibid., p. 139.

97. *GBCI*, vol. I, p. 98.

98. PDD, 15/1741–42, p. 428.

99. Ibid.

100. Ibid., pp. 438-9.

101. *GBCI*, vol. I, p. 99.

102. Ibid., fn.2, p. 99.

103. Burnell, *Bombay in the days of Queen Anne Introduction and Notes By T. Sheppard*, London,1933, pp. xvi–xvii.

104. Ibid., p. xvii.

105. Anderson, *English in Western India*, p. 172.

106. Burnell, op. cit., p. xviii.

107. Ibid., p. 24.

108. Ibid., pp. 24-5.

109. Hamilton, *A New Account*, p. 184.

110. *GBCI*, vol. III, pp. 162-3.

111. Ovington, *A Voyage to Surat*, p. 88.

112. Burnell, op. cit., p. 5.

113. Ives, *A voyage from England to India in the year MDCCLIV and an historical narrative*, pp. 42-3.

114. Anderson, *English in Western India*, p. 62.

115. Ibid., p. 63

116. *GBCI*, vol. III, p. 56.

117. Burnell, op.cit., p. xix.

118. Ibid., p. 70.

119. Ibid., p. 80.

120. Ovington, *A Voyage of Surat*, p. 86.

121. EFI, 1670-7, p. 19.

122. This is also corroborated by Fryer: 'To propagate their colony the Company sent out English women, but they beget the sickly generation and as the Dutch will observe, those thrive better that come of an European father and Indian mothers, which (not reflect on what creatures are sent abroad) may be attributed to their living at large, not debarring themselves wine and strong drink, which immoderately used, inflames the blood and spoils the milk in these hot countries as Aristotle long ago said...' He further says 'not withstanding this mortality country people and naturalised Portugal's live to good age, supposed to be reward of their temperance, indulging themselves neither in strong drinks, nor devouring flesh as we do. But I believe rather we are here, as exotic plants brought home to us, not agreeable to the soil; for to the loftier and fresher, and often times the temperate, the climate more unkind; but to old men and women it seems to be more suitable.' Fryer, op.cit., p. 69.

123. EFI, 1668-9, pp. 69, 241.

124. EFI, 1670-7, p. 19.

125. Ibid., p. 73. 'The previous experiment of this kind had not been altogether successful as in the year 1669, ...' we do not only send you English soldiers and their wives but also single women, that may become wives to our soldiers and others there. And the entertaining of the person of the quality above the soldier, who intended to proceed with his wife, had also encouraged some gentlewomen, who though we did not invite thereto, yet being engaged therein we do recommend them unto you, that they may be there civilly treated and have that respect from you as their virtues shall deserve. *EFI*, 1668-69, pp. 240-1.

126. EFI, 1670-7, p. 139.

127. EFI, 1668-9, p. 247.

128. EFI, 1670-7, p. 139.

129. Ibid., p. 140.

130. Ibid., p. 140. Anderson criticises Aungier for showing in this matter 'much protestant zeal, but little Christian love'. Against this is Giffard's statement that they were going 'pretty fast' towards matrimony, and would probably have reported the matter if Aungier's orders had the ill effects suggested by Anderson. The extant correspondence contains no further reference to this subject. Anderson, *English in Western India*, pp. 217-8.

131. Ibid.

132. Ibid.

133. Ibid., p. 177.

134. Malabari, *The Making of Bombay*, p. 227.

135. EFI, 1670-7, p. 111.

136. Ibid., p. 28.

137. Neibure as cited from Percival Spear's *The Naboobs*, New Delhi, 1963, pp. 66-7.

138. EFI, 1670-77, p. 73.

139. Ibid., p. 153.

140. Anderson, *English in Western India,* pp. 213-4.

141. The women of this class after marriage were compelled to continue wearing their native dress except in the cases, where they paid to the Company for the right to wear the Company's attire.

142. Court to the Directors to Bombay, 14 July 1680, *GBCI*, vol. II, p. 259.

143. Fryer, pp. 68-9; Hamilton, vol. I, p. 181; the fish used is the small one known as bummelo, which in its dried state is famous under the name of Bombay duck. EFI, 1670-7, p. 28, fn.3.

144. EFI, 1670-7, p. 28.

145. Spear, Percival, *The Naboobs*, p. 63.

146. Cf. David. *The History of Bombay, 1661-1708*, Bombay, 1973, p. 406.

147. Grose as Cf. Douglas James,*English and Western India*, pp. 134-5.

148. Grose as Cf. *GBCI*, vol. III, p. 165.

149. Ives, *A Voyage from England to India*, p. 31.

150. Campbell, vol. III, p. 554.

151. *GBCI*, vol. III, p. 166.

From Fishermen to Boatmen:
The *Mucquas* of Madras, 1650–1750

Jangkhomang Guite

The colonial construction of urban space was not simply a physical entity; it also connotes a relational identity, socially arranged and functionally specific, and, a change, both in social and economic terms.[1] Colonial Madras was also spatially and socially divided into 'white' and 'black' towns, 'left' and 'right' hand caste settlements, and within which was segregations on the basis of functional specialization marked with competition for power and privileges leading to intense struggle over space, idioms, symbols, and colour.[2] Much has been done on these aspects of urbanism and more refined studies are underway. This essay also outlines the redefinition and stabilization of caste structure and social demarcation within the domain of the right hand caste group, which was most ostensible in the fishing community popularly known as *Mucquas* or boatmen who were mostly Catholics by faith. Further, this essay underlines the fact that under the colonial scheme of space, caste, religion and political economy this erstwhile insignificant social group not only gained social and economic importance but also resorted to gain more and more of them, predicated with various forms of resistance to further their demands.

SOCIAL AND SPATIAL DIVISIONS

John Shortt was able to trace thirteen fishing 'tribes' in southern India out of eighteen 'who make fishing their calling' according to the legend.[3] Of these, Putteenevens, Curreans and Shembadavens relate to the fishing community of

*This paper is based on my Ph.D. thesis and was presented partly at the seminar on Coastal Histories at the Centre for Historical Studies, Jawaharlal Nehru University. I am grateful to my supervisor Prof. Yogesh Sharma and other participants of the seminar for their comments. I use the term *Mucquas* as it appears in English records and interchangeably used with boatmen to mean the entire fishing community of Madras.

Madras respectively known as *Pattnawarrs*,[4] *Cariallwarrs*,[5] and *Chumbudoo* or *Chomboddewar*.[6] He said, Shembadaven was the chief among the fishermen who fished only in the inland rivers and never took to sea fishing as it was 'against their caste to do so'.[7] On the other hand, Putteenevans and Curreans took to fishing on the sea.[8] In an assessment for contribution towards the construction of Black Town Wall in 1706 the three fishing castes were mentioned, viz., 'Chomboddeewar caste, alias Fishermen', 'Correala wars alias Muckquas', and 'Pottanapwarr alias Cattamaran men'.[9] However, this division seems to have been artificial if not invented as such traditional caste division within the fishing community seem to have been not much in vogue if not disuse in Madras. This is said because all of them were known to have been involved both in fishing and shipping activities depending upon the season. In fact, the term 'boatmen' was interchangeably used in the factory records to refer to all the fishing community of Madras who were involved in the shipping activities of the port. Besides, most of them had already been converted to Christianity probably in St. Thome from where most of these people came.[10] Hence, such division seems to have been redefined as given by custom to suit the interest of the English merchant government.

The social divisions of the erstwhile fishermen were further intensified by the spatial arrangement of their settlements. They were consciously located in a separate zone wherever possible. Usually, the fishermen were allotted the lands by the seaside. John Shortt has noted that these people live all along the Coromandel Coast and the village they occupy was called *Coopum*. They claimed propriety right to the soil extending from the sea to 120 yards inland and their villages were 2 to 3 miles apart divided by certain landmarks in between, in the coast and on the sea.[11] When the English first settle down in Madras there were already some fishermen's houses there.[12] But it is not known where they have been resettled.

The *Mucquas* or boatmen were first known to have been located along the seaside of the 'Black Town'. The caste dispute settlement of 1652 allocated separate streets for them for their wedding and burial processions, starting from their settlement at the seaside of the 'Black Town' until the Portuguese Church in the 'White Town'.[13] Some boatmen had also settled down 'by the sea side in Mootell Patta'.[14] By the 1670s when their population increased in numbers due to migration or otherwise; many of them were regrouped in a newly created 'town', mentioned again and again in the records as 'Mucquaw Town'.[15] Salmon (1699/1700) succinctly puts that 'there is a little suburb to the southward of the White Town, inhabited only by the black watermen and fishermen.'[16] This place is also shown in the map of Thomas Pitt as *Coupong* probably a corruption of *Coopum* (a fishermen's village).[17] Lockyer (1711) also noted that the 'English Town' (i.e. 'White Town') was joined in the southward by a 'Maqua Town, where the Boat-men live.'[18] This 'town' was probably inhabited between 1673 and 1679 as the first reference of it came only in 1679, and had escaped the description of John Fryer (1673) and others.[19] However, during the French occupation of Madras (1746-9) they were again removed from this 'town' to the nearby place, on the other side of the river, called Chepauk where they erected a church there, in the 'Mile end'.[20] Besides, many of them have also now settled down in and around the areas of Luz Church adjacent to Chepauk.[21]

The catamaran men were located near the seaside in the sandy tract between the sea and the 'Black Town'. On 21 November 1695 a heavy storm devastated the city of Madras where 'from York Point (the north-east corner bastion of the White Town Wall) to the Mud Point (the north-east corner bastion of the Black Town Wall) the sea hath washed down a considerable number of houses inhabited by Cattamaran men and poor people'.[22] We have also noted that they, along with the boatmen, were allotted separate streets in the 'Black Town' in 1652 caste dispute settlement.[23]

The fishermen were allotted the ground near the seaside of Muttial Petta where they lived together with the boatmen and lascars. The 1707 caste dispute settlement had stated that the right hand caste 'Boatmen, Laskars and Fishermen that have their houses by the sea side in Mootell Patta' were to remain undisturbed, but they should not give 'any molestation to the left hand caste'.[24]

Besides, Thomas Pitt's map had also identified some 'fishing huts' by the seaside on the other side of the river Coum, opposite to the 'Mucqua Town'. This place later came to be known as 'Chepauk' where many Christian fisher folks were relocated after the French occupation. Bowrey's fisherman, who survived the storm after being continuously on the sea without a boat hailed from Egmore indicating some fishermen settlement in that place.

Then, how do we explain this? Patrick Roche has shown that the social and spatial arrangement was mainly because of the stated policy of the British Merchant government to intensify the caste division and other demarcation in order to further their colonial interests.[25] The same view is also taken here. It was the stated policy for the merchant government to encourage the migration of as many people to settle down in Madras. The Directors often asked Madras government to encourage such migration but warned that it 'must never let the people find they are too many for you, for if you once throw the reins on their neck the next step may be they will throw you out of the saddle'. It directed that the best way to preserve the Company's government is to deal its subjects 'with a gentle hand'.[26] The 'gentle hand' it referred to was to divide the people as small as possible so that they could be controlled most effectively. For instance, in 1708 the heads of the weavers and oilmen castes were summoned in the Council to declare their caste as 'they were very fickle in their cast which troubles both castes'. The weavers declared that they belonged to the left hand caste and the 'Oylmen' to the right hand caste. They were then ordered to retain their declaration strictly or 'be severely punished'.[27] Control over the caste groups was made effective through their caste heads who were appointed by the government through whom it entered into the social network of its subjected people. It was the tradition in Madras to appoint 'heads' for each caste 'for the better keeping of those people in order and fast to the place'.[28]

In the case of the fishermen we have seen that they were divided into three groups on the basis of their traditional caste distinction despite the fact that most of them had already become Christians and that they all took to fishing and shipping. Locating them separately in different settlements further intensified this. Besides, they were located along the seaside adjacent to the left hand caste habitation in the 'Black Town' and Mutial Petta which separated them physically from their fellow right hand caste group of the 'Black Town' and Peddanaigue Petta.[29] Some boatmen were later transferred or regrouped in a separate location in the 'Mucqua Town', south of the 'White Town'.

The reasons for choosing the seaside for the settlement of the fisher folks can be twofold. We have seen that the fishermen traditionally claimed propriety rights to the soil extending from the sea to 120 yards inland.[30] They cannot be divorced from this traditional right as their predominant activities lie on the sea, i.e. fishing and shipping and dock work in the port. As it was the merchant government's scheme to redefine and intensify the traditional forms of division, and rights, accordingly that the fishermen were located near the seaside although they belonged to the right hand caste. This was the first reason, but there is more explanation for this. We know that most of the boatmen of Madras were Christians and by virtue of this they were closer to the merchant government. But under the colonial scheme of urban settlement the Indian Christians were not permitted to live within the 'Christian town' which was an exclusive reserve for the 'White' people, hence 'White Town.' At the same time, they were to be located separately from their fellow right hand caste group. Hence, both the schemes of religion and caste met at the seaside of Madras. It was because of this that the fishermen's settlements were firmly protected by the government whenever there was any regrouping done among the various castes, such as during the caste conflict settlement in 1652 and 1707.[31] The indispensable role of the boatmen at the port of Madras may be added as another important factor in locating them at the seaside. This will be dealt with in detail later. Thus, the unique spatial and social arrangement invented by the Company government in Madras was mainly to foster their colonial interests.

FROM FISHERMEN TO BOATMEN

Initially people generally took up employment in the urban space of Madras which was to their traditional 'calling'. However, this gradually gave way to the diversification of vocation for the upcoming urbanites especially due to the expanding opportunities. This is most ostensible in the shipping activities of Madras which were mainly performed by the *Mucquas* or boatmen as a full-time or part-time employment alongside their traditional fishing occupation.

However, before taking up this it would be pertinent to examine the two country boats expertly used by the boatmen in their new employment.

The 'Extraordinary Crafts': Catamarans and Masulas

The *Mucquas* of Madras mainly used two types of country boats for ferrying goods and men at the port, as well as for fishing: catamaran and *masula*. Catamarans are 'nothing more than three or four little planks or beams joined together and fastened securely like a raft … covered with mat and could carry a little sail made of bark of trees … the owner sat partly in the water rowing with his feet, and also with oar which had flat ends'.[32] According to John Fryer, a catamaran is a 'log lashed to that advantage that they wast off all their goods, only having a sail in the midst and paddles to guide them'.[33] Francois Martin said that 'depending upon their size, the catamaran is made of five or six pieces of wood which are bound and fastened together'.[34] To Abbe Carre it is 'three pieces of wood tied together'.[35] Colonel W.Campbell (1864) wrote that the catamaran 'is merely three rough logs of wood, firmly lashed together with ropes formed from the inner bark of the coconut tree. Upon this one, two, or three men, according to the size of catamaran, sit on their heels in a kneeling posture, and, defying wind and weather, make their way through the raging surf which beats upon the coast, and paddle out to sea at times, when no other craft can venture to face it.'[36] Lady Dufferin on her state arrival at Madras noted that a catamaran is 'two logs of wood lashed together, forming a very small and narrow raft'.[37]

According to John Shortt, catamarans or 'floats', consist of 'two to five logs of wood with curved extremities'.[38] He also said that the wood used for the purpose is that of 'Peruvian lilac, or *Melia Azadarach,* the Tamil name is Mullay bamboo, sometime the wood of the *Erythrina Indica,* or Moochee wood, is brought into use' and 'the dimensions of the logs are 20 feet long and 2 feet wide';'single logs are sometimes used by men who fish with a hook and line'.[39] He noted that catamarans had small sails, which are fastened to the log by an upright pole, and brought into play when the wind is favourable; otherwise each man uses a short oar or paddle and the anchor consist of a forked stick, which is weighted with stones.[40] Thus, we see the whole structure of catamarans, which were made of logs, fastened together, having curved extremities, small sails at the middle and rowed with oars or paddles.

Catamarans were used both for fishing as well as for ferrying heavy goods. Thomas Bowrey had succinctly put it that 'when any great Ordinance, Anchors, butts of water or the like ponderous ladeinge is carried off or on, they seize 4, 5, or 6 large pieces of boyant timber together, and this they call a cattamaran, upon which they can lade 3 or 4 tunns weight. When they go on fishinge, they are ready with very small ones of the like kind.'[41] These catamarans were extremely useful in Madras port especially for the heavy items such as elephants, horses, teak and timbers or other heavy materials that came into this port in large number.

The other means of transportation was the *masula*. This was a large surf boat built of planks roped together. This country boat was variously labelled by European accounts as *mussulas, mussoolas, massolas, massoolah, macule, masuris,* etc. But it was later largely standardized as *masula*. Thomas Bowrey described the *masula* as a boat 'built very sleight, having noe timbers in them, save shafts to hold their sides together, theire planks are very broad and thinne, sowed together with cayre,[42] being flat bottomed and every way much deformed, as on the other side demonstrated.'[43] John Fryer had also noted as 'a boat wherein ten men paddle, the two aftermost of whom are the steersmen, using their paddles instead of a rudder; the boat is not strengthened with Knee-Timber, as ours are; the bended planks are sowed together with rope-yarn of the cocoe, and calked with Dammar (a sort of rosin taken out of the sea) so artificially.'[44] Francois Martin called *masulas* as *chelingues* or *masuris*. According to him they were 'flate-bottomed boats made of several planks joined by rope yarn'.[45] *Masula* according to Lockyer is a 'large, flat-bottomed ill-shap'd Boats, not nail'd as ours, but sow'd together with coyr-twine, whence they are so pliable that the planks never start with the most violent shocks'.[46] To Lady Dufferin *masulas* are 'deep barges, the planks of which are sewn together to give elasticity, and the interstices stuffed with straw'.[47] John Shortt also noted that *masulas* are made of 'pieces being sewn together with cocoa-nut coir yarns, and from their yielding nature, resist the force of the waves better'.[48]

The *masulas* were mainly used to ferry lighter goods items. Bowrey said that *masulas* are 'for little use save carryinge of light goods (as bailes of Callicoes or silkes, not exceedinge 6 or 8 at one time)'.[49] Abbe Carre also said that *masulas*, which he called 'piphilis', are 'used to load merchandise on board ships'.[50] This boat was most commonly used at Madras port as textiles and spice bales were dominant items of export and import there.

These two boats were indispensable for Madras port due to the shallow sea and high surf along the coast. Bowrey said that these kind of boats 'are most proper for this coast, for all along the shore, the sea runneth high and breaketh, to which they doe buckle and also to the ground when they strike.'[51] Lockyer also said that 'here being a very high surf which sometimes breaks a great way from Shore, our English boats are of no use for landing or shipping of goods. For this end therefore they have Mussoolas.'[52] Fryer noted that *masulas* were built in such a way that 'it yields to every ambitious surf, otherwise we could not get ashore, the bar knocking in pieces all that are inflexible'.[53] According to Salmon, 'no large vessels can ride within two miles of the place, the sea is so very shallow; nor is there any landing but in the country boats, the surf runs so high and breaks so far from the shore'.[54] Francois Martin had also observed that 'descent is made difficult all along the Coromandel coast by the sea which is almost always rough here. If an attempt is made to board by launch, there is every danger not only of losing the boat but also of losing one's life....These light crafts are very responsive to the billowing waves which cast there ashore without the least damage.'[55] Gaspardo Baldi (sixteenth century) puts that 'the boats do not break, because they give to the wave; and because the beach is covered with sand, the boats stand upright on their bottoms'.[56]

In 1727 the Madras authority informed the directors at home that they did not provide as platform in those boats as they were too small and such a platform would make the 'boat so stiff that they would not bear our surf but either break to pieces or overset'. But to prevent the bales from damages they placed a large quantity of brushwood at the bottom of the boat, which kept the bale high enough to prevent the water.[57] It is interesting to note that when the French had threatened Madras in 1746, the directors 'at home' expressed confidence that it would be quite difficult for the French to approach the town as they would find difficult to persuade the boatmen under gunfire.[58] But to their surprise the French did it with the help of their own French boatmen. Thus, *masulas* and catamarans were the most efficient and perhaps the only means of transportation from and to the ships at Madras road which anchored about 2 miles away from the shore.

The 'Amphibious' Boatmen of Madras

We have already shown the structure and uses of catamarans and *masulas* and how they were indispensable for the transportation of goods at Madras port. Here it would be pertinent of look at how they were handled by the boatmen most expertly. Campbell said that 'of all the extraordinary craft which the ingenuity of man has ever invented, a Madras catamaran is the most extraordinary, the most simple, and yet, in proper hands, the most efficient'.[59] These 'extraordinary crafts' being in a proper hand of the boatmen were the most efficient means of running the Madras shipping activities. Their special expertise was shown by the account of the early Indian voyager's log-book which said: 'It is very curious to watch these catamarans putting out to sea. They get through the fiercest surf, sometimes dancing at their ease on the top of the waters, sometimes hidden under the waters; sometimes the man completely washed off his catamaran, and man floating one way and catamaran another, till they seem to catch each other again by magic.'[60]

Gaspardo Baldi's account of the sixteenth century is also interesting in this regard. He said: 'merchandise and passengers are transported from shipboard to the town by certain boats which are sewn with fine cords; and when they approach the beach, where the sea breaks with great violence, they wait till the perilous wave has passed and then, in the interval between one wave and the next, those boatmen pull with great force and so run ashore; and being there overtaken by the (next) waves they are carried still further up the beach.'[61] Lady Dufferin said that the rower of catamarans, which she called 'mosquito fleet', wears a 'fools' cap' and when he 'encounters a big wave, he leaves his boat, slips through the wave himself, and picks up his catamaran on the other side of it'.[62] The boatmen of Madras by virtue of their age-old tradition in using such boats for fishing on the high seas sustained to pass on such important expertise from generation to generation.

It was because of this special expertise that the boatmen of Madras enjoyed a worldwide reputation. John Shortt noted that the boatmen of Madras got a worldwide reputation in 'aquatic feats', 'for courage and daring in times of danger and difficulty'. He said that they are like 'amphibious animals, for they can live on land or sea, and are from morn to night naked and exposed to cold and wet in plying their boats to and from the roads … and appear to glory in the sea, and are altogether a very hard-working set'.[63] Their aquatic or amphibious life at sea makes them most adaptable and ultimately most efficient person to handle shipping tasks most expertly. In this respect, it would be worthwhile to recall the account of Thomas Bowrey who himself had seen one fisherman who retuned to the land after four days on the sea without his boat which was taken over by the storm while fishing.[64] Such resilience on sea was an extraordinary qualification that the boatmen of Madras possessed through generations.

The expertise of the boatmen of Madras, and the competence of their crafts were recommended in other English settlements. For instance, in 1713 the directors in England wrote to the Deputy Governor and Council of Bencoolen that some slaves from Madagascar were dispatched for the West Coast of Sumatra which they would have to 'train severall of them up to manage the boats for unlading and relading' ships. These slaves were to be trained by the boatmen of Madras for which the Madras government had promised to send some *masulas* and were trying to prevail over the boatmen of Madras to go to the West Coast and 'stay for a time till they have taught the Slaves'.[65] Again in 1726, the Governor and Council of St. Helena wrote to the Madras government requesting 'some catamarans and people to manage them' to ease their growing shipping activities in that place.[66] It was possible that some expert boatmen must have been sent to various places occasionally and marked the beginning of international naval centres gaining knowledge of the expertise and technology of Indian fishermen, which gradually spread to other countries. In the process the boatmen of Madras and their 'extraordinary crafts' became an indispensable part of networks of trade and commerce without which Madras trade could not have survived.

The Stirred Water: Mucquas on Their 'Calling'

We know that about eighteen castes of the Coromandel coast took to fishing as 'their calling'. The high seas and rivers were open for every fishermen, except that they had to fish within their own borders. However, since the coming of the English, fishing within the bounds of Madras, in the river and on the sea, was controlled in one form or another. The Company rented out the fishing rights generally to a person known as the 'fishing farmer'. A *cowle* or licence was issued to the highest bidder. Thus, the erstwhile communally owned sea, river and fishes now became the exclusive prerogative of the Company. Consequently, all fishermen were required to take permission to fish from the licence holder, this renter, the fishing farmer. Hence, an attempt is made here to show how fishing activity became more important by now and how it was implicated in the Company's notion of control.

Initially, the government collected fish at the seaside as tax gathered by a peon. The Company then started renting out this right to the highest bidder, which was withdrawn in the year 1680, when there was a strong opposition from the fishermen.[67] In 1694, it was again rented out to Mrs Anne Delgardner for 30 *pagodas* a year who held it for two years.[68] She was granted 'the sole liberty and privileges of fishing in the river adjoining unto the towne, forbidding any other person to catch fish in said river with any manner of nets whatsoever, without leave first obtained from said Anne Delgardner, under penalty of forfeiting their netts'. She was also given 'full power and authority to take and received custome fish of all Catamarans and others, as has been usually received by the Rt. Honble Company' from fishes brought in from the sea.[69] However, the Company reserved for itself the liberty of fishing in the river once a month. But from 1696 the fishing *cowle* was rented out to the heads of the *Mucquas* for 50 *pagodas* for certain reasons which will be explained later.[70] This *cowle* was usually granted for five years at a time and the payment was made half yearly. In 1697 the *cowle* was granted to Deigo Pascquall;[71] in 1707 to Pasquall and Joan;[72] in 1710 to Pasquall, Errlepaw, Andee Percoun;[73] in 1716 to Thome and Joanne;[74] in 1724 to Pasquall, Deigo, Ershiah, etc.,[75] and so on.

Besides, the fishermen had to pay the Town Concopoly (i.e. *Kanakpillai* or accountant) certain duties: 325 *fanams* per annum; 10 *fanams* per annum for fishing in the river; out of every draught of fish, ten fish; and to those that carried out from the town shall pay out of every *fanam*, 4 *cash*.[76]

Fishing seems to have been generally an individual activity although there was a possibility of group fishing in Madras. While the river fishing had already been contracted out to the 'Company of *Mucquas*' who had of the sole liberty and privilege fishing rights in the river, all others being forbidden, sea fishing continued to remain open to everyone. Although many of the erstwhile fisher folks took up various new jobs under the new colonial urban set up, especially as boatmen for loading and unloading goods in the port as their main occupation, now their erstwhile fishing profession continued to remain predominantly in their hands. This was mainly because sea fishing, like the shipping, required the same expertise over the deadly surf, and because shipping jobs were seasonal. So during the off shipping season many boatmen took to fishing as stopgap employment. In fact, there were many more people who could not be employed in the port. These people continued to practice fishing even more vigorously than before due to the ever growing population of Madras and its increasing demands. This increasing interest in fishing was pointed out by the fact that the erstwhile inland fishermen, the *chomboddeewar cast* now took to sea fishing.

The urban environment as well as their conversion to Christianity must have probably helped the inland fishermen to do away with their traditional caste inhibitions in taking to sea fishing.

Thomas Bowrey said that the fishermen or 'Moquaes' used catamarans for fishing at sea and had often seen them 'one league or more off shore, when the westerly winds have blown very hard … they have made sleight of it, onely let fall theire line with a stone fast thereto, … and they swimme on shore both against wind and sea'.[77] Abbe Carre had also noted that the Patanavars and Macouas used catamarans and the masulas in which 'two nude men spend all the day in the open sea to catch fish, which abound on this coast'.[78]

Nets were usually used for catching fish: a circular form, about 10 feet in diameter, and from 8 to 12 feet in length, pointed cone, loaded iron or leaden rings at the end; besides this, they used fish traps made of bamboo and hooks, for inland fishing.[79] On the sea, 'drag' and 'oblong' nets 'extending to a couple or three hundred feet in length, and from 3 to 4 feet in breadth were used.[80] The nets were made from the fibre of the Crotalaria Juncia, Janapha, coconut fibre and cotton wool.[81] The kind of nets used by the fishermen included seines which appear to have been known during this time. Other evidence supports this view. In October 1719, when the warehouse keeper asked for a supply of fish from the Mucquas for salting, he was told that enough fish could not be supplied till January 1720.[82] This evidence probably confirms the duration of the seining season when large catches were made, beginning in January till April. This season brings the pelagic shoals of anchovies, sardines, whitebait, silver bellies and Caranx towards the land where large catches were made.[83]

Thus, the increased demand for fishes in the town as well as for salting, not only demanded sufficient and regular supply but also necessitated more people to take up the task. The new circumstances made the fishing business more lucrative than before and hence improved the earnings of the fishermen to a great extend, despite certain restrictions.

The Civilizing Port: Mucquas as Boatmen in Madras Overseas Trade

It is difficult to say precisely when the fishermen actually become boatmen. It may be noted that due to their special expertise many fishermen must had already taken up such jobs under the Portuguese earlier. But in Madras, under the Merchant government, the Mucquas seem to have been better organized and hence they became boatmen. We know that the main task of the Madras boatmen was to ferry men and goods between Madras and ships anchored at the road. Lockyer noted that they could take two or three round trips in a day but later evidence suggested that a boat could take four trips.[84] Further, we have noted that the volume of goods carried by these boats was quite light: 6 to 8 bales of textiles by a masula, and 3 or 4 tonnes by a catamaran. This indicated that the Madras port required a large numbers of boatmen to cater for the ever-increasing ships and goods traded. The boatmen of Madras were also used for other purposes at sea, such as securing scattered goods whenever there was shipwreck, as messengers and postmen between several English port settlements, and so on.[85] This was possible with a well-organized body of boatmen and hence a new social group was gradually formed around the 'boatmen'. They were even considered as 'servants' of the Company.

As far as the organization of boatmen was concerned we have references of systematizing through the appointment of head boatmen by the merchant government. Through this headman the authority could organized and govern the entire activities of the boatmen. We have already noted that Christians were preferred for this important task. Therefore, the head of the boatmen would also usually be Christian. But it is not clear when such a system of appointing a head or chief came into force, and for that matter, why a Christian chief had to be appointed. It was possible that the Christian boatmen had already dominated the shipping activities since the inception of Madras, and perhaps such activities were carried out initially through the traditional caste heads. However, seeing that the government had little control over the existing system it must have decided to effect such appointments themselves.

In this respect, the Company, in 1674 directed that those persons to be employed in the 'command of the Boats to bring goods to and from the ships' were required 'to enter into covenants to render you true and just account and be bound to make oath to the truth of such account soe often as you shall require it'.[86] From this time onward the head of the boatmen was required to subscribe before the government for loyalty and truthfulness through 'covenants' or *tasheriffs*. Hence, the headmen had to function hereafter at the pleasure of the government who had arrogated the right of appointment and removal. This was followed by another order from the Directors in 1677 which specified that the persons to be employed in the boats must be Christians and 'if any natives be imployed they must be but under their [i.e. Christians] direction and at their command'.[87]

On the basis of such new power of appointment the Madras government appointed one Black Tom as the new headman or 'Muckadum or master' of the boatmen in November 1680 'for the better keeping those people in order and fast to the place'. His salary was fixed at 70 *fanams* per mensem as he was 'formerly in that imploy with the same charge' and 'the Boatmen being Christians as he is'.[88] The circumstance in which he was appointed is also interesting. In early November of that year many of the *Mucquas* and catamaran men joined the painters 'mutiny' and deserted Madras for Mylapore until they were forcibly brought back.[89] So in order to avoid such problems in future the former heads were dismissed and Black Tom was appointed. He was one of the former head boatmen and the choice of him is significant in that the appointment could be better legitimized although he was supposed to imbibe a new outlook under the new circumstances.

In 1696 the boatmen quarrelled with their chief and presented a petition to the government that he may be discharged and another person appointed in his place.[90] An enquiry was commissioned on the basis of this complaint but it was not clear whether a new chief was appointed. However, more chiefs were appointed later as evidence shows that there were four chiefs in 1708. In this year they were brought before the Council in which one of them was appointed as chief boatman for discovering the 'villainy' of the boatmen, whereas other three were whipped around the town. After seven months in prison they were released on payment of 500 *pagodas* as penalty.[91] However, disapproving this ruling, Frasers reinstated them later.[92]

By 1722 two new heads were appointed in place of two old ones with the help of the French padres.[93] This step was also taken when there was a conflict between the 'boat cooleys' and the two heads.[94] Interestingly, it was also known that whenever new headmen were to be appointed the boatmen of Madras and those of St. Thome were consulted. The President Nathaniel Elwick informed the Council that he had appointed the two headmen after taking all all concerned persons into confidence, including the *Mucquas* of St. Thome who were 'of the same cast' and who sent their approval of the alteration by signing on a *Cajan* sent to them.[95] This implies that the head boatmen of Madras was also to the boatmen of Mylapore, and their prior acceptance was thought essential for the smooth running of Madras shipping. The names of the two new headmen were not mentioned in the record but one of them must have been one 'Tomes (Head Boatmen)' whose name appeared in the petition of 'the right and left hand casts in behalf of themselves and all the inhabitants of the city of Madras' in December 1725.[96]

In this way the heads of *Mucquas* were appointed by the English government from time to time. But as far as the choice for the headmen were concerned we have no specific evidence. One later account pointed out that most of the headmen were from the same family. In 1795 Droomo Nursoo (Dharma Narasu), the Boat Overseer of the time, requested the government for the grant of a pension to his blind father. In his petition he mentioned that from his great grandfather who had been first appointed by Governor Pitt in 1705 his family continued to hold the office of the head boatman till date. His great grandfather was commissioned to Vizagapatnam to engage the boatmen of that place who would not be subservient to the Right Hand Caste people of Madras. For this valuable services he was granted 'in him and his heirs the employ of sea-side *dubash*, or broker and headmen over the Massoola and catamaran people' by a *cowle* (i.e. decree) on 12 December 1705. His father succeeded his great grandfather and when he become blind since when the said Droomo had took over the headship. The government agreed to grant a pension of one and a half *pagodas* per month to his father.[97]

We can assess the number of boatmen and boats employed in Madras port from some indirect evidence. Given the volume of trade and number of ships anchoring at Madras port each year, the need for boatmen to run the shipping business appeared to have been very large. Generally, more than 60 ships frequented Madras every year. For instance, in the year 1710 about 74 ships came and went with full loaded goods.[98] Besides, there were many more sloops, ketches and brigantines plying between the coastal entrepôts. These smaller vessels came and went more frequently. Even the large ships plying to Bengal could go two round trips in a year; for example, the ships *Chalton* and the *Morning Star* in 1710. The carrying capacity of the overseas bound ship was usually 400 to 600 tonnes. Given the hefty volume of goods traded vis-à-vis the maximum number of three trips a day by a boat the need for a large number of boatmen can be established.[99] No wonder, it was said that when President Pitt left Madras for England in June 1709 there were 50 trading ships and above 200 small crafts at one time in Madras road.[100] Further, the number of ships coming to Madras in 1738 was said to be 120 'sail of ships'.[101] Given this figure of 200 boats with that of Fryer's account of ten men for each boat the number of boatmen required at Madras port probably shot up to over 2,000 persons.

In fact, because of the ever-growing trade and commerce there was a perennial shortage of boatmen in Madras as the business was seasonal.[102] This is mainly due to the specialty of the task which can be handled only by the boatmen. Therefore, many boatmen from different places were usually invited during the peak season; the Mylapore boatmen probably constituted the largest in number mainly because they were also Christians and of the 'same caste' as the Madras boatmen. We have also noted that 'Mucqua town' was created in the 1670s to cater for the increasing numbers of boatmen who had migrated from other places. In spite of such measures, the shortage of boatmen was felt quite considerably throughout this period. This is apparent in the reports of President Nathaniel Elwick in 1722, in which he lamented that many ships were usually forced to wait for the next season due to shortage of boatmen although they worked hard as usual.[103] His solution for this problem included the building of thirty boats at Company's expenses, appointment of two new heads, procuring more boatmen, changing the system to the boatmen's satisfaction, and finally building a shed or 'Banksall' for the workmen to 'repair or build boats under fifty foot long and fifteen foot broad'.[104] The Council accepted all the proposals and it was accordingly carried out.[105] It seems that such measures greatly eased the problems of shipping in Madras.

As far as their wages were concerned we have some important accounts during this period. In 1654 the boatmen's wages ranged from 2-3 *fanams* per *masula*.[106] In 1678, after a strong protest from the boatmen, the government increased it to 5 *fannams* 'per Mussulas lading of 6 bales'.[107] By 1711 Lockyer said that the 'hire' of *masula* was 'six fannams or eighteen pence a trip; but the Company has seven Boats per *pagoda*, which is money dearly earned.'[108] By 1727 the charge for 1 *masula* was still 5 *fanams*.[109] Given the enormity of the work, the earning of boatmen was much below what might be expected but enough for their subsistence. We know that one *masula* can go two to three trips a day and more than ten persons have to run the boat to cross the deadly surf.[110] This meant that a boat can earn a maximum of 15 *fanams* per day, which was to be divided among them. By computation each boatman could earn more than 40 *fanams* a month, which was reasonable. However, such 'good penny' seems to have been quite insufficient considering the crucial role they played and more so in sustaining and gaining social importance in the local society. This may be the reason why the *Mucquas* were seen to be quite restless.

However, the government compensated for this in other forms so that the boatmen would remain serviceable to the Company at all times. Witsen (1690) said that *masulas* are 'useful boats but have to be taken to parts and resewn often'.[111] Seeing that the boats required annual refurbishing to prevent them from damage, the Company helped in lending advance money annually to the boatmen. Initially, 100 *pagodas* and later 200 *pagodas* were annually rendered as advance money to the boatmen for repairing their boats before the onset of the new shipping season, around the month of October or November.[112] This money used to be deducted later from their accounts of earning. Even when they could not repay, the government usually wrote off or paid out of the Company's account as stopping such advance would only upset the shipping business.[113] But non-repayment of such advance money rarely happened; it only occurred during bad times, say during the famine of 1728-9 for instance.[114] In most cases boatmen earned several times more than the advances. For instance, in 1733 they earned 171.19 *pagodas* over and above 200 *pagodas* of the advances.[115] As an additional incentive, the boatmen were also taken care of by the government during difficult times such as when their boats were blown out to sea by storms,[116] during famines,[117] and so on.

It was because of such indispensable services to the Company that the Merchant government was always compelled to forge good relations with the boatmen: 'the business of the place is not to be done as it should be unless those fellows are continued in their employs'.[118] The question of boatmen came up again and again in the consultations since the early part of the settlement. We know that they were always protected whenever there was regrouping of settlements done. In 1749, after returning to Madras, the English authority was also not only compelled to let the Capuchin priests dwell within Madras but also reinstated the Capuchin priests to their former office and given possession of the Fishermen's Church in the 'Mile End' as the Boatmen 'who are of that communion, may probably be induced to leave us should we expel them [the French Padres] our limits'.[119] Luz church was also returned to St. Thome priests as the Madras boatmen were 'all settled near it'.[120] Thus, the importance of the boatmen in the business of the Company was well established.

RESISTANCE FOR SOCIAL UPLIFT: THE *MUCQUAS* IN TRANSITION

We have already established the importance of boatmen in the Company's business in Madras and how much the Company valued them. It follows that this experience would also certainly have had some social repercussion too. The notion of being close to the rulers, of gaining a strong reputation, of being indispensable to the Madras overseas trade, and gaining economic status, would have transformed the social outlook of the boatmen. This new outlook was expressed in two forms. First, they participated in various caste activities to redefine or revitalize their social standing in local society. Second, they persevered to gain more and more economic benefits to reinforce their gaining social importance by demanding more and more from the government through various forms of protest. It is pertinent to begin with some aspects of their social life as background information.

We have noted that most of the boatmen were Christians and for their various ceremonies like marriages, burials, festivals, etc., they went to St. Andrew Church in the 'White Town' a new status that would have a lasting social ramification.[121] There was also a separate Church in the Fishermen's town to cater their daily spiritual needs.[122] Hence, the Capuchin priests of Madras had great command over the boatmen, in both religious and temporal affairs.[123] Besides, the boatmen would have been one of the earliest beneficiaries of the various schools ran by their Capuchin fathers. Abbe Carre said that the Capuchin priests of Madras have 'three schools of different sorts of Christians whom they teach there with fruitful results, viz., Portuguese, Hindus and Malabars'.[124] At the same time one can also see the practice of many old traditions among them such as they believed in the old women 'saints' like one old blind women Adroza,[125] the faith in traditional 'astrologers for weathers',[126] and so on. John Shortt has noted that as a body, the fisher folks were 'dissipated set, and, whenever they can afford it, partake freely of toddy or arrack. They are also particularly partial to tobacco, which they chew and smoke'.[127]

The coming of Europeans brought them many changes. Life at the seaside of Madras during the shipping season appeared like a modern bazaar; the sea-gate was always thronged with people, some laying wagers, others waiting for Masters, and the rest to satisfy their curiosities.[128] The boatmen sailed out till the ships with people and goods or bring them to the shore. Their women-folks assisted them in cleaning, repairing or making the fishing nets or in selling the fishes or other items by the seaside.

On the sea, they were cheerful lot. In early morning they set out for fishing excursion on the sea as far as forty to fifty miles away from the shore and returned with their catches in the evening.[129] Lockyer said that the boatmen, despite their long journeys from the shore to the ships amidst the high surf, were never seen to be sad but 'they are merry birds, howling out a *Ela, Yela* as chorus to their songs at almost every stroke'.[130] Bowrey had also seen them in a cheerful mood.[131] Thus, for the brave fisher folks sea life had offered a source of pleasure, for more than just an occupation. And the new changing circumstances added more salt to their life, as they became an indispensable part of it. All these changes make them a most vibrant social group in the town.

Participation in Caste Affairs

As far as their participation in the caste activities were concerned, we have three important instances on record. First, the fisher folks joined the Painter's Mutiny (1680) in large numbers. Many boatmen and catamaran men deserted Madras along with other right hand caste group under the leadership of their respective caste heads. They joined the mutineers in stopping all provisions and firewood to Madras, threw off the textile packages of the company from the oxen that carried them, and burnt some houses.[132] They returned only when their wives and children were 'taken out of their houses and driven into the Pagodae (Temple)'.[133] The heads of the boatmen who led the mutiny were dismissed, and in their place one Black Tom was appointed.[134]

Again, in 1686, the Madras Government proposed to raise taxes for the construction of the Black Town Wall. However, the entire inhabitants of Madras 'in contempt of the government ... desist from their labours and services' to the Company, closed all shops, stopped grains to be brought into town 'insolently declaring that they would continue their rebellion, till they were freed from the said present and all future taxes'. But when the government clamped down heavily they ceased the protest.[135] The Company's servants, especially the 'Washermen, Muckwaes, cattamaran men, and Cooleys' were in the forefront who 'returned to their several businesses' only after their caste heads called off the protest.[136]

In 1707 an unusual quarrel broke out between the right and left hand castes, which lasted in an acute form for upward of six months and was not easily settled during the remaining year.[137] In the ensuing disputes many right hand caste groups deserted Madras and repaired themselves at St. Thome from where they threatened Madras, while others shut themselves in their houses. The deserters consisted chiefly of 'Boatmen, Washermen, Fishermen and other necessary handicrafts'.[138] Even after the intervention of the Capuchin Fathers the deserters, which included (head of the watchmen and peons) many Christian boatmen and fishermen, did not return to the town. The *Peddanaik* of St. Thome finally brought them back to Madras on the night of 4 October 1707.[139]

The participation of the boatmen in such caste affairs would be seen in much better light if we take into account the changing socio-economic circumstances of the fishing community in Madras. It should be noted that the general social demarcation of the larger part of the Madras population was made on the basis of left and right hand castes. If any group found difficulty in accommodating themselves within these folds it was apparent—from the spatial division—that it would seem difficult to get any foothold not only in the physical space but also in the social space of Madras. Under such circumstances it seemed unwise to sever one's caste linkage from either of these two broad divisions. This became more prudent for the upcoming social group such as the boatmen due to their gaining economic importance under colonial urbanism. Therefore, the participation of the *Mucquas* in such caste affairs may be seen as a form of assertion to gain a higher social standing in local society on the one hand, and to enlarge their social base on the other. By joining the mutinies the fisher folk might have aspired to enter the social network of the right hand caste although they were already Christians.

We have seen that such a new consciousness was mediated by their caste leaders who acted as 'elite' and 'intelligentsia' for the fishing people. The dismissal of their caste headman after the Painter's Mutiny was one case in point. Seeing such consequences, the head boatmen, Pasquall and Joan, had submitted a petition to the Governor after joining the caste disputes in 1707 claiming they had joined the conspiracy inadvertently and 'by the instigation and ill advice of some designing people'. They also told the Governor that they realized they belonged to neither of the two castes after becoming Christians only after returning from St. Thome to joined the right hand caste group and asked for his protection from the caste exaction in future.[140] This was but a mere excuse as many other castes had also already done the same before the boatmen citing their inadvertent behaviours, all blaming their caste heads. In fact the boatmen were the last to do so.[141] Therefore, we can say that they had joined the caste conspiracies with a clear resolution in mind to elevate their social standing in local society. This new social outlook must made them very assertive in gaining more and more economic benefits from the prolific trade of Madras.

Resistance for Higher Economic Benefits to Reinforce Social Status

The apparent complicit relationship between the Company and the boatmen was not always smooth sailing. This was mainly because of the government's low attitude towards the welfare of the fisher folk and also towards their gaining social importance. The grievances of the boatmen over the new system were manifold but three areas become discernible: low wages, control of their traditional fishing rights, and the various odd taxes. Usually, led by their headmen or chiefs, the poor fisher folk consciously resorted to various forms of resistance to further their demands. Some important methods employed by them include: petitions, ceasing of work of stoppage of their supplies fish, avoidance of payment of taxes, stealing Company goods, and finally desertion or migration. These forms of resistance constituted what James Scott called 'weapons of the weak'.[142]

As early as in 1678 the *Mucquas* pleaded to the Government to increase their pay to 5 *fanams* per boat.[143] To make the demand acceptable they claimed that their stealing habit was due to the low wages. When the petition was not considered many boatmen deserted Madras; some even 'forsacking it' and settling elsewhere.[144] On 7 January 1678 the boatmen had demanded 'five *fanams* of Madrass per Mussulla'. Having witnessed the seriousness of the demand the government agreed to pay them 4 *fanams* per boat upon condition that if any person steal, the head of the caste shall make good the loss, or else deliver them to be transported to St. Hellena or elsewhere. Dissatisfied with the ruling, all the *Mucquas* again deserted Madras on the night of 10 January, 'carrying their Oars with them'.[145] Seeing this, the government had finally conceded to their demands and accordingly sent one Cassa Verona with a letter of pardon to bring them back to Madras. They were brought back on 12 January 1678.[146]

This instance exemplified their mode of protest: first petition and appeal, and then migration and desertion if their first plea was not accepted. Such 'weapons' were cleverly employed by the boatmen with a predetermined state of mind, not just a spontaneous response to government policy. They knew that the best time for such agitation was when their labour was most required, i.e. January, the peak season for shipping at Madras port. We know that Madras shipping could be carried only with the valuable service of the boatmen. At this juncture any en masse migration or desertion by the boatmen would put the Company business on standstill. The Company was thus left with no option. This is clearly seen in the Council minutes which said 'there being no other remedy nor possibility of prevention, for if bigger guards were sett upon them it would but make them the more subtle in the attempt, and difficult to reclaim'.[147] However, to prevent another desertion in future the government contemplated to construct a curtain next to the Mucqua Town from St. Thomas point by the riverside. This did not happen anyway but later one 'black outguard' was posted at this point 'to give intelligence to the fort'. Thus, by prevailing upon the necessities of the Company the boatmen pushed through their demands.

Realizing their bargaining power during the previous year, the boatmen in 1680 launched another protest against 'the fellow that hath rented the custome of the fish'. To ensure their demand be heard they stopped supplying fish to Madras, which 'gives great offence to many people'. When there was strong pressure from the 'townes people' the Governor and Council finally withdrew the licence from the renter.[148] If they could not stop the Company from controlling their traditional rights over fishing in the river, the *Mucquas* strived to take for themselves the fishing *cowle* so that the rivers remained in their hands. This, they got in 1696 after a long struggle. We know that the Company finally rented out the fishing *cowle* to *Mucqua* headmen as 'it being much more convenient for them than any other farmers who meet with great trouble from the Mucquaw men in the collection of the custom fish'.[149] They resisted the fishing farmer by not paying him taxes as another form of protest. Naturally, the fishermen would resist the appropriation of the communally owned river or any other natural resources by someone exclusively.

In 1681 the infamous Quit Rent was imposed upon all the inhabitants of Madras. Although it was of nominal charge the townsmen strongly protested, as it was a new legislation. The *Mucquas* or boatmen were especially reluctant to pay this tax. In 1693, when the arrears of quit rent from the boatmen kept on increasing, the Company was 'forced to accept' their labour in filling up a ditch near St. Thomas Gate in lieu of the payment of areas of rent. Again in

1695 the Company insisted that the *Mucquas* repair their boats themselves in lieu of the quit rent.[150] As it became a troublesome affair for the Rental General to collect quit rent from the boatmen, from 1697 the government began to pay for them and later deducted from their accounts of earnings in shipping.[151] This was not done only to the *Mucquas* but the poor *talliars*, washermen, and pariahs. Soldiers and gunroom crew were later exempted.[152] When the collectors of quit rent became cruel and corrupted the task was assigned to the caste heads from 1695.[153]

The *Mucquas* also resorted to stealing the Company's goods while shipping in the port. For this the company got the name of 'all notorious thieves'. For John Shortt (1867) the boatmen of Madras were 'great rogues, and would rob one to his face'.[154] But it would be interesting to see how such discourses began to take place in the writing of the later English ethnographers like Shortt, and look into whether such things really occurred. In fact, the case of stealing Company's goods by the boatmen came up again and again in the records. In 1687, the President when he was informed of 'robbery committed by the Muckwaes of a parcel of corall stolen out of chests bringing ashore' he ordered an immediate enquiry. Upon hearing these orders all the *Mucquas* 'left their dwellings to secure themselves' for Pulicat and Sadraspatam. However, they were recalled with the assurance that 'only the guilty shall suffer for this crime'.[155]

Again, in 1697 the boatmen were suspected of stealing pepper while bringing it ashore. This came about when the weight of pepper brought from York Fort was found to be less than was in the account. No action was taken in the absence of substantive evidence of theft. To prevent such future occupancies the ship master, searcher of the sea custom house and the warehouse keepers were ordered to keep records, and to send a note with every boat for verification: furthermore, a packer should be sent beforehand to 'sew up the mouths of all the bags of pepper, and mend them if defective to prevent the Mucquaws stealing'.[156]

In 1708, the boatmen opened the Company's bales and many pieces were stolen. One of the boatmen informed the Peddanaik who immediately secured the head boatmen in whose house was discovered 'four pieces of Beteelaes buryed under ground'. Upon this evidence an order was issued to apprehend all the boatmen. When some of the boatmen returned 'their associates and this villainy made them some private sign, so that they returned as far as the Paddy boats'. The Governor sent two boats of Vizakapatnam boatmen with soldiers to prevent them escaping. Seeing this, ten boats slipped away southwards upon which the Governor ordered the gunners 'to fire shot from the battery and sink them if possible, several of which fell very near'. It was decided that once the losses were known from England where it was to be unloaded, the warehouse and sea gate '*Concopolys*' (i.e. accountants) seaside peons and boatmen should equally pay for the shortfall.[157] The Company's warehouse keepers in England found that 80 pieces of cloth, including 16 pieces of longcloth, 3 pieces of

Salampores, and 61 pieces of *Bettellees*, were stolen from the bales. This signified the participation of all the boatmen. The 'carelessness in not sending Peons with every boat of goods' was blamed by the Directors and warned that it 'must never again be practiced'.[158] While the boatman who informed the 'villainy' was rewarded the chief boatmen's post, the other three heads were whipped round the town, stood three days in the pillory, and sentenced to seven months imprisonment.[159] This action was also highly approved by the Company 'at home' as it would act as an example to the public.[160] But this instance must have brought about lots of disconcerted feeling among the boatmen. So much so, that the three heads, whom Pitt's had deposed were soon reinstated to their previous position by Frasers so that they could again work for the Company more 'cheerfully'.[161]

No further instance of stealing came after this incident. But these few instances created the notion that the *Mucquas* were 'notorious thieves' and that it became a common practice to suspect or accuse them whenever there was any loss or stealing occurred. For instance, in June 1738 'near a fifth part of the rice' brought from Moco Moco was found deficient 'where upon a new tub was ordered to be sent thither for a fixed measure'. The head of the boatmen was summoned in this regard as the government suspected 'it was stole by the Boats people'. They wrote home that they would 'watch them more narrowly'. The Directors not only appreciated the action but also asked the President to proceed with a further enquiry.[162] Again, in 1742, several pieces of clothes were missing while loading a ship. The government again suspected the boatmen but as there was no evidence they reported the loss to the Company at home. The Directors ordered that 'a Serjeant and soldiers or two' be placed on each boat while loading or unloading the ships so that 'such Roguerys may be prevented'.[163] The various instances of stealing even before the 1708 incident were of doubtful condition as there was no clear proof for such thefts.

Apart from the 1708 incident all the other incidents of stealing by the boatmen were based on mere suspicion. Even the incidence of 1708 needs deeper examination. We have seen that the 1708 incident of stealing was done with the participation of all the boatmen. This suggest that it was something more than mere theft. Hence, the best explanation for the cause of stealing Company's goods is but another form of protest against the low wages paid to boatmen of Madras. This was also earlier spelt out in their petition to the government in 1678, in which they said that they should be paid higher wages as their low wages prompted them to resort to stealing.[164]

It should also be noted that the demand for higher pay was more to improve their economic status than to ease their poverty. We have calculated that their monthly earnings was more than 40 *fanams* per head which could go up to 60 *fanams* if they could make four trips a day. Compared to the earnings of these heads boatmen (70 *fanams* per month), this was quite a reasonable amount for subsistence. Salmon's account of their settlement might lead us to believe that they were poverty-ridden lot. But it should be noted that Salmon drew a similar picture of the 'Black Town', which was supplemented by the government's accounts.[165] So it can be said that the *Mucquas* were much better off being in the employment of the Company and other merchants as boatmen, from which they earned 'good penny'. For instance, in 1696 the catamaran men were seen to have bought several of the houses from the pariahs who were transferred to the 'Pariars Town'.[166] Therefore, it can be said that when the boatmen avoided paying taxes, demanded high pay, and stole the Company's goods it should be read as forms of protest against the attitude of the government for their higher social aspirations which mainly hinged upon their economic viability. Thus, the boatmen really wished to enrich themselves for an elevated economic status and ultimately reinforce their social standing in the local society.

In the foregoing account we see that the *Mucquas* were determined to expand their social base by joining the caste affairs and aspired to improve their social standing in the local society by improving their economic status. They adopted various forms of resistance which include pleas and petitions, avoidance of tax payment, ceasing of works or non-cooperation, stealing and finally desertions and migration to other places which consciously materialized. Therefore, it can be said that when the poor boatmen revolted against the government they did not do so in haste; they reacted within the compass of their consciousness against the colonial set up. If they could not confront the well-armed Company government openly in the battlefield they did not remain subjugated without showing any sense of resentment and resistance. In the ensuing protests, their headmen or chiefs (whom the government had appointed, and, who were expected to mediate between the Company and the fisher folk) led them. Thus, the *Mucquas* or boatmen of Madras became the first champions of a passive resistance movement against the colonial government in India. And in fact because of such dynamism they gradually became one of the most vibrant social groups in later period. In 1867 John Shortt found 'men of wealth and influence, and many well-educated individuals' among the Christian boatmen of Madras.[167] This transformation process began during the period of this study.

CONCLUSION

In conclusion, it can be said that the fishermen or *Mucquas* of Madras were also subjected to the colonial policy of divisions: socially and spatially. It also underlines the intensification of divisions within the larger right hand caste demarcation in Madras. However, within all these subjections the *Mucquas* strived to become one of the most vibrant social groups in the town. This is mainly because of their proximity to the Company government on the one hand and their indispensable shipping services in supporting the Company's business, on the other. Under the colonial urban set-up they took up several jobs hitherto unknown to them. Of this the most important was in the shipping activities where the erstwhile fishermen mostly took up the tasks of ferrying men and goods to and from the port with the help of their 'extraordinary' boats and became an indispensable part of Madras global networks of trade and commerce. Besides, fishing had now become a more profitable job than before due to the increasing demands in the markets. By virtue of such economic privileges under the Company's government the fishermen now not only gained some social importance but they strove to gain more of them by participating in the caste affairs although they were already converted to Christianity. They also demanded more economic power to reinforce their social standing local society, which made them so restless. They demanded higher pay, exemption from taxes, withdrawal of controls over their fishing rights, etc. Their forms of resistance included petitions, ceasing of work, stopping supply of their goods, non-cooperation, stealing of Company's goods and finally desertion and migration. They employed these methods to bargain for benefits from the flourishing trades in which they became an important part. In fact, due to their attitude of resistance and application, the *Mucquas* emerged as one of the most dynamic social groups during the later period.

NOTES

1. See for example, P.A. Roche, 'Caste and the British Merchant Government in Madras, 1639-1749', *Indian Economic and Social History Review (IESHR)*, 1975, pp. 381-407; T.G. Mcgee, *The South East Asian City: A Social Geography of the Primate Cities of South East Asia,* London, 1967; R. Redfield and M. Singer, 'The Cultural Role of Cities,' *Man in India,* September 1956; and M. Singer, *When a Great Tradition Modernizes,* Vikas, Delhi, 1972.

2. A. Appadurai, 'Right and Left Hand castes in South India' *IESHR*, 1974, pp. 216-59; Roche, 'Caste and the British Merchant Government', *IESHR,* 1975, pp. 381-407; S.J. Lewandowski, 'Changing form and function in the Ceremonial and the colonial Port City in India: An Historical Analysis of Madurai and Madras', *Modern Asian Studies (MAS),* 1977, pp. 183-212; and M. Satish Kumar, 'Idioms, Symbolism and Divisions: Beyond the Black and White Towns in Madras, 1652-1850', in S. Raju (ed.), *Colonial and Post-colonial Geographies of India,* Sage, 2006, pp. 23-48.

3. John Shortt, 'The Fishermen of Southern India', *Transactions of the Ethnological Society of London,* vol. V (1867), pp. 193-201, *www.jstor.org.*

4. Pattanavan: the fishermen on the east coast from Kistna to the Tanjore District are popularly called Karaiyan, or seashore people. Lit: dwellers in a town or *pattanam*. E. Thurston, *Castes and Tribes in South India,* vol. vi, pp. 177-8.

5. Tamil Nadu State Archives (TNSA), Records of Fort St. George (RFSG), *Diary and Consultation Books* (henceforth *DCB*), 30 October 1707, p. 75.

6. *DCB*, 28 November 1695, pp. 152-3.

7. John Shortt, 'The Fishermen of Southern India', pp. 193-5.

8. Ibid., p. 196.

9. *DCB*, 6 July 1706, p. 55.

10. *DCB*, 30 October 1707, p. 75; 27 November 1680, p. 87; and 28 November 1695, pp. 152-3; C. Fawcett, *English Factories in India,* n.s., 1670-1677, vol. II, p. 146.

11. John Shortt, 'The Fishermen of Southern India,'p. 196.

12. The Dutch sources noted 15 to 20 fishermen's huts there. See H.D. Love, *Vestiges of Old Madras, 1640-1800,* Asian Education Services, New Delhi, 1996, I, p. 35.

13. The '*Pattnawarrs* and *Cariallwarrs* are to pass with their weddings and burials from the back side of Mr. Porters House to the middle of the Quarter Porters house and soe to proceed to the Portuguez Church, they may likewise goe through the great street.' See *DCB*, 30 October 1707, p. 75.

14. *DCB*, 25 August, 1707, p. 56.

15. *DCB*, 8 October 1687; 21 November 1695, p. 151; 25 May 1721.

16. Thomas Salmon, *Modern History or the present state of all nations,*1724, cited in Love, *Vestiges,* II, p. 75

17. See Pitt's Map (1710) in Love, *Vestiges,* vol.II.

18. Charles Lockyer, *An Account of the Trade in India, containing rules for good government in trade, price courants, and tables* (1711), cited in Love, *Vestiges,* vol. II, p. 80.

19. H.Yule, *Hobson-Jobson,* Mucoa, where he cited *Factory Record,* Fort St. George, no.1, *N. and E.* p. 2, 29 January 1679; John Fryer, *A New Account of East India and Persia, 1672-1681,* 1985, p. 24.

20. TNSA, RFSG, *Fort St. David Consultations* (henceforth *FDC*), 11 December 1749, p. 277.

21. *FDC*, 11 September 1749, p. 183.

22. *DCB*, 21 November 1695, p. 151.

23. *DCB*, 30 October 1707, p. 75.

24. *DCB*, 25 August 1707, p. 56.

25. Roche, 'Caste and the British Merchant Government', *IESHR,* 1975, pp. 381–407. Such conscious policy can also be seen from the work of Mcgee, *The South East Asian City,* 1967.

26. TNSA, RFSG, *Dispatches from England* (henceforth *DFE*) 1706-1710, 4 February 1708, p. 38.

27. *DCB*, 15 January 1708.

28. *DCB*, 27 November 1680, p. 87.

29. It should be noted that the right and left hand castes had their separate settlements and streets in the 'Black Town': left hand on the eastern and right hand on the western side. Peddanaigue petta, on the inland side (west of 'Black Town'), was an exclusive settlement of the right hand castes, whereas, Muttialpetta near the sea was an exclusive settlement of the left hand castes. See for example, Appadurai, 'Right and Left Hand castes' *IESHR,* 1974, pp. 216–59; Roche, 'Caste and the British Merchant Government,' *IESHR,* 1975, pp. 381–407; Refer also to the maps of Madras for details.

30. John Short, 'The Fishermen of Southern India,' p. 196.

31. For details of the two settlements see, *DCB*, 30 October 1707, p. 75; and 25 August 1707, p. 56.

32. Yule, *Hobson-Jobson,* Catamaran: *Kuttumaram* in Tamil is a 'raft, consisting of three logs of very buoyant timber,' *Madras Manual of Administration,* vol. iii, p. 137.

33. Fryer, *A New Account of East India and Persia,* p. 24.

34. Lotika Varadarajan (ed. & trans.), *Memoirs of Francois Martin (1670-1694): India in the seventeenth Century: Social, Economic and Political,* vol. I, part I, p. 84.

35. Charles Fawcett, *The Travels of the Abbe Carre in India and the Near East, 1672 to 1674,* Asian Educational Services, New Delhi, 1990, vol. ii, p. 594.

36. W.Campbell, *My Indian Journal,* 1864, cited in Thurston, *Caste and Tribes,* vol. vi, p. 180.

37. Ibid., p. 180.

38. John Shortt, 'The Fishermen of Southern India', p. 197.

39. Ibid., p. 197.

40. Ibid., pp. 197-8.

41. Thomas Bowrey, *A Geographical Account of countries rounds the Bay of Bengal, 1669 to 1679,* ed. by R.C. Temple, Manohar, 1997, p. 43.

42. *Coir:* rope made from coconut husks.

43. Bowrey, *A Geographical Account,* p. 42.

44. Fryer, *A New Account of East India and Persia,* p. 37.

45. *Memoirs of Francois Martin (1670-1694),* vol. I, part I, p. 84.

46. Lockyer's accounts cited in Love, *Vestiges,* II, p. 81.

47. Thurston, *Castes and Tribes,* vol. vi. pp. 180-81.

48. John Shortt, 'The Fishermen of Southern India', p. 197.

49. Bowrey, *A Geographical Account,* pp. 42-3.

50. Fawcett, *The Travels of the Abbe Carre,* vol. ii, p. 594.

51. Ibid., pp. 42-3.

52. Lockyer's account cited in Love, *Vestiges,* vol. II, p. 81.

53. Fryer, *A New Account of East India and Persia,* p. 37.

54. Salmon's account cited in Love, Vestiges, vol. II, p. 75.

55. *Memoirs of Francois Martin (1670-1694),* vol. I, part I, p. 84.

56. Cited in Eric Kentley, 'The *Masula*—A Sewn Plank surf boat of India's Eastern Coast', in Sean McGrail, *Boats of South Asia,* Routledge Curzon, London, 2003, pp. 122-3.

57. TNSA, RFSG, *Dispatches to England* (henceforth *DTE*), 22 September 1727, para. 22.

58. *DFE,* 14 November 1746, paras. 22, 23.

59. Campbell, *My Indian Journal,* 1864, cited in Thurston, *Caste and Tribes,* vol. vi, p. 180.

60. Thurston, *Caste and Tribes,* vol. vi. pp. 179-80.

61. Kentley, 'The *Masula*', pp. 122-3.

62. Thurston, *Caste and Tribes,* vol. vi, p. 180.

63. John Shortt, 'The Fishermen of the Southern India', pp. 199-200.

64. Bowrey, *A Geographical Account,* p. 44.

65. *DFE,* 1713-14, 15 January 1713, pp. 50-1.

66. *DCB,* 12 April 1726, p. 41.

67. *DCB,* 1 July 1680, pp. 43-4.

68. *DCB,* 21 May 1695, p. 67.

69. *DCB,* 3 May 1694.

70. *DCB,* 4 May 1696, p. 63; 14 November 1695, p. 148.

71. *DCB,* 24 September 1696, p. 112; 2 May 1698, p. 49; 16 May 1699, p. 41.

72. *DCB,* 2 December 1707, pp. 86-7.

73. *DCB,* 4 April, 1710, p. 35.

74. *DCB,* 12 March, 1716, p. 36; 23 June 1720, p. 102.

75. *DCB,* 10 April 1724, p. 45.

76. *DCB,* 24 December 1701, pp. 109-10.

77. Bowrey, *A Geographical Account,* pp. 43-4.

78. Fawcett, *The Travels of the Abbe Carre,* vol. ii, pp. 594-5.

79. John Shortt, 'The Fishermen of the Southern India', p. 194.

80. Ibid., p. 198.

81. Ibid.

82. *DCB*, 19 October 1719, p. 190.

83. Kentley, 'The *Masula*', p. 131.

84. *Lockyer's Account* in Love, *Vestiges*, vol. II, pp. 81-2.

85. See for instance, *DCB*, 19 April 1696, p. 55; 21 September 1718, p. 168.

86. *DFE*, 8 January 1674, p. 29.

87. *DFE*, 1670-77, 12 December 1677, p. 123.

88. *DCB*, 27 November 1680, p. 87.

89. *DCB*, 1 November 1680, p. 75; 6 November 1680, p. 76; 8 November 1680, p. 76.

90. *DCB*, 24 September 1696, p. 112

91. *DCB*, 21 January 1708, pp. 10-11; 20 January 1708, pp. 7-8; 19 August 1708, p. 46.

92. *DCB*, 9 November 1711, p. 156.

93. *DCB*, 6 October 1722, p. 126; 17 October 1722, p. 131.

94. *DCB*, 17 October 1722, p. 131; 30 October 1722, p. 134.

95. *DCB*, 30 October 1722, p. 134.

96. *DCB*, 27 December 1725, p. 198.

97. *Public Consultation*, 11 December 1795, as reproduced in Love, *Vestiges*, vol. II, p. 539.

98. As computed from the shipping lists of *DCB*, 1710.

99. It should be noted that a *Masula* can ply two or three turns/trips a day 'being the most fellow can make.' See Lockyer's account in Love, *Vestiges*, vol. II, p. 81.

100. *DFE*, 1706-10, 7 July 1710, p. 153.

101. *DFE*, 21 March 1739, para. 41.

102. Shipping season mainly stretched from November till August of the following year.

103. *DCB*, 6 October 1722, p. 126.

104. Ibid.

105. *DCB*, 17 October 1722, p. 131; 30 October 1722, p. 134.

106. Love, *Vestiges*, vol. I, p. 143.

107. *DCB*, 12 January 1678, p. 128.

108. Love, *Vestiges*, vol. II, p. 81.

109. *DCB*, 25 July 1727, p. 93.

110. Ibid., p. 81.

111. Cited in Kentley, 'The *Masula*', p. 122.

112. For instance, in 1711, the warehouse keeper was ordered to advance the *Mucquas* 'two hundred *pagodas* according to custome to repair their boats during this vacation that they be in good condition against the arrival of the Averilla.' See *DCB*, 1 November 1711, p. 144.

113. See for example, *DCB*, 27 January 1724, p. 16; *Dispatches from England*, 2 February 1724, paras.79; 31 October 1730, p. 141.

114. See for instance, *DCB*, 4 December 1728, p. 161.

115. *DCB*, 29 October 1733, p. 151.

116. For instance, on 25 November 1695, the paymaster was ordered to advance 100 *pagodas* to the boatmen 'for building new Mussulas and repairing the old ones, there having been blown away and broke to pieces in the late storme (on 21 November) seven or eight.' See *DCB*, 25 November 1695, p. 151.

117. In February 1718, 300 *pagodas* and 40 *garse* of rice were provided to boatmen when provisions were scarce. During the great famine of 1728-9 the government also provide them money and rice. See *DCB*, 20 February 1718, p. 33; 19 March 1718, p. 52; 4 December 1728, p. 161.

118. In 1708 the relationship between boatmen and govt. was strained after Pitt disposed their three headmen, which caused the Company dearly so that Mr. Frasers immediately reinstated them. It was disapprovingly taken up again in 1711 in the Council asking the President to 'reconcile them' so that they may be ready for the arrival of ships. See *DCB*, 12 November 1711, p. 159.

119. *FDC*, 2 September 1749, pp. 174-75; 11 December 1749, p. 277.

120. *FDC*, 11 September 1749, p. 183.

121. For these separate streets were reserved for them from their settlement St. Andrew's Church; one in 'Black Town' and one from 'Mucqua Town' to the said church. See *DCB*, 30 October 1707, p. 75; 29 February 1676, p. 89.

122. *FDC*, 11 December 1749, p. 277.

123. The government had also used to 'discoursed with the Portugueeze Padrees about regulating them (boatmen) … they being CHRISTIANS but very mean and numerous, over whom the CLERGY have a Great Command.' See *DCB*, 17 October 1722, p. 131.

124. Fawcett, *The Travels of the Abbe Carre*, vol. II, p. 553.

125. Niccolao Manucci, *Storia do Mogor or Mugul India*, III, Low Price Publication, New Delhi, 1996 (1st pub. 1907-8), p. 200. She lived with reputation of a saint where 'a whole village of fisher-folk had recourse to her' so that they stop coming to the church because of which she 'was admonished, expelled from Madras,' only to find her new rendezvous at the foothills of St. Thomas Mount.

126. *DCB*, 3 November 1684.

127. John Shortt, 'The Fishermen of Southern India,' p. 196.

128. Love, *Vestiges*, II, p. 85.

129. John Shortt, 'The Fishermen of Southern India,' p. 197.

130. Love, *Vestiges*, II, pp. 81-2.

131. Bowrey, *A Geographical Account*, p. 43.

132. *DCB*, 6 November 1680, p. 76.

133. Ibid.

134. *DCB*, 27 November 1680, p. 87.

135. *DCB*, 3 January 1686, p. 3.

136. *DCB*, 4 January 1686, p. 5.

137. For details of the caste disputes in 1707 see Roche, 'Caste and the British Merchant Government,' *IESHR*, 1975, pp. 381-407; Appadurai, 'Right and Left Hand castes', *IESHR*, 1974, pp. 216-59; and Kumar, 'Idioms, Symbolism and Divisions', pp. 23-48.

138. *DCB*, 21 August 1707, p. 54.

139. *DCB*, 26 June 1707, p. 36; 21 August 1707, p. 54; 25 August 1707, pp. 55-6; 24 September, p. 66.

140. *DCB*, 2 December 1707, pp. 86-87; see also *DTE*, 1707-08, p. 79.

141. On 20 October 1707 the Washermen and other Right Hand caste groups also appealed to the Governor to be pardoned on the next day (21 October); see *DCB*, 20 and 21 October 1707, pp. 71, 73.

142. J.C.Scott, *Weapons of the Weak: Everyday Forms of Peasant Resistance,* New Haven, 1986. His penetrating study on the rural Malaysia shows the poor peasants daily forms of resistance like giving false information to officials, non-co-operation with the imposed rules and regulations, migrations, etc.

143. Till then they received 3 *fanams* per *masula* of 6 bales that required 10 to 12 men to transport.

144. *DCB*, 12 January, 1678, p. 129.

145. Ibid.

146. *DCB*, 12 January, 1678, pp. 129-30.

147. *DCB*, 12 January, 1678, p. 129.

148. *DCB*, 1 July 1680, pp. 43-4.

149. *DCB*, 4 May 1696, p. 63.

150. *DCB*, 14 November 1695, p. 148.

151. *DCB*, 19 April 1697, p. 34. The *Mucquas* share of the Quit Rent was only *pagodas* 8:18 *fanams.*

152. *DCB*, 30 March 1699, p. 27; 8 January 1700, p. 3.

153. *DCB*, 14 November 1695, p. 148.

154. Shortt, 'The Fishermen of Southern India', p. 199.

155. *DCB*, 29 September 1687.

156. *DCB*, 7 March 1698, p. 26.

157. *DCB*, 20 January 1708, pp. 7-8; see also *DTE*, 1707-08, p. 93.

158. *DFE*, 1706-1710, 4 February 1708, pp. 20 & 39.

159. *DCB*, 21 January 1708, pp. 10-11; 19 August 1708, p. 46.

160. *DFE*, 1706-10, 9 January 1709, p. 93.

161. *DCB*, 12 November 1711, p. 159.

162. *DFE*, 2 January 1739, para. 24.

163. *DFE*, 21 March 1743, para. 52.

164. *DCB*, 12 January 1678, pp. 129-30.

165. Salmon's account cited in Love, *Vestiges,* vol. II, p. 75; see also *DCB*, 14 November 1695, p. 148.

166. *DCB*, 26 November 1696, p. 143.

167. Shortt, 'The Fishermen of Southern India', p. 199.

Editor and Contributors

ANIRUDH DESHPANDE is Associate Professor at the Department of History, University of Delhi, and the author of *British Military Policy in India 1900-1945: Colonial Constraints and Declining Power* (2005); *Class, Power and Consciousness in Indian Cinema and Television* (2009); and *The British Raj and its Indian Armed Forces, 1857-1939* (jointly edited with Partha Sarathi Gupta, 2002).

JANGKHOMANG GUITE is Assistant Professor at the Department of History, Assam University, Silchar, Assam. His doctoral work is entitled 'Catholics in a Protestant Enclave: The Catholic Community and the English Company in Madras, 1640-1750'. He has also written several research articles.

PIUS MALEKANDATHIL is currently Associate Professor, Centre for Historical Studies, Jawaharlal Nehru University, New Delhi. Earlier, he was Reader in History at Goa University (2000-3) and Sree Sankaracharya University of Sanskrit at Kalady, Kerala (2003-6). Dr Malekandathil has published numerous papers in national and international journals. Some of his publications are: *The Germans, the Portuguese and India* (1999); *Portuguese Cochin and the Maritime Trade of India, 1500-1663* (2001); *Jornada of D. Alexis Menezes: A Portuguese Account of the Sixteenth Century Malabar* (2003); *The Portuguese, Indian Ocean and European Bridgehead: Festschrift in Honour of Prof. K.S. Mathew* jointly edited with T. Jamal Mohammed (2001); *The Portuguese and the Socio-Cultural Changes in India: 1500-1800* jointly edited with K.S. Mathew and Teotonio R. de Souza (2001); *The Kerala Economy and European Trade* jointly edited with K.S. Mathew (2003); *Goa in the Twentieth Century: History and Culture* jointly edited with Remy Dias (2008).

TILOTTAMA MUKHERJEE is Lecturer at the Department of History, Jadavpur University, Kolkata. Her doctoral dissertation is entitled 'Markets, Transport and the State in the Bengal Economy, 1750-1800'. Her recent publications include articles entitled 'Commodities, Trade and the Economy of Bengal: A Reassessment of the Early English East India Company State' in S. Jeyaseela Stephen, ed., *The Indian Trade at the Asian Frontier*, and 'The Co-ordinating State and the Economy: The Nizamat in Eighteenth-century Bengal' in *Modern Asian Studies*.

JOY L. K. PACHUAU is Associate Professor at the Centre for Historical Studies, Jawaharlal Nehru University, New Delhi. Her research interest includes formation of regional identities and the role of religion in this construction. Issues relating to the role of Christianity are also of primary interest to her. She has research publications in these areas in the contexts of coastal India in the sixteenth and seventeenth centuries as well as north–east India in the nineteenth and twentieth centuries

MICHAEL PEARSON is Emeritus Professor of History at the University of New South Wales, and Adjunct Professor of Humanities at the University of Technology, Sydney. Professor Pearson has authored *Merchants and Rulers in Gujarat* (1976); *Coastal Western India* (1981); *The Portuguese in India* (1988); *Port Cities and Intruders: The Swahili Coast, India, and Portugal in the Early Modern Era* (1999) and, *The World of the Indian Ocean 1500-1800* (2005) among many others.

VAIBHAV SHARMA has completed his Ph.D. from the Centre for Historical Studies, Jawaharlal Nehru University, New Delhi. His dissertation is entitled 'The Growth of Bombay, 1660–1760'.

YOGESH SHARMA is Associate Professor at the Centre for Historical Studies, Jawaharlal Nehru University, New Delhi. He has co-edited *Portuguese Presence in India in the Sixteenth and Seventeenth Centuries* (2008) and *Biography as History: Indian Perspectives* (2009), and has written several research articles. His research interest focuses on maritime studies, urbanization and trade history.

ABHAY KUMAR SINGH is Reader in History, Gujarat University, Ahmedabad. His areas of interest are maritime, medical and epidemiological history, Indian Ocean environmental history, and the technological history of India. Besides several research papers, his major publications include *Modern World System and Indian Proto-Industrialization: Bengal 1650-1800* (2006) in two volumes, and his forthcoming publication is entitled *Maritime Morbidity and Mortality: Intra-Indian Ocean Transmission of Diseases 1500-1800* in three volumes.

ARVIND SINHA is Associate Professor at the Centre for Historical Studies, Jawaharlal Nehru University, New Delhi. He was an Associate in Eurindia Project (an EC project) from 2003 and a Fellow at Nehru Memorial Museum and Library from 2005. His publications include *The Politics of Trade: Anglo-French Commerce on the Coromandel Coast, 1763-1793* (2002), *Sankranti Kaleen Europe* (2009), *Europe in Transition: From Feudalism to Industrialization* (2010), besides several research articles in journals and edited volumes.

Index